Objective
Seed Science and Technology

(2nd Revised Edition)

Related Books

Objective Seed Science and Technology

(2nd *Revised Edition*)

K. Vanangamudi

S. Kavitha

K. Raja

Published by:

Jodhpur
5 A, New Pali Road, P.O. Box 91
Jodhpur - 342 001 Rajasthan (India)
E-mail: info@scientificpub.com
Website: www.scientificpubonline.com

New Delhi
4806/24, Ansari Road, Daryaganj,
New Delhi - 110 002 (India)

Print: 2025

1st Edition 2018

2nd Edition: 2020

ISBN: 978-93-89412-60-4
eISBN: 978-93-89412-61-1

Printed in India

Preface

Continuous advancements in ***Seed Science and Technology*** ensure seeds are produced, processed and maintained with high quality. If seed supply is not cared properly, it will lose its ability to sprout, resulting in empty fields. Seed security is food security.

Knowing the importance of Seed Science in Agriculture, all the State Agricultural Universities, ICAR Institutes and other educational Institutions are actively involved in research and teaching of Seed Science and Technology. Job opportunities are also very encouraging for the agricultural and horticultural graduates who are specializing in Seed Science and Technology. Many agricultural and horticultural graduates are interested in higher education, JRF and SRF fellowships, ARS scientists, NET examinations, agricultural extension officers in banks, agricultural and horticultural officers in the State agriculture, horticulture and seed certification departments, public and private sector seed corporations and companies and NGO's.

Keeping the above needs of graduates and aspirants, the book entitled "Objective Seed Science and Technology" is revised by incorporating all the errors pointed out by readers.

I acknowledge the readers, students, graduates, scientists and teachers for their support and suggestions to improve and update this edition. Special heartfelt and sincere thanks to Mr. Tanay Sharma of Scientific Publishers for his continued support, encouragement and love.

K. VANANGAMUDI

January, 2020

Contents

PART 2 : ADVANCED IN SEED SCIENCE AND TECHNOLOGY : SEED PHYSILOGY AND BIOCHEMISTRY

Part 1

Seed Science and Technology

(As per ICAR Syllabus)

Unit 1

SEED BIOLOGY

1.1. FLORAL BIOLOGY

1.1.1. Choose the correct answer

1. Tassel is the staminate inflorescence of
(a) Maize (b) Paddy
(c) Cotton (d) Sunflower

2. The mode of pollination in maize is
(a) Cross pollination (b) Self pollination
(c) Cleistogamy (d) Chasmogamy

3. The male inflorescence of maize is
(a) Tassel (b) Silk
(c) Cyme (d) Racemose

4. Pollination of maize is by
(a) Insect (b) Wind
(c) Water (d) Machine

5. Cross pollination in radish is favoured by
(a) Insects (b) Wind
(c) Water (d) Beetles

6. Inflorescence in carrot flower is called as
(a) Protectants (b) Compound umbel
(c) Protogynous (d) Raceme

7. Pollination in cauliflower is done by
(a) Wind (b) Honeybee
(c) Insect (d) Water

8. Harvestable maturity stage for bhendi is
(a) Yellow colour (b) Hair line crack formation
(c) Dried pod (d) Matured pod

9. Example for dioecious fruit crop
(a) Papaya (b) Jack
(c) Mango (d) Sapota

10. Pollination in tea is effected by
(a) Insects (b) Water
(c) Wind (d) Animals

11. The only cone bearing angiosperm is
(a) *Casuarina equisetifolia* (b) *Pinus elliotii*
(c) *Pinus roxburghii* (d) *Santalum album*

12. The monoecious plant normally having unisexual flowers
(a) Blackgram (b) Groundnut
(c) Maize (d) Paddy

13. Example for self pollination
(a) Tomato (b) Brinjal
(c) Chilli (d) Cucurbits

14. The crop with longest style is
(a) Sunflower (b) Maize
(c) Cumbu (d) Wheat

15. Self incompatible crop is
(a) Carrot (b) Onion
(c) Cabbage (d) Spinach

16. Heterostyly favours
(a) High seed yield (b) Genetically pure seed
(c) Self pollination (d) Cross pollination

17. Maturity of stamens and pistils at different time is called as
(a) Cleistogamy (b) Dichogamy
(c) Dioecy (d) Dicliny

18. Flower colour is a
(a) Truly qualitative characteristic (b) Pseudo qualitative characteristic
(c) Standard characteristic (d) Special characteristic

19. Hollow style is common in
(a) Maize (b) Cotton
(c) Groundnut (d) Blackgram

20. Flowering of inflorescence from bottom to top
(a) Acropetal (b) Basipetal
(c) Centripetal (d) Peripetal

21. King of coarse grain is
(a) Cumbu (b) Jowar
(c) Ragi (d) Maize

22. Prevention of self pollination by flower structure is
(a) Autogamy (b) Allogamy
(c) Herkogamy (d) Dichogamy

23. Rain favours pollination in
(a) Cardamom (b) Cinnamum
(c) Pepper (d) Nutmeg

24. If the flower is having all the floral parts, the flower is called as
(a) Incomplete flower (b) Complete flower
(c) Imperfect flower (d) Pistillate flower

25. If the flower is having only male and female organs, the flower is called as
(a) Incomplete flower (b) Imperfect flower
(c) Perfect flower (d) Pistillate flower

26. If the flower has gynoecium only that flower is named as
(a) Pistillate (b) Staminate
(c) Pistillode (d) Staminode

27. If the flower has androecium only that flower is named as
(a) Pistillate (b) Staminate
(c) Pistillode (d) Staminode

28. If the flower has rudimentary stamen that flower is named as
(a) Pistillate (b) Staminate
(c) Pistillode (d) Staminode

29. Self pollination is otherwise called as
(a) Allogamy (b) Homogamy
(c) Autogamy (d) Chasmogamy

30. The cross pollination is otherwise called as
(a) Allogamy (b) Homogamy
(c) Autogamy (d) Chasmogamy

31. The flowers do not open and self pollination is the rule
(a) Allogamy (b) Homogamy
(c) Autogamy (d) Cleistogamy

32. The insect pollination mechanism is called as
(a) Anemophily (b) Hydrophily
(c) Entomophily (d) Ornithophily

33. The water pollination mechanism is called as
(a) Anemophily (b) Hydrophily
(c) Entomophily (d) Ornithophily

34. The wind pollination mechanism is called as
(a) Anemophily (b) Hydrophily
(c) Entomophily (d) Ornithophily

35. The differential maturation of male and female flowers is known as
(a) Portandry (b) Protogyny
(c) Dichogamy (d) Dicliny

36. The pollination by birds is usually named as
(a) Chrotophily (b) Ornithophily
(c) Psycophily (d) Palenophily

37. The pollination by bats is usually known as
(a) Chrotophily (b) Ornithophily
(c) Psycophily (d) Palenophily

38. The pollination by beetle is called as
(a) Canthophily (b) Symphysophily
(c) Waspophily (d) Mycophily

39. The pollination by sawfly is called as
(a) Canthophily (b) Symphysophily
(c) Waspophily (d) Mycophily

40. Site of double fertilization is
(a) Embryo sac (b) Micropyle
(c) Ovary (d) Ovule

41. The pollination by flies is named as
(a) Myophily (b) Psycophily
(c) Palenophily (d) Ornithophily

42. The pollination by butterflies is called as
(a) Myophily (b) Psycophily
(c) Palenophily (d) Ornithophily

43. The pollination by moths is called as
(a) Myophily (b) Psycophily
(c) Palenophily (d) Ornithophily

44. Perisperm is
(a) Haploid (b) Diploid
(c) Tetrapolid (d) Hexaploid

45. The transfer of pollen to stigma of different flowers of same plant is called as
(a) Chasmogamy (b) Cleistogamy
(c) Geitenogamy (d) Xenogamy

46. The transfer of pollen to stigma of flowers of two different plant is known as
(a) Chasmogamy (b) Cleistogamy
(c) Geitenogamy (d) Xenogamy

47. If the male and female organs present in the same flower, the flower is called as
(a) Hermaphrodite (b) Unisexual
(c) Andromonoecious (d) Gynomonoecious

48. If the male and female organ is present separately in two different flowers, the flower is named
(a) Hermaphrodite (b) Unisexual
(c) Andromonoecious (d) Gynomonoecious

49. The plant that possess any one of the following flower type is called as andromonoecious
(a) ♀ (b) ♂
(c) ♂ + ♂ (d) ♂ + ⚥

50. The plant that possess any one of the following flower type is called as gynomonoecious
(a) ♂ + ♂ (b) ♂
(c) ♀ (d) ♀ + ♀

51. The period of dehiscence of anthers and pollen distribution is called as
(a) Flowering (b) Pollination
(c) Fertilization (d) Anthesis

52. Occurrence of pollination and fertilization in an unopened flower bud and 100% self fertilization will occur which is called as
(a) Chasmogamy (b) Cheiropterophily
(c) Anemophily (d) Cleistogamy

53. Maturation of male and female reproductive organs of a hermaphrodite flower at different time is called as
(a) Dichogamy (b) Plasmogamy
(c) Cleistogamy (d) Herchogamy

54. The union of male and female gametes in sexual reproduction is known as
(a) Fertilization (b) Pollination
(c) Infertility (d) Incompatibility

55. The outer bract of the flowers of grasses sometimes referred to as flowering glume is called as
(a) Lemma (b) Glume
(c) Palea (d) Anther

56. The study of pollen is termed as
(a) Cartilinology (b) Palentology
(c) Ornithology (d) Palenology

57. The inflorescence of wheat is called as
(a) Racemose (b) Sporodix
(c) Spike (d) Cyme

58. The individual flower stalk in an inflorescence is known as
(a) Spikelet (b) Pedicel
(c) Auricle (d) Pellicle

59. The main axis of an inflorescence or, in the case of single flowers, the flower or fruit stalk is known as
(a) Peduncle (b) Spikelet
(c) Auricle (d) Pellicle

60. Collective term for the outer part of flower comprising calyx and corolla is known as
(a) Spike (b) Perianth
(c) Auricle (d) Spikelet

61. Study of the relations between seasonal climatic changes (e.g. temperature, day length and precipitation) and periodic biological phenomena such as flowering, fruiting, and dormancy is known as
(a) Chronology (b) Genology
(c) Physiology (d) Phenology

62. Region in the ovary where the ovules originate and are attached to the carpel is known as
(a) Placenta (b) Ligule
(c) Planta (d) Auricle

63. Many seeded fruit derived from a compound pistil embedded in a fleshy hypanthium or floral tube of epigynous flowers is known as
(a) Pod (b) Caryopsis
(c) Capsule (d) Pome

64. A spiral arrangement of modified leaves bearing the reproductive organs in conifers and certain other plants, functionally equivalent to angiosperm flowers is called as
(a) Achene (b) Strobilus
(c) Pome (d) Caryopsis

65. The stalk of the pistil between stigma and ovary is called as
(a) Style (b) Peduncle
(c) Ligule (d) Auricle

66. A centrifugal inflorescence on which the secondary or lateral branches continue to grow and extend beyond the main axis is called as
(a) Bolt (b) Curd
(c) Cyme (d) Caryopsis

67. The inflorescence of Allium cepa is
(a) Umbel (b) Corymb
(c) Spadix (d) Spike

68. Zygomorphic symmetry or bilateral symmetry is common in flowers of the family
(a) Solanaceae (b) Malvaceae
(c) Papilionaceae (d) Apocyanaceae

69. The components of perianth are called as
(a) Petals (b) Sepals
(c) Tepals (d) Stamens

70. Unbranched inflorescence with elongated axis with sessile florets is known as
(a) Raceme (b) Cyme
(c) Corymb (d) Spike

71. Characteristics of inflorescence in Asteraceae family is
(a) Capitulum (b) Catkin
(c) Spadix (d) Fascicle

72. Temperature plays a major role in sex expression in
(a) Castor (b) Cotton
(c) Cabbage (d) Cauliflower

73. In bajra, heavy wind causes
(a) Floral abortion (b) Tip sterility
(c) Abortive seeds (d) Under developed embryo

74. Anemophily indicates
(a) Pollination of water (b) Pollination by wind
(c) Pollination by insects (d) Pollination by birds

75. Inability of flowering plant to produce functional pollen is
(a) Male sterility (b) Self incompatibility
(c) Mutation (d) Inbreeding depression

76. Pin type flowers of Primula have
(a) Long style and short stamen (b) Short style and long stamen
(c) Long style and long stamen (d) Short style and short stamen

77. Barnase Barstar model is in
(a) Genetic male sterility (b) CGMS
(c) CMS (d) Transgenic male sterility

78. Self incompatibility is very much utilized for
(a) Production of hybrids (b) Production of varieties
(c) Production of composites (d) Production of synthetics

79. The pollination behaviour of coconut is
(a) Self pollination (b) Cross pollination
(c) Often cross pollination (d) Self and cross pollination

80. Vegetative apomixis can also be known as
(a) Non recurrent apomixis (b) Agamospermy
(c) Agamogony (d) Pseudogamy

81. Epigynous flowers are more common in the members of
(a) Leguminosae (b) Poaceae
(c) Asteraceae (d) Umbelliferae

82. Tapetum is present in
(a) Placenta (b) Carpel
(c) Locule (d) Anther

83. The flowers are protandry in
(a) Pearl millet (b) Sunflower
(c) Maize (d) Castor

84. Flowering behaviour of castor is
(a) Monoecious (b) Dioecious
(c) Andromonoecious (d) Gynomonoecious

85. Pistil, stamen and petals are the common structural components of
(a) Flower (b) Inflorescence
(c) Apical bud (d) Vegetative bud

86. In which family, the flower is anomaly of very small fruits
(a) Compositae (b) Umbelliferae
(c) Graminae (d) Liliaceae

87. A single flower of gymnosperm is known as
(a) Strobilus (b) Strobili
(c) Microspore (d) Megaspore

88. Gymnosperms do not have botanical flower, but have the following for regeneration
(a) Male and female strobili (b) Cones
(c) Megaspore (d) Microspore

89. Angiosperm flowers containing both stamens and pistil are known as
(a) Perfect (b) Simple
(c) Solitary (d) Raceme

90. The flower with stamen, pistil, sepals and petals are known as
(a) Complete (b) Monoecious
(c) Hermaphrodite (d) Cyme

91. In angiosperm, flowers in which male sex organ mature first is known as
(a) Protandry (b) Protogynous
(c) Perfect (d) Monoclinous

92. In angiosperms, the functions which are to minimize the chances of self fertilization is known as
(a) Dichogamy (b) Monogamy
(c) Dioecious (d) Cross pollination

93. Both male and female reproductive organs occur on the same individual plant, but separate from each other, the plant is known as
(a) Monoecious (b) Imperfect
(c) Dioecious (d) Dicot

94. In gymnosperm, flowering denotes
(a) Period of pollen shedding and receptivity of the female organ
(b) Fusing of megaspore to microspore
(c) Opening of cones
(d) Formation of fruiting bodies

95. The largest species (30,00,000) and genera (1,21,000) of plant kingdom are grouped under
(a) Angiosperm (b) Gymnosperm
(c) Dicot (d) Monocot

96. Monocot and dicot are the sub divisions of
(a) Angiosperm (b) Gymnosperm
(c) Orthodox species (d) Recalcitrant species

97. Most of the woody plants are grouped as
(a) Dicot (b) Monocot
(c) Gymnosperm (d) Angiosperm

98. The pollen enclosed within the anther is released at maturity through pores in
(a) Ericaceae (b) Lauraceae
(c) Fagaceae (d) Febaceae

99. In Erythrina, cross pollination occurs through
(a) Birds/Honey eaters (b) Insects
(c) Wind (d) Water

100. The transfer of pollen to stigma of different flowers of same plant is called as
(a) Chasmogamy (b) Cleistogamy
(c) Geitenogamy (d) Xenogamy

101. The transfer of pollen to stigma of different flowers of two different plant is known as
(a) Chasmogamy (b) Cleistogamy
(c) Geitenogamy (d) Xenogamy

102. The pollen is produced in
(a) Ovule (b) Anther
(c) Stigma (d) Style

103. Naked ovules are found in which one of the following group?
(a) Angiosperm (b) Microsperm
(c) Gymnosperm (d) Macrosperm

104. Which solution is used for testing the pollen viability?
(a) 2% potassium permanganate (b) 2% acetic acid
(c) 2% aceto-carmine (d) 2% naphthalene acetic acid

♦ Answer Keys ♦

Q.No.	Ans.	Q.No.	Ans.	Q.No.	Ans.	Q.No.	Ans.	Q.No.	Ans.	Q.No.	Ans.
1	a	2	a	3	a	4	b	5	a	6	b
7	b	8	b	9	a	10	a	11	a	12	c
13	a	14	b	15	c	16	d	17	b	18	a
19	a	20	a	21	b	22	c	23	c	24	b
25	c	26	a	27	b	28	d	29	c	30	a
31	d	32	c	33	b	34	a	35	c	36	b
37	a	38	a	39	b	40	a	41	a	42	b
43	c	44	b	45	c	46	d	47	a	48	b
49	d	50	d	51	d	52	d	53	a	54	a
55	a	56	d	57	c	58	b	59	a	60	b

Q.No.	Ans.	Q.No.	Ans.	Q.No.	Ans.	Q.No.	Ans.	Q.No.	Ans.	Q.No.	Ans.
61	d	62	a	63	d	64	b	65	a	66	c
67	a	68	c	69	c	70	d	71	a	72	a
73	b	74	b	75	a	76	a	77	d	78	a
79	b	80	b	81	c	82	d	83	c	84	a
85	a	86	a	87	a	88	a	89	a	90	a
91	a	92	a	93	a	94	a	95	a	96	a
97	a	98	a	99	a	100	c	101	d	102	b
103	c	104	c								

1.1.2. Fill in the blanks

1. Flower pistil mature before stamen is called as ....
2. .... is an example for protandry
3. .... is an example for protogyny
4. Naked ovules are found in ....
5. Sexual reproduction involves fusion of .... gametes
6. Pollination refers to the transfer of pollen grains from anthers to ....
7. Inflorescence of carrot is ....
8. High temperature at flowering affect the .... of pollen grains
9. .... is a dioecious plant
10. Dispersal of seeds by ant is known as ....
11. Self incompatibility in crops can be broken by ....
12. Triple waxy bloom is noticed in .... crop
13. Heterostyly is common in ....
14. Androecium in sunflower is ....
15. Sorghum is an .... crop
16. Mechanism that aid cross pollination in bajra is ....
17. Mechanism that aid cross pollination in maize is ....
18. Cowpea is .... pollinated crop
19. Fruit of cruciferous family is called as ....
20. Cross pollination is called as ....
21. Self pollination is named as ....
22. Style of the flower fertilized by the stamen of another flower of same plant is known as ....
23. Complete self pollination is called as ....
24. Presence of pistillate and staminate flowers in same flower is ....
25. Pistil matures first is ....
26. Anther matures first is ....
27. Flowers with different styles are called as ....

28. Long style and short stamen flower is ....
29. Short style and long stamen flower is ....
30. Maize is a .... pollinated crop
31. .... reproductive part of a plant
32. Lack of any one of the reproductive parts is called as ....
33. In .... flowers, gynoecium is present
34. In .... flowers, androecium is present
35. Pistil rudimentary is called ....
36. Stamen rudimentary is called as ....
37. True to type will be maintained by .... pollinated crops
38. Synchronised maturation of ♂ & ♀ in the same flower is called as ....
39. Self pollination is a rule in ....
40. Cross pollination is effected by external agents like ....
41. Cross pollination is effected by biotic agents like ....
42. An example for anemophilous pollination is ....
43. An example for hydrophilous pollination is ....
44. Union of male and female gametes in sexual reproduction is called as ....
45. Site of double fertilization ....
46. An example for trimonoecious (♂, ♂ , ♀) is ....
47. An example for gynomonoecious is ....
48. An example for andromonoecious is ....
49. Carrot is a .... pollinated crop
50. .... is known as angiosperms
51. Heterostyly is the common feature of .... crop
52. A special kind of spike covered by spathe is called ....
53. Pollen is produced in the ....
54. Stamens are the specialized ....
55. Pollen grain germination occurs in ....
56. Sperms made their way to ovule through ....
57. Pollination by wind is called ....
58. Pollination after flower opening is called ....
59. An example for often cross pollination is ....

♦ Answer Keys ♦

Q.No.	Ans.	Q.No.	Ans.	Q.No.	Ans.
1	Protogyny	2	Maize	3	Bajra
4	Gymnosperm	5	Female and Male	6	Stigma
7	Umbel	8	Fertility	9	Papaya

Q.No.	Ans.	Q.No.	Ans.	Q.No.	Ans.
10	Myrmecochory	11	Supplementary pollination	12	Castor
13	Brinjal	14	Syngenecious	15	Often Cross Pollinated
16	Protogyny	17	Protandry	18	Self
19	Siliqua	20	Allogamy	21	Autogamy
22	Geitonogamy	23	Cleistogamy	24	Monoecious
25	Protogyny	26	Protandry	27	Heterostyly
28	Pin flower	29	Thrum flower	30	Cross
31	Flower	32	Incomplete flower	33	Pistillate
34	Staminate	35	Pistillode	36	Staminode
37	Self	38	Homogamy	39	Bengalgram/Horsegram
40	Wind/Insect	41	Insect/Animal	42	Bajra
43	Water hyacinth	44	Fertilization	45	Embryo sac
46	Castor/banana	47	Ailanthus	48	Sorghum/Sunflower
49	Cross	50	Flowering plant	51	Brinjal
52	Spadix	53	Anther	54	Microsporophyte
55	Stigma	56	Pollen tube	57	Anemophily
58	Chasmogamy(or) Allogamy	59	Redgram/Cotton		

1.1.3. State True or False

1. Flowers of bhendi varies based on style length
2. When receptivity of stigma precedes the dehiscence of anther in a bisexual flower, the condition is termed protandry
3. Genetic contamination is common in self-pollinated crops
4. Bajra is a protogynous plant
5. Bajra is typically a protandry
6. Maize is a monoecious plant
7. Chilli is an often cross pollinated crop
8. Natural crossing is a serious problem in self pollinated crops when it is a male sterile line
9. Heterostyly character is common in brinjal
10. Heterostyly favours cross pollination in bhendi
11. Pigeon pea is an often cross pollinated crop
12. Heterostyly in brinjal aids cross pollination
13. Sunflower is pollinated by insects
14. The reproductive structure of angiosperms is flowers
15. Gynoecium is the male reproductive part in the flowers
16. The arrangement of floral parts like petals and sepals on the thalamus is called placentation

17. Maize is a typical example for monoecious type of plant
18. The cucurbits have hypogynous flowers with superior ovary
19. Dichogamy nature in flower favours the cross pollination
20. Effect of foreign pollen on seed coat of maize seed is called as meta xenia
21. Flowers with differential length of the style and stigma is called herkogamy
22. A special kind of spike covered by a spathe is called spadix
23. Heterostyly is more common in tomato crop
24. Peduncle is the main axis of flower
25. Radicle is a individual flower stalk in a inflorescence
26. Receptacle is found at the tip of floral structures
27. Sepals are the small outgrowth above the petal
28. Stamen is the male organ of a plant
29. Pistil is the male organ of a plant
30. Calyx is the union of petals
31. Corolla is otherwise called as calyx
32. Calyx and corolla together makes the perianth
33. Androecium is the female part of flower
34. Gynoecium is the female part of the flower
35. Autogamy is the fertilization of ovule by the pollen of the same flower
36. In cleistogamy, the flower do not open at all
37. Maize is an often cross pollinated crop
38. Sorghum is a self pollinated crop
39. Megaspore mother cell undergoes a meiotic division to form four megaspores
40. Three cells present at chalazal end are called as antipodal cells
41. Exine is thin and composed of pseudo cellulose
42. Fusion of one male gamete with egg cell is called as syngamy
43. Fusion of 2nd male gamete with two egg cells of central part of female gametophyte to form PEN (3N) (primary endosperm nucleus)
44. Nuclear divisions unaccompanied by cell wall formation is called as nuclear endosperm development
45. Pistil is otherwise called as androecium

♦ Answer Keys ♦

Q.No.	Ans.	Q.No.	Ans.	Q.No.	Ans.	Q.No.	Ans.	Q.No.	Ans.	Q.No.	Ans.
1	False	2	False	3	False	4	True	5	False	6	True
7	True	8	True	9	True	10	False	11	True	12	True
13	True	14	True	15	False	16	False	17	True	18	False
19	True	20	True	21	False	22	True	23	False	24	True

25	False	26	False	27	False	28	True	29	False	30	False
31	False	32	True	33	False	34	True	35	True	36	True
37	False	38	False	39	False	40	True	41	False	42	True
43	False	44	True	45	False						

1.2. SEED DEVELOPMENT AND MATURATION

1.2.1. Choose the correct answer

1. Maturation of ovule leads to formation of
   (a) Fruit (b) Embryo
   (c) Endosperm (d) Seed
2. Methods of endosperm development are
   (a) 2 (b) 3
   (c) 4 (d) 1
3. The last phase of seed development is
   (a) Fertilization (b) Dehydration
   (c) Desiccation (d) Pollination
4. Carbohydrates are transported from .... through the sieve tubes of the phloem
   (a) Source only (b) Sink only
   (c) Source to sink (d) Sink to source
5. The fully developed leaf export percentage of its assimilate to sink is
   (a) 50-60 (b) 40-50
   (c) 80-100 (d) 60-80
6. In leaf to root translocation, water recycles from sink to source through
   (a) Xylem (b) Phloem
   (c) Osmosis (d) Plasmolysis
7. Reactions found during photosynthesis is
   (a) $6\ CO_2 + H_2O + \text{Sunlight} \rightarrow C_6H_{12}O_6 + 6\ O_2$
   (b) $6\ CO_2 + 6\ H_2O + \text{Sunlight} \rightarrow C_6H_{12}O_6 + O_2$
   (c) $6\ CO_2 + 6\ H_2O + \text{Sunlight} \rightarrow C_6H_{12}O_6 + 6\ O_2$
   (d) $6\ CO_2 + 6\ H_2O + \text{Sunlight} \rightarrow C_6H_{12}O_6 + 8\ O_2$
8. In photosynthesis, the energy released by the protein when they diffuse across the thylakoid membrane into the stroma is used to produce
   (a) $NADH_2$ (b) ATP
   (c) Photolysis (d) CO_2 fixation
9. How many turns of Calvin cycle are taken to produce one hexose molecule
   (a) 1 (b) 3
   (c) 6 (d) 12

10. The structural unit of photosynthesis
    (a) Thylakoid (b) Stroma
    (c) Chlorophyll membrane (d) Nucleus

11. The substance for starch synthesis in plants
    (a) ATP (b) ADP-Glucose
    (c) UDPH (d) FADP

12. The key regulatory enzyme in starch synthesis is
    (a) ATP-Glucose (b) $NADPH_2$
    (c) FADH (d) ADP-Glucose phytophosphorylase

13. In the final stage of seed development accumulation of triglycerol and oleosins takes place in
    (a) Embryo (b) Endosperm
    (c) Seed coat (d) Perisperm

14. Fatty acid is released from
    (a) ATP (b) ACP
    (c) AMP (d) ADP

15. Palmitic acid has carbon atoms of
    (a) 12 (b) 14
    (c) 16 (d) 18

16. Tryptophan is produced in
    (a) Plastid (b) Golgi apparatus
    (c) Endoplasmic reticulum (d) Mitochondria

17. During seed development, auxin accumulate in
    (a) Embryo (b) Endosperm
    (c) Nucleus (d) Perisperm

18. Gibberellin is first extracted from
    (a) *Fusarium oxysporum* (b) *Gibberella fujikuroi*
    (c) *Fusarium moniliforme* (d) *Fusarium graminerium*

19. Who first isolated gibberellin from the fungus
    (a) Yabuta and Sumuki (b) Mac Millan and Suter
    (c) Skoog and Saltza (d) Black and Mac Millan

20. Naturally occuring cytokinin
    (a) 6 Benzyl adenine (b) Zeatin
    (c) Benzimadazole (d) 1-Benzyl adenine

21. Which one of the following is not the role of secondary metabolites
    (a) Protection against microbial infections
    (b) Repair of damaged cell wall
    (c) Attract pollinators and seed dispersal agent
    (d) Serve as allelopathy agents

22. Which one of the following secondary metabolite do not contain nitrogen
    (a) Steroids (b) Alkaloid
    (c) Amines (d) Glucosinolates
23. Among the following secondary metabolites combinations, which one contains nitrogen
    (a) Diterpenes and Steroids (b) Alkaloids and Flavanoids
    (c) Alkaloids and Glucosinolates (d) Monoterpenes and Polyacetylenes
24. Among the following facts, which one is not true about terpenes
    (a) Soluble in water (b) Fusion of 5 carbon isoprene unit
    (c) Originate from turpine (d) Decompose to give isoprene
25. Gibberellin is a
    (a) Flavanoid (b) Diterpene
    (c) Phenolic compound (d) Tetraterpene
26. A group of organic compounds that contain sulphur and nitrogen and derived from glucose and an amino acid
    (a) Hydrocyanic acid (b) Glucosinolate
    (c) *Saponis* (d) Phenolic compound
27. When mutation occur naturally that is called as
    (a) Somatic mutation (b) Germinal mutation
    (c) Induced mutation (d) Spontaneous mutation
28. Example for physical mutagen is
    (a) EMS (b) Ethidium bromide
    (c) X-rays (d) Nitrous acid
29. Example for mutation agent is
    (a) X-rays (b) EMS
    (c) β-rays (d) UV radiation
30. Example for non ionizing radiation is
    (a) UV-rays (b) X-rays
    (c) β-rays (d) α-rays
31. Mutations occurring in bready cells which do not transmit to next generation is called as
    (a) Germinal mutation (b) Somatic mutation
    (c) Induced mutation (d) Reverse mutation
32. Development of a new mutant type from a wild type called as
    (a) Forward mutation (b) Reverse mutation
    (c) Delinked mutation (d) Somatic mutation
33. Gene mutation is otherwise called as
    (a) Chromosomal aberration (b) Chromosomal variation
    (c) Point mutation (d) Endogenous mutation

34. Which one of the following events takes place in the last phase of seed development
(a) Fertilization (b) Embryo development
(c) Endosperm development (d) Dehydration

35. The starch is transported to the developing and maturing reproductive organs of a plant from
(a) Source (b) Sink
(c) Source to sink (d) Sink to source

36. Which one of the following parts recycles water from sink to source during the export and import of photo assimilate?
(a) Xylem (b) Phloem
(c) Sieve cells (d) Plasmadesmata

37. For export of photo assimilate from source to sink, protoplasmic streaming theory is one of the method of export
(a) De Vries (1885) (b) Thaine (1964)
(c) Munch (1926) (d) Fenson (1971)

38. Which one of the following parts translocate both solute and solvent
(a) Phloem vessels (b) Xylem vessels
(c) Vascular bundle (d) Plasmadesmata

39. The export of photo assimilate from phloem to the sink is called as
(a) Phloem unloading (b) Phloem loading
(c) Xylem unloading (d) Xylem loading

40. Physiological maturity symptom of rice is
(a) Golden yellow panicle (b) Complete drying of plant
(c) Shedding of seeds (d) Shattering of seeds

41. Harvestable maturity stage for bhendi is
(a) Yellow colour (b) Hair line crack formation
(c) Dried pod (d) Mature pod

42. The seed nut of coconut should be matured by
(a) 9 months (b) 10 months
(c) 12 months (d) 14 months

43. Under normal conditions, fruit set in coffee is affected by
(a) Floral abnormalities (b) Failure of anther dehiscence
(c) Non-fertilization (d) Embryo abortion

44. The seeds of *Azadirachta indica* reaches physiological maturity when the fruits turn
(a) Green (b) Yellow
(c) Brown (d) Orange

45. The mother palm selected in coconut seed garden should have atleast .... no. of fully opened leaves
(a) 20 (b) 30
(c) 40 (d) 50

46. The seeds are formed without fertilization is known as
   (a) Apogamy (b) Apospory
   (c) Anagenesis (d) Apomixis
47. The term that represent the tissue of seeds that develops from sexual fusion of the polar nuclei of the ovule and the second male sperm cell is called as
   (a) Embryo (b) Radicle
   (c) Endosperm (d) Plumule
48. Production of large number of vegetative progeny (initially, of very small size) through tissue culture is called
   (a) Micro propagation (b) Mega propagation
   (c) Apospory (d) Explant propagation
49. The process in which an egg cell of an embryo sac derived from megaspore develop into an embryo without fertilization is called
   (a) Perfect fertilization (b) Parthenogenesis
   (c) Endocarpy (d) Double fertilization
50. Occurrence of more than one embryo in seed is known as
   (a) Hyperembryony (b) Hypoembryony
   (c) Heteroembryony (d) Polyembryony
51. Ovule orientation in which the ovule is inverted with respect to its funiculus is known as
   (a) Exotropous ovule (b) Anatropous ovule
   (c) Inverted ovule (d) Exerted ovule
52. Fusion of one of the two sperms with the secondary nucleus leading to the formation of a triploid primary endosperm nucleus is called
   (a) Double fusion (b) Instigated fusion
   (c) Triple fusion (d) Direct fusion
53. The plant generation which produces gametes from which, after fertilization, the sporophyte develops is called as
   (a) Gametophyte (b) Microspore
   (c) Ligule (d) Corolla
54. Line of cells that differentiates in the early development of the embryo and anchors it to the parental tissue are known as
   (a) Ligule (b) Rachis
   (c) Peduncle (d) Suspensor
55. Arrangement of ovules in the ovary is
   (a) Placentation (b) Aestivation
   (c) Insertion (d) Differentiation
56. The type of ovule in which the ovule is erect with micropyle and funicle lie on the same axis (vertical line)
   (a) Orthotropous (b) Amphitropous
   (c) Anatropous (d) Campylotropous

57. The placenta with a single ovule more or less at the base of the ovary
(a) Axile (b) Parietal
(c) Marginal (d) Basal

58. Microgametogenesis refers to
(a) Production of ovules
(b) Development of embryo from egg cell
(c) Production of sperm nuclei
(d) Endosperm development

59. During embryo development, the apical cell divides by longitudinal wall in
(a) Asterad type (b) Solanad type
(c) Caryophyllad type (d) Chenopodial type

60. Coconut endosperm is an example for
(a) Nuclear endosperm (b) Cellular endosperm
(c) Helobial endosperm (d) Liquid endosperm

61. During seed formation, the stage of development of eight celled embryo sac from the functional megaspore is known as
(a) Microsporogenesis (b) Megagametogenesis
(c) Microgametogenesis (d) Megasporogenesis

62. Campylotropous ovule is common in
(a) Amaranthus (b) Onion
(c) Barley (d) Wheat

63. During the maturation phase of the seeds, the polysomes
(a) Will increase (b) Will disappear
(c) Remain intact (d) Will decrease

64. After physiological maturity, the quality of the seeds will be
(a) Increased (b) Decreased
(c) Same (d) Highly increased

65. During microsporogenesis, the following will be produced
(a) Microspore mother cell (b) Tetrads of haploid microspores
(c) Female gametophyte (d) Zygote

66. Calcium alginate is commonly used for
(a) Increasing seed vigour (b) Synthetic seed production
(c) Increasing seed germination (d) Increasing seed viability

67. At physiological maturity, the moisture content of the seed usually may range from
(a) 20 to 25 per cent (b) 10 to 15 per cent
(c) 50 to 60 per cent (d) 60 to 70 per cent

68. During seed development, phytin is formed in the
(a) Endoplasmic reticulum (b) Golgi apparatus
(c) Cytoplasm (d) Nucleus

69. During seed development, the .... of the seeds will be in the decreasing trend
    (a) Moisture content (b) Seed weight
    (c) Seed coat thickness (d) Seed consistency
70. In synthetic seed production, the callus is initiated and multiplied using
    (a) 2,4-D (b) GA_3
    (c) Amylose (d) ABA
71. Occurrence of adventive embryony is common in
    (a) Tomato (b) Chow-chow
    (c) Neem (d) Citrus
72. The protein accumulated in the embryo during the later stages of seed development
    (a) LEA proteins (b) Prolamines
    (c) Glutelins (d) Zeatin
73. For seed maturation, the process of drying is
    (a) Not essential (b) Not considered
    (c) Sometimes essential (d) Always essential
74. Harvestable maturity occurs
    (a) After physiological maturity (b) Before physiological maturity
    (c) During physiological maturity (d) After harvest
75. The maturity index for ashgourd is
    (a) Appearance of ashy coating on the fruit
    (b) Turning of yellow colour
    (c) Disappearance of ashy coating on the fruit
    (d) Drying of fruit stalk
76. The symptom of physiological maturity for brinjal is
    (a) Hair line crack (b) Dunken layer
    (c) Drying of leaves (d) Yellowing of fruit
77. Seed bearing organ in gymnosperm is
    (a) Megastrobili (b) Microstrobili
    (c) Receptacle (d) Perianth
78. Seed bearing organ in angiosperm is
    (a) Ovary (b) Ovule
    (c) Flower (d) Inflorescence
79. Megastrobili is also known as
    (a) Cones (b) Fruit
    (c) Capsule (d) Pome
80. In gymnosperm, fruit is
    (a) Matured ovule and surrounding freshly post
    (b) Matured ovule
    (c) Matured megastrobili (d) Matured scales

81. An example for naturally occuring auxin
    (a) 2,4-Dichlorophenoxy acetic acid
    (b) 1-Naphthalene acetic acid
    (c) Indole -3-acetic acid (IAA)
    (d) 2-Methoxy-3,6-dichlorobenzoic acid

82. An example for synthetic auxin
    (a) Phenyl acetic acid
    (b) Indole-3-acetic acid
    (c) Indole-3-butyric acid
    (d) 1-Naphthalene acetic acid-NAA

83. Role of auxin
    (a) Cell division
    (b) Cell elongation
    (c) Apical dominance
    (d) Cell differentiation

84. Precursor of auxin is
    (a) Tryptamine
    (b) Tryptophan
    (c) Indole-3-pyruvate
    (d) N-hydroxy tryptamine

85. The physiological maturity of rice reaches on
    (a) 18-21 DAF
    (b) 28-31 DAF
    (c) 38-41 DAF
    (d) 48-51 DAF

86. In pigeonpea, the seed attains maturity on
    (a) 30 days after anthesis
    (b) 40 days after anthesis
    (c) 50 days after anthesis
    (d) 60 days after anthesis

87. Greengram seed reaches physiological maturity on
    (a) 20 days after anthesis
    (b) 25 days after anthesis
    (c) 30 days after anthesis
    (d) 35 days after anthesis

88. Sunflower seed crop can be harvested at
    (a) 30 -35 days after anthesis
    (b) 35-40 days after anthesis
    (c) 40-45 days after anthesis
    (d) 45-50 days after anthesis

89. Bhendi seed crop can be harvested when the fruit exhibits
    (a) Shattering of seeds
    (b) Dehiscence of fruit
    (c) Hair line cracks
    (d) Complete craking of fruits

♦ Answer Keys ♦

Q.No.	Ans.	Q.No.	Ans.	Q.No.	Ans.	Q.No.	Ans.	Q.No.	Ans.	Q.No.	Ans.
1	d	2	b	3	b	4	c	5	d	6	a
7	c	8	b	9	c	10	a	11	b	12	d
13	a	14	b	15	c	16	a	17	b	18	b
19	a	20	b	21	a	22	a	23	c	24	a
25	b	26	b	27	d	28	c	29	b	30	a
31	b	32	a	33	c	34	d	35	c	36	d
37	a	38	b	39	a	40	a	41	b	42	d
43	c	44	b	45	b	46	d	47	c	48	a
49	b	50	d	51	b	52	c	53	a	54	d
55	a	56	a	57	d	58	c	59	a	60	d

61	d	62	a	63	b	64	b	65	a	66	b
67	a	68	a	69	a	70	a	71	d	72	a
73	d	74	a	75	a	76	d	77	a	78	a
79	a	80	a	81	c	82	d	83	b	84	b
85	b	86	c	87	b	88	a	89	c		

1.2.2. Fill in the blanks

1. Mutation in somatic tissues, usually affecting an axillary bud is called as ....
2. Leaf colour and .... are the common indices assessing the seed maturity in the field
3. Harvestable maturity stage has .... moisture content than the physiological maturity stage
4. The optimum stage recommended for harvesting of seed crop is ....
5. .... seeds are polyembryonic in nature
6. After fertilization ovule develop into ....
7. Hard seeds are developed due to high .... during seed development and maturation
8. Physiological maturity index for sorghum is ....
9. Physiological maturity index for sunflower is lemon colour in ....
10. Prior to ...., disorganization of membrane system/organelle occur
11. Physiological maturity is indicated by accumulation of maximum ....
12. Desiccated coated embryo is form of .... seed
13. Black layer formation in maize, sorghum and pearl millet is an indication of......................
14. Variations in seed quality within a single pod/fruit are called ....
15. Seed is .... in nature
16. Physiological maturity in bhendi is ....
17. Seed maturation period in Amaranthus ....
18. Development of well defined nucellar or integumentary tissue in the micropylar region is called as ....
19. The maximum accumulation of .... indicates the physiological maturity of seeds
20. During syngamy, male gamete is fused with ....
21. The maturity index of neem fruit is change of fruit from ....
22. Improper development of seeds are called as ....
23. The fruit morphology is decided by ....

24. Double fertilization occur only in ....
25. In double fertilization, sperms cell fuses with central cell (Polar nuclei) while other with the ....
26. The central cell (polar nuclei) after fuses with sperm nuclei form ....
27. The sperm fuses with egg cell which develops into .... that is diploid
28. The maturity index of jatropha fruit is change of fruit colour from green to ....
29. Physiological maturity index for groundnut is .... of pods
30. The appearance of .... beneath the embryo of wheat seed indicate harvesting stage of seed
31. The pearl millet seed reaches physiological maturity on .... after anthesis
32. In soybean, change of seed from green to .... shows the physiological maturity.
33. The safflower seed can be harvested on .... days after anthesis.
34. In cotton, the seed can be harvested at 52 days after anthesis with ....
35. Stylosanthus seed may be harvested when pod colour changes from green to ....

♦ Answer Keys ♦

Q.No.	Ans.	Q.No.	Ans.	Q.No.	Ans.
1	Bud mutation	2	Fruit colour	3	Less
4	Physiological maturity	5	Citrus	6	Seed
7	Temperature	8	Dunken layer formation	9	Thalamus region
10	Physiological maturity	11	Dry matter production	12	Synthetic
13	Physiological maturity	14	Polymorphism	15	Hygroscopic
16	Fruit cracking	17	30 days	18	Epistase
19	Dry matter	20	Female gamete	21	Green to yellow
22	Peps	23	Fusion of carpels	24	Angiosperms
25	Egg	26	Triploid cell (3 N)	27	Zygote
28	Yellow	29	Ratling	30	Dark pigment
31	30-40 days	32	Yellow	33	45
34	Craking of bolls	35	Light brown		

1.2.3. State True or False

1. Embryo is not in spherical stage during the globular stage of embryo development
2. Hard wall cells transport water and mineral nutrients in a kind of tissue called xylem
3. The xylem transports a sucrose residue
4. Protoplasmic streaming theory was supported by Curtis (1935)
5. Phloem vessel is one way flow
6. Electro osmotic mechanism may account for the translocation of sugars in sieve tubes
7. Decrease in the temperature of soil increases the rate of movement of materials towards root
8. Synthesis of sucrose occur in plastids
9. Synthesis of starch occur in plastids
10. Carbon fixation reaction is also known as dark reaction
11. Site of Calvin cycle is grana
12. Glucose is the product of photosynthesis
13. Light reaction occur in the grana
14. Dry weight accumulation phase synthesis small quantities of protein
15. Thiamin binding protein gene in sesame is found only in seed maturity stage
16. The amount of mRNA is maximum in mature seeds
17. Double fertilization occurs only in angiosperm
18. Most RNA is single stranded
19. ACP is part of a multi enzyme complex
20. During the initial or lag phase of seed development more storage triglyceride is present
21. Mac Millan and Suter first isolated gibberellins from immature seeds of *Phaseolus cocinous*
22. MEP (Methyl Erythritol Phosphate) biosynthetic pathway for production of terpenoid precursor occurs in cytoplasm
23. MVA and MEP pathways are 2 biosynthetic pathways for production of terpenoid precursor
24. Skoog, Saltza and Strong first isolated kinetin
25. Zeatin is the most abundant and widely distributed natural cytokinin in higher plants
26. 6 Benzyl adenine is a naturally occuring cytokinin
27. Cytokinins are synthesized from Adenosine monophosphate and isopentenyl pyrophosphate
28. Isopentenyl diphosphate is precursor of ABA
29. Specific fragrance in roses is due to terpenes

30. Isopentenyl diphospate and its isomer is the activated five carbon building blocks of terpene biosynthesis
31. Phenylalanine is the important intermediate in the biosynthesis of most plant phenolics
32. Malenic acid pathway involved in the biosynthesis of most plant phenolics of higher plants
33. Phenylalanine Ammonia Lynse (PAL) is the most studied enzyme in plant secondary metabolism
34. Phenyl propanoids are important building block of the more complex phenolic compounds
35. Alkaloids are usually synthesized from the amino acids like lysine, tyrosine and tryptophan
36. Systemin is the first and so far only polypeptide hormone discovered in plants
37. Hugo de Vries introduced the term mutation in 1980
38. Mutations are not generally harmful to the organism
39. Pleiotropy occurs when a single gene will not influence multiple phenotypic traits
40. Primary effect of radiation is ionization
41. X-rays can cause DNA and protein damage
42. Alternate pathway of heterosis is influenced by environment
43. In hybrid substances of heterosis, the new product is having under dominance
44. DICER is an enzyme complex not similar to RNase III
45. siRNA (small interfering RNAs) is double stranded RNA molecules
46. ABA content of seeds decrease during development and increase during maturation
47. Hard wall cells transport water and mineral nutrients in a kind of tissue called xylem
48. Transcellular streaming is long distance translocation theory of photo assimilate translocation
49. Electro osmotic mechanism may account for the translocation of sugars in sieve tubes
50. Decrease in the temperature of soil increases the rate of movement of materials towards roots
51. Formation of dunken layer is the physiological maturity symptom of paddy
52. Self incompatibility is the reason for poor seed set in sunflower
53. Existence of seed polymorphism in the same plant is referred as genetic polymorphism
54. Endosperm of a seed is 2n in nature
55. Seed quality is better in harvestable maturity stage than the physiological maturity stage

56. Angiosperms produce only vertical ovules
57. As far as pre-embryo development is concerned, it is one and the same in both dicot and monocot seeds
58. Physiological maturity precedes harvestable maturity
59. Viability and vigour are maximum at physiological maturity
60. Archesporium is the basic cell or starting point towards the development of pollen grain/egg cell
61. 80% of angiosperms produce anatropous ovule
62. Facultative apomixis produces 5 – 10% of sexual seeds
63. Sunflower seeds attain harvestable maturity when the thalamus turns pale yellow
64. Moisture absorption is through micropyle in cotton while, it is surface absorption in greengram
65. During early phase of seed development, protein accumulation will be maximum
66. After double fertilization, the central cells in the embryosac become 3 n endosperm nucleus
67. The outer integument of the ovule develops into the tegmen of seed coat
68. The scar on the ovule made where the funiculus detaches at maturity of seed is known as hilum
69. Seeds are not influenced by memories of their place of birth
70. The endosperm is 2/3 maternal and 1/3 paternal in origin
71. Seed becomes physiologically matured only when it attains harvestable maturity
72. More than one embryo from a single zygote present in this one embryo sac is polyembryony
73 Harvestable maturity precedes physiological maturity in peas
74. Asteroid type of embryo are very common in the family of compositae
75. In synthetic seed production, sodium alginate is used for encapsulation
76. Dunken layer is associated with sorghum
77. The seed formation without sexual fusion is called as pseudogamy
78. The development of dunken layer is an indication of physiological maturity in maize

♦ Answer Keys ♦

Q.No.	Ans.	Q.No.	Ans.	Q.No.	Ans.	Q.No.	Ans.	Q.No.	Ans.	Q.No.	Ans.
1	False	2	True	3	False	4	True	5	False	6	True
7	False	8	False	9	False	10	True	11	False	12	True
13	True	14	False	15	True	16	True	17	True	18	True
19	True	20	False	21	True	22	False	23	True	24	True
25	True	26	False	27	True	28	True	29	True	30	True

Q.No.	Ans.	Q.No.	Ans.	Q.No.	Ans.	Q.No.	Ans.	Q.No.	Ans.	Q.No.	Ans.
31	True	32	False	33	True	34	True	35	True	36	True
37	False	38	False	39	False	40	True	41	True	42	True
43	False	44	False	45	True	46	False	47	True	48	False
49	True	50	False	51	False	52	True	53	False	54	False
55	False	56	False	57	True	58	True	59	True	60	True
61	False	62	True	63	True	64	True	65	False	66	True
67	False	68	True	69	False	70	True	71	False	72	True
73	False	74	True	75	False	76	True	77	False	78	False

1.3. APOMIXIS, PARTHENOCARPY, POLYEMBRYONY AND SYNTHETIC SEED

1.3.1. Choose the correct answer

1. The phenomenon of a single major gene affecting more than one character is known as
   (a) Pathogenecity (b) Inheritance
   (c) Plasticity (d) Pleiotrophy
2. An individual having more than two copies of a single genome or more than one genome is called as
   (a) Monosomic (b) Polyploidy
   (c) Nulliploidy (d) Disomic
3. An embryo arises from unfertilized egg cell is called as
   (a) Sporophytic apomixis (b) Gametophytic apomixis
   (c) Nucelluar embryony (d) Adventitious embryony
4. An embryo formed directly from nucellus or integument tissues is called as
   (a) Sporophytic apomixis (b) Gametophytic apomixis
   (c) Nucelluar embryony (d) Parthenogenesis
5. Parthenogenesis is due to incomplete meiosis of gametophytic cell
   (a) Mitosis (b) Cell division
   (c) Meiosis (d) Nuclear division
6. An example for saprophytic apomixis is
   (a) Apple (b) Citrus
   (c) Sapota (d) Plum
7. An example for vegetative apomixis is
   (a) Agave (b) Opuntia
   (c) Vettiver (d) Colocasia
8. A seedless fruits produced through
   (a) Parthenocarpy (b) Apomixis
   (c) Parthenogenesis (d) Polyembryony

9. An example for parthenocarpy fruit is
   (a) Mango (b) Seedless watermelon
   (c) Sapota (d) Papaya
10. A seedless watermelon plants are grown from
   (a) Cuttings (b) Seeds
   (c) Tissue culture (d) Graftings
11. Artificial parthenocarpy is developed by
   (a) Plant hormones (b) Crossing
   (c) Vegetative propagation (d) Genetic engineering
12. Fruit growth without reproduction is termed
   (a) Drupe (b) Naked seed
   (c) Parthenocarpy (d) Physiological anomaly
13. Seedless banana is
   (a) Parthenocarpic fruit (b) Multiple fruit
   (c) Drupe fruit (d) True fruit
14. The phenomenon wherein, the ovary develops into a fruit without fertilization is called
   (a) Parthenocarpy (b) Apomixis
   (c) Asexual reproduction (d) Sexual reproduction
15. A phenomenon in which two or more embryos developing from a single fertilized egg is called as
   (a) Apomixis (b) Parthenocarpy
   (c) Polyembryony (d) Twins
16. A phenomenon in which two seedlings emerge from one seed is called as
   (a) Apomixis (b) Parthenocarpy
   (c) Polyembryony (d) Twins
17. In tissue culture, undifferentiated mass of cells are called as
   (a) Tissue (b) Cell
   (c) Callus (d) Plantlet
18. In tissue culture, an embryo that develops from an explant tissue is known as
   (a) Direct embryogenesis (b) Indirect embryogenesis
   (c) Somatic embryogenesis (d) Synthetic seed
19. The use of somatic embryogenesis is
   (a) Clonal propagation (b) Micro propagation
   (c) Macro propagation (d) Mass propagation
20. Who has coined the word synthetic seed?
   (a) Toshio Murashige (b) Chin and Chin
   (c) Skoog (d) Roberts
21. In which year, the term synthetic seed has been coined?
   (a) 1950 (b) 1960
   (c) 1970 (d) 1980

22. The hydrated synthetic seed is produced by
    (a) Sodium alginate (b) Potassium alginate
    (c) Polyoxyethylene glycol (d) Carragenan
23. Which one of the following chemical is used for coating desiccated synthetic seed?
    (a) Polyoxyethylene glycol (b) Potassium nitrate
    (c) Sodium nitrate (d) Sodium alginate
24. Synthetic seeds are
    (a) Artificially synthesized seeds
    (b) Somatic embryos encapsulated in suitable matrix
    (c) Seeds of plants modified genetically
    (d) Artificially synthesized plants
25. Somatic embryoids are
    (a) Identical zygotic embryos and without seed coats
    (b) Identical with zygotic embryos and with seed coats
    (c) Non-identical with zygotic embryos and without seed coats
    (d) Non-identical with zygotic embryos and with seed coats
26. The preserved embryos are termed as
    (a) Synthetic seeds (b) Semi- synthetic seeds
    (c) Natural seeds (d) Fermented seeds
27. Encapsulation is necessary to produce and protect synthetic seeds. The encapsulation is carried out by various types of hydrogels, which are
    (a) Soluble in water (b) Soluble in organic solvents
    (c) Insoluble in water (d) Insoluble in organic solvents
28. Which of the following is not true about synthetic seeds?
    (a) Can be stored for a year without the loss of valuables
    (b) Easy to handle
    (c) Can be directly sown in the soil like natural seeds
    (d) Need hardening in the green house
29. Synthetic seed is produced by encapsulating somatic embryo with
    (a) Sodium chloride (b) Sodium alginate
    (c) Sodium acetate (d) Sodium nitrate

♦ Answer Keys ♦

Q.No.	Ans.	Q.No.	Ans.	Q.No.	Ans.	Q.No.	Ans.	Q.No.	Ans.	Q.No.	Ans.
1	d	2	b	3	b	4	a	5	c	6	b
7	a	8	a	9	b	10	b	11	a	12	c
13	a	14	a	15	c	16	c	17	c	18	a
19	d	20	a	21	c	22	c	23	d	24	b
25	a	26	a	27	a	28	d	29	b		

1.3.2. Fill in the blanks

1. The mechanism underlying the polyembryony is ....
2. Two seedlings emerge from one seed is called ....
3. .... is an natural or artificially induced production of fruit without fertilization of ovules
4. Plants that don't require pollination or other stimulation to produce parthenocarpic fruits have ....
5. An example for vegetative parthenocarpy is ....
6. The development of parthenocarpic fruit by application of plant hormone is termed as ....
7. Parthenogenesis means development of an embryo from ....
8. .... are formed from plant cell that are not involved in the development of embryos
9. .... is defined as the artificial encapsulation of somatic embryo

♦ Answer Keys ♦

Q.No.	Ans.	Q.No.	Ans.
1	Programme cell death	2	Polyembryony
3	Parthenocarpy	4	Vegetative parthenocarpy
5	Seedless cucumber	6	Artificial Parthenocarpy
7	Unfertilized egg cells	8	Somatic embryos
9	Synthetic seed		

1.3.3. State True or False

1. Apomixis is a type of reproduction that does involve fertilization
2. Parthenocarpy occasionally occurs as mutation in nature
3. A seedless fruit can be produced through apomixis
4. Parthenogenesis is component process of apomixis

♦ Answer Keys ♦

Q.No.	Ans.	Q.No.	Ans.	Q.No.	Ans.	Q.No.	Ans.
1	False	2	True	3	False	4	True

1.4. CELL BIOLOGY

1.4.1. Choose the correct answer

1. Non living secretion of cell membrane
   (a) Cell wall (b) Endoplasmic reticulum
   (c) Cytoplasm (d) Nuclear membrane
2. Internal component of cell is
   (a) Cytoplasm (b) Protoplasm
   (c) Nucleus (d) Nucleolus

3. Cytoplasm acts as a
(a) Boundary (b) Shock absorber
(c) Storage organ (d) pH medium

4. The cell organelle which secrets and stores chemicals
(a) Ribosome (b) Plastids
(c) Golgi apparatus (d) Endoplasmic reticulum

5. The cell organelle which transports chemicals between and within cells
(a) Ribosome (b) Cell wall
(c) Cell membrane (d) Endoplasmic reticulum

6. Ribosomes are called as
(a) Site of protein synthesis (b) Digestive plants
(c) Shock absorber (d) Site of hydrolysis

7. Site of photosynthesis
(a) Plastids (b) Grana
(c) Chloroplast (d) Thylakoids

8. Inner membrane of mitochondria unfolded to form
(a) Cristae (b) Cisternae
(c) Grana (d) Vacuoles

9. Site of protein synthesis
(a) Ribsome (b) Endoplasmic reticulum
(c) Chloroplast (d) Lysosomes

10. Golgi apparatus consists a stacks of flattened sacs of unit membrane called
(a) Cristae (b) Cisternae
(c) Grana (d) Vacuoles

11. Net ATP production in glycolysis
(a) 4 (b) 2
(c) 32 (d) 36

12. The water is taken through the cell wall
(a) Symplastically (b) Apoplastically
(c) Transcelluar (d) Stachyose

13. Protoplasmic streaming theory was proposed by
(a) De Vries (1885) (b) Thaine (1964)
(c) Munch (1926) (d) Fenson (1971)

14. Specific enzyme found in the stroma of the chloroplast is
(a) Catalase (b) Peroxidase
(c) RUDP carboxylase (d) Pyruvate

15. The process of breakdown of organic molecules to harvest energy in cellular respiration is called
(a) Anabolism (b) Respiration
(c) Catabolism (d) Diffusion

16. Chlorophyll is found in
(a) Eukaryotes (b) Prokaryotes
(c) Virus (d) Bacteria

17. The golgi apparatus was discovered by
(a) Camila golgi (b) Juliano
(c) Cruz (d) Burr

18. How many type of RNA present in protein synthesis
(a) 5 (b) 2
(c) 4 (d) 3

19. Which golgi is closest to the ER
(a) Medial golgi (b) Cis golgi
(c) Golgi (d) S. Golgi

20. Which enzyme is soluble in water
(a) Globlin (b) Prolamin
(c) Albumin (d) Glutelin

21. The chemical detected in thiamine binding protein only in albumen cells
(a) Immano gold (b) Immano silver
(c) Gammano platinum (d) Immano gold-silver

22. Change in number of chromosomes per cell is called as
(a) Chromosomal variation (b) Chromosomal aberration
(c) Gene mutation (d) Exogenous mutation

23. The cell membrane contains .... in addition to lipid and cellulose
(a) Carbohydrate (b) Protein
(c) Fat (d) Sucrose

24. The commonly used reporter gene in plant transformation
(a) GUS (b) Kanamycin
(c) Hygromycin (d) DREB

25. Plants having two sets of chromosomes are known as
(a) Dioecious (d) Diploid
(c) Dichrom (d) Disome

♦ Answer Keys ♦

Q.No.	Ans.	Q.No.	Ans.	Q.No.	Ans.	Q.No.	Ans.	Q.No.	Ans.	Q.No.	Ans.
1	a	2	b	3	b	4	c	5	d	6	a
7	a	8	a	9	a	10	b	11	b	12	b
13	a	14	c	15	c	16	a	17	a	18	d
19	b	20	c	21	d	22	a	23	b	24	a
25	d										

1.4.2. Fill in the blanks

1. Cell theory was first developed by ....
2. Plant cell differs from animal cell by the presence of cell wall, cytoxysome and ....
3. Cell membrane consists of .... lipid and cellulose
4. EST stands for ....
5. .... reveals gene sequence and structure of DNA elements
6. In a cell, site of protein synthesis is ....

♦ Answer Keys ♦

Q.No.	Ans.	Q.No.	Ans.
1	Sehwann and Schleiden	2	Plastids
3	Protein	4	Essential Sequence Tags
5	Genome	6	Ribosomes

1.4.3. State True or False

1. Cell theory was developed in the year 1835
2. Animal cells contain more vacuoles
3. Glyoxysome present in animal cell
4. Cell membrane is selectively permeable
5. Cell wall is selectively permeable
6. Chloroplast is a type of plastid
7. Site of cellular respiration is peroxisome
8. Vacuole consists of double layer unit membrane
9. Centriole forms spindle fibre during cell division
10. Site of photorespiration is peroxisome
11. Glycolysis occurs in cytoplasm
12. Glycolysis occurs only in cellular respiration
13. Glycolysis occurs in aerobic and anaerobic respiration
14. Ribosome contains (30s) small sub units and (60s) large sub units

♦ Answer Keys ♦

Q.No.	Ans.	Q.No.	Ans.	Q.No.	Ans.	Q.No.	Ans.	Q.No.	Ans.
1	False	2	True	3	False	4	True	5	False
6	True	7	False	8	False	9	True	10	True
11	True	12	False	13	True	14	False		

1.5. SEED STRUCTURE

1.5.1. Choose the correct answer

1. The scar on the ovule made when the funicular detaches at maturity is known as
   (a) Raphe (b) Chalaza
   (c) Hilum (d) Funicle
2. The area between chalaza and the hilum of many species is known as
   (a) Chalaza (b) Hilum
   (c) Funicle (d) Raphe
3. Starch is stored in
   (a) Nucleus (b) Mitochondria
   (c) Golgi body (d) Amyloplast and Cotyledon
4. The percentage of oryzenin in rice is
   (a) 46 (b) 55
   (c) 80 (d) 56
5. Spherosomes are
   (a) Protein bodies (b) Fat bodies
   (c) Enzymes (d) Hormones
6. Myristic acid is present in
   (a) Milk (b) Coconut
   (c) Cocoa butter (d) Safflower
7. Saturated fat is high in
   (a) Pumpkin (b) Coconut
   (c) Almond (d) Sesame
8. Erucic acid is present in
   (a) Castor bean (b) Rape
   (c) Sorghum (d) Cotton
9. Ricinoleic acid is present in
   (a) Rape (b) Tomato
   (c) Sorghum (d) Castor bean
10. Olive oils are the richest in
    (a) Diene (b) Triene
    (c) Monoene (d) Tetraene
11. Red, orange and yellow carotenoids belong to which group of terpene compounds
    (a) Monoterpene (10 carbons) (b) Diterpene (20 carbons)
    (c) Triterpene (30 carbons) (d) Tetraterpene (40 carbon)
12. Which part of the seed mainly contains tannin
    (a) Embryo (b) Endosperm
    (c) Seed coat (d) Cotyledons

13. The most widespread group of flavanoid pigment is
(a) Anthocyanin (b) Xanthophyll
(c) Yellow carotenoid (d) Orange carotenoid

14. The starch is converted into simple sugars by α-amylase and the sucrose is transported to which part of the seed
(a) Aleurone layer (b) Coleoptile
(c) Coleorhiza (d) Scutellum

15. A crop that is commonly used for seed research is
(a) Lettuce (b) Arabidopsis
(c) Rice (d) Groundnut

16. The storage structure in seed is
(a) Embryo (b) Endosperm
(c) Seed coat (d) Aleurone layer

17. Caryopsis is the seed type of
(a) Paddy (b) Sunflower
(c) Groundnut (d) Carrot

18. The perisperm is present in
(a) Coffee (b) Maize
(c) Groundnut (d) Cabbage

19. The protein present in soybean is
(a) Glycine (b) Arachin
(c) Zein (d) Zeatin

20. Botanically seed is a
(a) Matured ovule (b) Matured seed coat
(c) Matured endosperm (d) Matured ovary

21. Awn is present in
(a) Rice (b) Maize
(c) Tomato (d) Castor

22. The storage structure of coffee seed is
(a) Endosperm (b) Perisperm
(c) Embryo (d) Aleurone layer

23. The shape of cotyledon in cotton seeds is
(a) Coiled (b) Folded
(c) Curved (d) Twisted

24. Transduction process is the production of
(a) Secondary messengers (b) Lipids
(c) mRNA (d) Genes

25. Response to signal transduction takes place in
(a) Cytoplasm (b) Plasma membrane
(c) Nucleus (d) Perisome

26. Is covalently bounded to a non protein pigments
(a) Linage (b) Cytochrome
(c) Photoreceptor (d) G-protein

27. Protein linkage is molecule
(a) Response (b) Release
(c) Receptor (d) Effector

28. Aleurone layer cells are induced to synthesize and secrete into endosperm
(a) Peroxidase (b) α-amylase
(c) Lipase (d) Kinase

29. Involve in signal transduction pathway
(a) Chloroplast (b) Perisome
(c) Phytochrome (d) Ribosome

30. Seed is defined as
(a) Matured ovule (b) Embryo enclosed in a seed coat
(c) Matured endosperm (d) Matured ovary

31. Give example for recalcitrant seed
(a) Guava (b) Teak
(c) Papaya (d) Coffee

32. During the initial phase of tea seed development, the seed volume occupied by liquid endosperm is
(a) 23% (b) 52%
(c) 98% (d) 100%

33. The beautiful, brilliant, scarlet net like membrane in nutmeg, known as mace is nothing but
(a) Raphae (b) Aril
(c) Strophiole (d) Caruncle

34. Arecanut selected for seed purpose should float in water
(a) Vertically with calyx end upwards
(b) Horizontally
(c) Vertically with calyx end downwards
(d) Diagonally

35. Elephant bean of coffee will contain
(a) One seed per locule
(b) More than one seed per locule
(c) 1-2 seeds per locule
(d) 3 seeds per locule

36. The unit of handling is actually a fruit but for conventional reasons it is termed as seed in
(a) *Hardwickia binata* (b) *Prosopis julioflora*
(c) *Tectona grandis* (d) *Casuarina equisetifolia*

37. The pods of this tree mature during January - March and majority of the pods keep hanging on the tree for a long duration
(a) *Acacia nilotica* (b) *Tectona grandis*
(c) *Santalum album* (d) *Albizia lebeck*

38. The seeds of this species are called samaras
(a) *Ailanthus excela* (b) *Azadirachta indica*
(c) *Terminalia chebula* (d) *Tectona grandis*

39. Seeds of *Azadirachta indica* can be dried in shade, but cannot be stored for long. The seed belongs to
(a) Orthodox (b) Recalcitrant
(c) Intermediate (d) Difficult recalcitrant

40. The fruit of this species is an indehiscent pod
(a) *Prosopis juliflora* (b) *Albizia lebeck*
(c) Pongamia *pinnata* (d) Acacia *nilotica*

41. The fruit is dry and dehiscent in nature
(a) *Santalum album* (b) *Azadirachta indica*
(c) *Mangifera indica* (d) *Acacia nilotica*

42. Cocoa seeds being recalcitrant require high seed moisture of
(a) 33-35 per cent (b) 55-65 per cent
(c) 40-50 per cent (d) 70-80 per cent

43. Halum cutting in potato is adopted to avoid
(a) Viral disease (b) Bacterial disease
(c) Genetic contamination (d) Physical contamination

44. Multiple carpels fused into a single structure known as
(a) Capitesta stigma (b) Syncarpous style
(c) Hypogynous (d) Gamosepalous

45. Miniature plant of the seed is
(a) Plumule (b) Radicle
(c) Cotyledons (d) Embryo

46. One of the following is not recalcitrant seed
(a) Jack (b) Jamum
(c) Amla (d) Citrus

47. Monocot seed
(a) Redgram (b) Cotton
(c) Sunflower (d) Onion

48. Appendage in paddy seed is
(a) Aril (b) Hairs
(c) Hilum (d) Awn

49. Food material is supplied to embryo in monocot through
(a) Scutellum (b) Plumule
(c) Radicle (d) Coleoptile

50. Endospermic dicot seed
(a) Chickpea (b) Bean
(c) Groundnut (d) Fenugreek

51. Ragi seed is called
(a) Caryopsis (b) Achene
(c) Utricle (d) Nut

52. The cover enclosing the plumule is
(a) Coleorhiza (b) Scutellum
(c) Coleoptile (d) Aleurone layer

53. Cell grows and become larger in which phase of cell cycle activity
(a) G0 (b) G1
(c) S (d) M

54. DNA synthesis is associated with
(a) Accumulation of ß-tubulin (b) Degradation of ß-tubulin
(c) Accumulation of α-tubulin (d) Degradation of α-tubulin

55. RNA and protein synthesis stimulated in stage of
(a) Pre-germination (b) Germination
(c) Post germination (d) Both pre and post germination

56. H_2O permeability is greater in the area of
(a) Microphyle (b) Hilum
(c) Xylem (d) Phloem

57. Which variety of soybean has the highest protein content responsible for imbibition
(a) SJ 5 (b) CM 2
(c) CM 60 (d) CO 1

58. The swelling force on the cell wall when H_2O enters is called
(a) Osmotic presence (b) Turgor pressure
(c) Diffusion (d) Imbibition

59. Auxin comes under which hormone group
(a) Growth hormones (b) Stress hormones
(c) Hormones (d) Shock hormones

60. Which one of the following hormone used for fruit ripening
(a) Auxin (b) Cytokinin
(c) ABA (d) Ethylene

61. Which one of the following hormone used for bud activation, cell division, embryo development
(a) GAs (b) CKs
(c) ABA (d) SAs

62. Building block of protein is
(a) Fatty acid (b) Amino acid
(c) Glucose (d) Phosphorous

63. Uracil is present in
(a) RNA (b) DNA
(c) Nucleus (d) Cytoplasm

64. Nucleotide without phosphate group is called as
(a) RNA (b) DNA
(c) Nitrogenous base (d) Nucleoside

65. Which of the following is not a nitrogenous base
(a) Adenine (b) Guanine
(c) Uracil (d) Lysine

66. Site of protein synthesis is
(a) Rough endoplasmic reticulum (b) Nucleus
(c) Cytoplasm (d) Ribosome

67. Transcription take place in
(a) Cytoplasm (b) Nucleus
(c) Ribosome (d) Endoplasmic reticulum

68. Enzyme involved in transcription
(a) RNA polymerase (b) DNA polymerase
(c) Kinase (d) Endonuclease

69. The major protein content of wheat seed is
(a) Albumin (b) Globulin
(c) Glutelin (d) Prolamin

70. Pyrimidines are
(a) Cytosine (b) Thymine
(c) Adenine (d) Cytosine and thymine

71. Hydroxyl groups on sugar represent
(a) Alcohol (b) Ethylene
(c) Starch (d) Fat

72. In cell with nucleus (Eukaryotes), the cell cycle consists of how many phases
(a) Two (b) Three
(c) Four (d) One

73. The phase which is collectively known as inter phase
(a) G1 phase (b) S phase
(c) G2 phase (d) M phase

74. For synthesis of new strand, a short fragment of DNA or RNA called as
(a) Primers (b) Gene
(c) Polymerase (d) Marker

75. How many possible replication for DNA is available
(a) 3 (b) 4
(c) 6 (d) 2

76. Which enzyme is specially used in joining okazaki fragment synthesized in DNA synthesis
(a) Ligase (b) Polymerase
(c) Helicase (d) Synthatase

77. Fats are degraded by
(a) Amylase (b) Protease
(c) Lipase (d) Endonuclease

78. Product of oxidation of fatty acids is
(a) Acetyl COA (b) Mono glycerides
(c) Glycerol (d) Triglycerides

79. Oxidation takes place in
(a) Mitochondria (b) Cytoplasm
(c) Glyoxysome (d) Golgi apparatus

80. Who determined the location of key glyoxylate cycle enzymes ie citrate lyase and malate synthetase
(a) Breiden Batch and Beevers (b) El-Sebaiy and El-Mahydy
(c) Wanasundara (d) Mayber

81. Key enzymes involved in glyoxylate of lipid breakdown during germination are
(a) Phospo fructokinase
(b) Malate synthetase and isocitrate lyase
(c) Pyruvate kinase
(d) Pyruvate carboxylase

82. Glyoxylate cycle of lipid breakdown takes place in
(a) Glyoxysome (b) Mitochondria
(c) Cytoplasm (d) Ribosome

83. Final product of glyoxylate cycle is
(a) Malate (b) Citrate
(c) Glyoxylate (d) Succinate

84. Gluconeogenesis of lipid break down takes place in
(a) Cytoplasm (b) Mitochondria
(c) Glyoxysome (d) Polysomes

85. Which is the isomer (s) of glucose
(a) Fructose (b) Galactose
(c) Fructose and galactose (d) Sucrose

86. Which one of the following is a disacharide
(a) Fructose (b) Raffinose
(c) Starch (d) Sucrose

87. The major form of carbohydrate stored in seeds is
(a) Sucrose (b) Starch
(c) Cellulose (d) Glucose

88. About polysacharides, which of the following fact is not correct?
(a) Reducing (b) Water insoluble
(c) Monosacharides joined by glucosidic bond
(d) Hydrolyzed by enzymes/acids

89. The major storage tissue of coffee
(a) Embryo (b) Cotyledon
(c) Perisperm (d) Scutellum

90. What is the major storage organ in plant
(a) Endosperm (b) Cotyledon
(c) Aleurone layer (d) Perisperm

91. In rice, the predominant starch catabolic pathway is
(a) Hydrolytic (b) Phosphorylytic
(c) α-oxidation (d) β-oxidation

92. What is the major form of carbohydrate involved in translocation in plants
(a) Starch (b) Glucose
(c) Hemi-cellulose (d) Sucrose

93. Paddy seeds lack in
(a) Zein (b) Albumin
(c) Globulin (d) Prolamin

94. Small integument out growth is called as
(a) Caruncle (b) Aril
(c) Hairs (d) Elaeiosome

95. Anatomically elaeiosome are
(a) Appendages (b) Seed coat
(c) Third cotyledon (d) Embryo

96. A one seeded achene and pappus indehiscent fruit type is common feature in
(a) Leguminaceae (b) Compositae
(c) Polygonaceae (d) Palmae

97. Seeds with little variability in size and shape are called as
(a) Euryopermous (b) Cleistogamous
(c) Apomixis (d) Stenospermous

98. In gymnospermous seeds, the food is stored in
(a) Cotyledon (b) Seed coat
(c) Embryo (d) Megagametophyte

99. Formation of fruits under the soil through gynophores is called
(a) Geocarpic (b) Hysterogeocarpic
(c) Hypocarpic (d) Epicarpic

100. Orchid seeds are devoid of
(a) Embryo (b) Seed coat
(c) Cotyledon (d) Endosperm

101. Perisperm is
(a) Haploid (b) Diploid
(c) Tetraploid (d) Hexaploid

102. The coloured flesh mass present on the outside of the seed in nutmeg is called
(a) Aril (b) Stigma
(c) Anther (d) Endosperm

103. A slender bristle shape, elongated appendage present in wheat, rye and rice is called as
(a) Spike (b) Auricles
(c) Awns (d) Spikelet

104. Fruit of castor is called as
(a) Grain (b) Capsules
(c) Bulb (d) Achene

105. The protective sheath that covers the young shoot of the embryo in plants of the family graminaceae is called as
(a) Endosperm (b) Coleoptile
(c) Protoplast (d) Coleorhiza

106. Radicle is enclosed in a protective cover called as
(a) Endosperm (b) Coleoptile
(c) Protoplast (d) Coleorhiza

107. Inflorescence of cauliflower is called
(a) Spike (b) Curd
(c) Bulb (d) Achene

108. A miniature plant which consists of plumule, radicle and cotyledon is called as
(a) Embryo (b) Aleurone layer
(c) Endosperm (d) Perisperm

109. The growing point of the embryo which gives rise to the shoot or the above ground part of the plant can be termed as
(a) Epigeal (b) Epicotyl
(c) Epistasis (d) Epigyny

110. Plants in which the cotyledons appear above the surface of the soil is called as
(a) Epigeal (b) Epicotyl
(c) Epistasis (d) Epigyny

111. Each ovule is attached to the placenta by a distinct stalk known as
(a) Embryo (b) Radicle
(c) Hilum (d) Plumule

112. The fruit derived from one carpel and splitting along one side is called as
(a) Spike (b) Achene
(c) Caryopsis (d) Folicle

113. The correct sequence of cell division phases on duration basis is
(a) Prophase - metaphase - anaphase - telophase
(b) Metaphase - prophase - telophase - anaphase
(c) Prophase - metaphase - telophase - anaphase
(d) Anaphase - prophase - metaphase – telophase

114. The part of the embryo or seedling below the cotyledonary node and above the root is called
(a) Radicle (b) Hypocotyl
(c) Hilum (d) Plumule

115. The point where the integuments meet at the nucellar apex has been known as
(a) Radicle (b) Hypocotyl
(c) Hilum (d) Microphyle

116. Per cent of oil content is highest in
(a) Sesame (b) Groundnut
(c) Castor (d) Soybean

117. The outermost wall of the ovary is called as
(a) Mesocarp (b) Metacarp
(c) Epicarp (d) Pericarp

118. The rudimentary root of the seed or seedling that forms the primary root of the young plant is known as
(a) Rachis (b) Recerne
(c) Rachilla (d) Radicle

119. A layer of nutritional tissue of a diploid maternal origin arising from the nucleus and often surrounding the endosperm is called as
(a) Mesoperm (b) Metasperm
(c) Episperm (d) Perisperm

120. The portion of the seed embryo, which produces the aerial portion, which develops into a plant is called as
(a) Embryo (b) Radicle
(c) Endosperm (d) Plumule

121. Dehiscent fruit developed from one carpel and split along the two sides is called
(a) Spike (b) Pod
(c) Achene (d) Caryopsis

122. Highest protein content is found in
(a) Rajmash (b) Mung bean
(c) Gram (d) Soybean

123. Wheat protein is called as
(a) Zein (b) Glutelin
(c) Ricin (d) Zeatin

124. Crop which has highest unsaturated fatty acid content is
(a) Castor (b) Coconut
(c) Linseed (d) Niger

125. Vitamin C is also known as
(a) Folic acid (b) Ascorbic acid
(c) Riboflavin (d) Tocopherol

126. The fruit derived from a compound pistil in which the one seeded carpels are separated from one another at maturity is called as
(a) Heliocarp (b) Schizocarp
(c) Epicarp (d) Plumule

127. The outer layer of the seed coat which is smooth and rough is known as
(a) Lemma (b) Palea
(c) Testa (d) Glume

128. Collective term for the nutritional tissue between the embryo and the seed coat, including perisperm and endosperm is called as
(a) Albumin (b) Protein
(c) Globulin (d) Glutenin

129. Outgrowth on the seed from other parts than the funiculus is termed as
(a) Arilloid (b) Axil
(c) Aril (d) Awn

130. Arilloid (type of outgrowth on the seed) occurring near the micropyle is named as
(a) Caruncle (b) Capsule
(c) Caryopsis (d) Catkin

131. Spike like inflorescence of unisexual flowers found in some angiosperms e.g. *Alnus* and *Betula* is termed as
(a) Caruncle (b) Capsule
(c) Caryopsis (d) Catkin

132. Region in the ovule opposite to micropyle where the integuments fuse with funiculus is called as
(a) Caruncle (b) Ligule
(c) Chalaza (d) Endosperm

133. The percentage of amylose present in a starch grain is nearly
(a) 1 – 2% (b) 40 – 45%
(c) 20 – 25% (d) 38 – 42%

134. The seed bearing structure of conifers consisting of a central stem, woody or fleshy scales, bracts and seeds is termed as
(a) Cone (b) Bract
(c) Corolla (d) Caryopsis

135. The stalk of an ovule or seed, attaching it to the ovary plancenta is called as

(a) Ligule (b) Peduncle
(c) Funicle (d) Auricle

136. The one or two layers (often fused) of tissue covering and surrounding the nucleus of an ovule is called as

(a) Integument (b) Radicle
(c) Capsule (d) Epicarp

137. The chemical changes within a cell that provide the energy required by a plant or animal is called as

(a) Catabolism (b) Respiration
(c) Metabolism (d) Anabolism

138. Minute opening in the integuments of an ovule through which the pollen grain or pollen tube possess to reach the embryo sac is called as

(a) Micropyle (b) Auricle
(c) Integument (d) Pericarp

139. Ridge formed on the seed coat if the funiculus is fused with the integument in part of its length in anatropous or campylotropous ovules is known as

(a) Raphae (b) Auricle
(c) Ligule (d) Scales

140. Common term for the hard endocarp of a drupe containing the seed is called as

(a) Achene (b) Caryopsis
(c) Pome (d) Stone

141. A bulky, short, vertical underground stem which stores food is generally called as

(a) Corm (b) Achene
(c) Bolt (d) Caryopsis

142. The area between chalaza and hilum in the ovule is known as

(a) Raphae (b) Nucleus
(c) Micropyle (d) Funicle

143. Ex-albuminous seed refers to

(a) Seed completely filled with endosperm
(b) Seed with little or no endosperm
(c) Seed with well developd chalazosperm
(d) Seed with endosperm and perisperm

144. Anemochory is seed dispersal by

(a) Wind (b) Water
(c) Animal (d) Bird

145. Seeds collected and stored by animals for nutrition is called
(a) Entozoochory (b) Epizoochory
(c) Synzoochory (d) Ornithochory

146. Longest seed viability period of lotus seeds in soil seed bank is
(a) 1 year (b) 10 years
(c) 100 years (d) 1000 years

147. Ants are attracted by seeds having
(a) Big size (b) Small size
(c) Elaeisome (d) Colour

148. Transient seed banks have viability of
(a) 1 year (b) 10 years
(c) 100 years (d) 1000 years

149. Parachute is mechanism for seed dispersal by
(a) Wind (b) Water
(c) Animals (d) Birds

150. Largest seed in the plant kingdom
(a) *Linum usitatissimum* (b) *Lactuca sativa*
(c) *Lodoicea maldivica* (d) *Lactuca scariola*

151. Single seeded fruit where pericarp can be removed from the mature seed is called as
(a) Caryopsis (b) Schizocarp
(c) Achene (d) Nut

152. An example for monocotyledonous exalbuminous seed
(a) Rice (b) Onion
(c) Sorghum (d) Maize

153. Schizocarp of fruit is found in
(a) Castor (b) Coriander
(c) Cabbage (d) Cowpea

154. The outgrowth of integument in castor is known as
(a) Hilum (b) Caruncle
(c) Raphae (d) Strophiole

155. Presence of mericarp is common feature of the family
(a) Compositae (b) Malvaceae
(c) Solanaceae (d) Umbelliferaceae

156. Niger is a
(a) Cereal (b) Pulses
(c) Oilseed (d) Fibre

157. A seed having little or no endosperm but, a well developed perisperm
(a) Albuminous seed (b) Exalbuminous seed
(c) Endospermous seed (d) Non endospermous seed

158. A seed which has a curved embryo
(a) Paddy (b) Wheat
(c) Tomato (d) Moringa

159. The single cotyledon of the seeds of poaceae family is reduced to
(a) Scutellum (b) Seed coat
(c) Endocarp (c) Mesocarp

160. A specialized thin tissue which covers the radicle is known as
(a) Root cap (b) Coleorhiza
(c) Coleoptile (d) Nucleus

161. The major storage protein of wheat and maize
(a) Prolamin (b) Albumin
(c) Viclin (d) Legumin

162. A pore present in the seed coat through which water enters into the seed is known as
(a) Gynaphore (b) Micropyle
(c) Chalaza (d) Synergids

163. The storage organ in gymnosperm is
(a) Archegonia (b) Pollen scale
(c) Megagametophyte (d) Cotyledons

164. Glutelins are soluble in
(a) Water (b) Alcohols
(c) Dilute acids and alkali solution (d) Methane

165. Presence of folded cotyledons are common in the seeds of the family of
(a) Leguminaceae (b) Malvaceae
(c) Euphorbiaceae (d) Liliaceae

166. A chemical needed for testing the viability of pollen is
(a) $CaCl_2$ (b) $CaCo_3$
(c) Iodine (d) Potassium

167. Sarcotesta is associated with the seed of
(a) Ber (b) Papaya
(c) Pomegranate (d) Guava

168. The major storage of food reserve of coffee is
(a) Endosperm (b) Cotyledon
(c) Perisperm (d) Scutellum

169. In castor, the major storage organ is
(a) Cotyledon (b) Endosperm
(c) Perisperm (d) Aleurone layer

170. The translocated form of sugar from the site of photosynthesis to the developing seed is
(a) Sucrose (b) Glucose
(c) Fructose (d) Mannitol

171. Paddy fruit is called as
    (a) Caryopsis (b) Folicle
    (c) Panicle (d) Ligule
172. A seed which contains both monogerm and multigerm are
    (a) Cauliflower (b) Tomato
    (c) Cabbage (d) Beet root
173. Siliqua is associated with
    (a) Thenai (b) Wheat
    (c) Cabbage (d) Palak
174. Liquid endosperm is the specialized character of
    (a) Oil palm (b) Sugarcane
    (c) Date palm (d) Coconut
175. Examples for tiny seed
    (a) Orchids (b) Eucalyptus
    (c) Tobacco (d) Amaranthus
176. The biggest seed is
    (a) *Lodoicea maldivica* (b) *Cocas nucifera*
    (c) Tongue oil tree (d) *Lagerstroemia* spp
177. The botanical name for double coconut which weighing more than a kilo is
    (a) *Lodoicea maldivica* (b) *Cocas nucifera*
    (c) *Lantana camera* (d) *Lolium spp*
178. In strict botanical sense, fruit is
    (a) Matured ovary (b) Fertilized ovule
    (c) Ripened ovule (d) Matured anther
179. The fruit produced by species whose ovary is made up of several carpel's but may separate and form separate fruits
    (a) Aggregate fruit (b) Pome
    (c) Pod (d) Capsule
180. The fruit of neem is known as
    (a) Capsule (b) Berry
    (c) Pome (d) Drupe

♦ Answer Keys ♦

Q.No.	Ans.	Q.No.	Ans.	Q.No.	Ans.	Q.No.	Ans.	Q.No.	Ans.	Q.No.	Ans.
1	c	2	d	3	d	4	c	5	b	6	a
7	b	8	b	9	d	10	c	11	d	12	c
13	a	14	d	15	b	16	b	17	a	18	a
19	a	20	a	21	a	22	b	23	b	24	c
25	c	26	c	27	d	28	b	29	c	30	a
31	d	32	c	33	b	34	a	35	a	36	c
37	d	38	a	39	c	40	a	41	d	42	a

Q.No.	Ans.	Q.No.	Ans.	Q.No.	Ans.	Q.No.	Ans.	Q.No.	Ans.	Q.No.	Ans.
43	a	44	b	45	d	46	c	47	d	48	d
49	a	50	d	51	c	52	c	53	b	54	a
55	a	56	a	57	c	58	b	59	a	60	d
61	b	62	b	63	a	64	d	65	c	66	d
67	c	68	a	69	c	70	d	71	a	72	c
73	c	74	a	75	a	76	a	77	c	78	a
79	c	80	a	81	b	82	a	83	d	84	a
85	c	86	d	87	b	88	a	89	c	90	b
91	a	92	d	93	d	94	c	95	c	96	b
97	a	98	d	99	b	100	d	101	b	102	a
103	c	104	b	105	b	106	d	107	b	108	a
109	b	110	a	111	c	112	d	113	c	114	b
115	d	116	c	117	d	118	d	119	d	120	d
121	b	122	d	123	b	124	a	125	b	126	b
127	c	128	a	129	a	130	a	131	d	132	a
133	c	134	a	135	c	136	a	137	c	138	a
139	a	140	d	141	a	142	a	143	b	144	a
145	c	146	d	147	c	148	a	149	a	150	c
151	a	152	b	153	b	154	b	155	d	156	c
157	a	158	c	159	a	160	b	161	a	162	b
163	c	164	c	165	b	166	c	167	b	168	c
169	b	170	a	171	a	172	d	173	c	174	d
175	a	176	a	177	a	178	a	179	a	180	d

1.5.2. Fill in the blanks

1. Matured ovule is ....
2. The first leaf of a plant ....
3. Single reduced cotyledon of monocot ....
4. .... is the carrier of new technologies
5. While seed development, ... changed into seed coat
6. Caruncle is the appendages of .... seeds
7. Botanically seed is a ....
8. .... seed has the caruncle at the micropylar end
9. .... in the cell has its own DNA
10. The sunflower seed is a fruit called ....
11. The genetic makeup of an endosperm is ....
12. In a monocot, food reserves are stored in ....
13. In a dicot seed, the food reserves is stored in.........................
14. In castor, a dicot seed, the food reserves are stored in ....
15. A seed, where there is no storage reserve available for germination is ....

16. In all the cereals, although we call it as seed, it is not a true seed. Botanically it is called ....
17. Non-endospermous seed is called as an ....seed
18. Largest seed is ....
19. Smallest seeds is ....
20. The seeds do not germinate when they have ....
21. During fatty acid synthesis elongation of fatty acids occur in ....
22. .... is responsible for desiccation tolerance of seeds
23. In the seeds, starch is stored in the form of ....
24. Perisperm is .... in origin
25. .... can be dried to safe moisture upto 10-15%
26. .... cannot be dried to 10% moisture
27. Any two dehiscent fruit types are ....
28. Seed coat is otherwise known as ....
29. The respiratory organelle of a cell is ....
30. α amylase and β amylase are called as ....
31. The hypocotyl becomes the .... part of the plant
32. Young method is used to calculate the............ content of seeds
33. The sugarcane seed is called as ....
34. .... is a seed appendage
35. Seed is the basic input of agriculture and it should be ....
36. .... is derived from the nucellus of the ovule
37. The ovules are attached to the placenta by ....
38. The minute opening of integument is ....
39. Ovule is surrounded by ....
40. In the ovary, micropyle is oriented opposite to placenta it is known as ....
41. The placentation in which micropyle is facing the placenta is known as ....
42. An example for anatrophus ovary is...........................
43. The orientation of the ovule from which the ovule is placed at an angle to the placenta is known as ....
44. For an orthodox seed, increase in the moisture content will reduce the ....
45. Membrane that protects the developing plumule in graminae is ....
46. The whole genetic material of any organism is known as...............
47. Two types of genomics are ....
48. Transcription means...........................
49. In a cell, the site of protein synthesis is ....

50. Palm oil is highly .... oil.
51. The nucleotide sequence used to detect the presence of complement array sequence in nucleic acid sample is called as ....
52. The .... of a sprouting seed develops into the lower stem and roots.
53. Most stored food in a bean seed is contained in their ....
54. The embryonic plant is called a .... before germination.
55. The .... of a sprouting seed develops into the upper stem and leaves.

♦ Answer Keys ♦

Q.No.	Ans.	Q.No.	Ans.	Q.No.	Ans.
1	Seed	2	Cotyledon	3	Scutellum
4	Seed	5	Integuments	6	Castor
7	Matured ovule	8	Castor, Cassava	9	Nucleus
10	Achene	11	3 n (triploid)	12	Endosperm
13	Cotyledon	14	Perisperm	15	Orchids
16	Caryopsis	17	Exalbuminous	18	Lodoicea
19	Orchid	20	Dormancy	21	Cytosol
22	ABA	23	Amylose and Amylopectin	24	Maternal
25	Orthodox seeds	26	Recalcitrant seeds	27	Follicles, capsules
28	Testa	29	Mitochondria	30	Enzymes
31	Lower stem	32	Protein	33	Pluff
34	Awn	35	Genetically pure	36	Perisperm
37	Funiculus	38	Micropyle	39	Integument
40	Orthotropous ovary	41	Anatrophus	42	Leguminaceae
43	Campylotropous	44	Storage life	45	Coleoptile
46	Genome	47	Structural genomics and functional genomics	48	DNA to RNA
49	Ribosome's	50	Saturated	51	Probes
52	Hypocotyl	53	Cotyledon (endosperm)	54	Seed
55	Epicotyl				

1.5.2. State True or False

1. The outermost layer of the endosperm is known as the aleurone layer
2. Starch is the polymer of glucose
3. The amylose content of seed is more than the amylopectin content
4. Cellulose is the polymer of glucose with β 1-6 linkage
5. Globulin content of oat is 80%
6. Hordein enzyme is present in prolamin
7. Albuminoids contains carbohydrates
8. Fats and oils are organic compounds
9. In the saturated fat, the bonds hold all the hydrogen bond

10. If the bond hold one additional hydrogen atom, the fat is monosaturated
11. Chemically fats are triesters of glycerol and fatty acids
12. Palm oil is a highly saturated oil
13. Saturated fats are usually lipids at room temperature
14. Lipids is used to refer both liquid and solid fats
15. Seed lipid reserves are stored in the form of spherosomes
16. Unsaturated fatty acids have single bonds
17. High erucic acid is available in rape seed oil
18. Peanut contains high monounsaturated fatty acid
19. Gibberellins are nothing but cyclic diterpene
20. So far 20 different types of gibberellins are known
21. Single seeded fruit with fused fruit and seed wall is known as achene
22. Perisperm is the storage structure of coffee
23. Seed is a living entity
24. Seed coat develops from the integument of the ovule
25. Single seeded fruit with fused fruit and seed wall is caryopsis
26. Storage protein present in wheat is glutelin
27. Protein linkage is receptor molecule
28. Auxin promote cell elongation
29. Adaptive enzyme formed or activated as a result of the presence of hormone
30. Acid growth hypothesis used to decrease the plasticity of cell
31. Constitutive enzyme always present but pathway, hormone can affect level
32. Hormones stimulate the signal transduction
33. Seed is a hygroscopic in nature
34. Endosperm is present in cucurbitaceous seeds
35. Protective cover of plumule is coleorhiza
36. Shape of the seed depends upon the shape of ovule
37. Perisperm is separated from the seed coat in papaya
38. Corny endosperm is due to more of amylopectin
39. Orchid seeds lack proper storage tissue
40. Seed is a ripened ovary
41. In a monocot seed, the cotyledon is represented by a structure called scutellum
42. Seed is a ripened ovary containing an embryo
43. Seed is a fertilized matured ovary
44. The entire testa of seed forms the sarcotesta in euphorbiaceae
45. Endomorphic rather than exomorphic characters are useful in varietal identification

46. Profile value of seed is nothing but the ratio between the seed length and breadth
47. Perisperm is the remenant of russel that is adhering to seed coat in black pepper
48. Length – breadth ratio is useful for classification and fixation of procurement price in rice cultivars
49. Cotton is an endospermous seed
50. Micropyle is formed by exostome and endostome
51. Imbibition depends on chemical composition of the seed
52. The imbibition distance over 10 mm water flows exceeds the seed
53. Enzyme activity is low in the radicle during imbibition
54. Nucleic acid includes DNA and RNA
55. Oligopeptides are short peptides
56. Polysomes are otherwise called as polyribosomes
57. Polysomes are absent in dry seeds
58. Polyadenylation protects the mRNA from degradation by exonucleases in the cytoplasm
59. Most of DNA synthesis is probably nuclear in orgin
60. DNA synthesis takes place continuously throughout dormancy of imbibed embryos
61. Flow cytometric analysis of embryos from dry seed showed a very low peak of 4 C nuclei content
62. Fats are degraded by lipase into glycerol and free fatty acids
63. ABA inhibit DNA repair in embryos of *Avena fatua*
64. X-ray radiation inhibits DNA synthesis in lettuce
65. DNA synthesis activated in cotyledon and higher towards radicle tip
66. Citric acid cycle of lipid breakdown will yield isocitrate (oxaloacetate)
67. Glyconeogenesis of lipid metabolism yields fumarate (sucrose)
68. Triglycerides are found in plants as liquid oils
69. In seeds of jojoba plants, lipids are stored as liquid wax
70. In sunflower seed, lipids accumulated in endosperm
71. In tobacco seed, lipids are stored in both embryo and endosperm
72. β-oxidation pathway relates to starch metabolism
73. Mitochondria is involved in lipid metabolism
74. In β-oxidation, fatty acids is converted into acetyl COA
75. α-amylase is present in air dry seed
76. α-amylase is synthesized in aleurone layer under stimulation of GA in cereals
77. Maltase breaksdown the maltose into fructose

78. α-amylase cannot hydrolyse α-1,6 glucosidic bonds in amylopectin
79. During hydrolysis of amylopectin by α-amylase, dextrin is produced in addition to glucose and maltose
80. α-amylase breaks down the amylose and yields glucose and maltose
81. Phosphorylytic starch catabolic pathway predominates during the early stages of seed germination in peas
82. In cereal seeds, GA is synthesized in embryo and then transported to aleurone layer
83. Starchy seeds are generally more tolerant to anoxia than fatty seeds
84. In monocot, the embryo includes endosperm
85. In monocot, the embryo excludes endosperm
86. In dicot, the embryonic axis does not include cotyledon
87. In dicot, the embryonic axis include cotyledon
88. Air resistance to seeds improves wind dispersal
89. Ballistic seed dispersal is self seed dispersal mechanism
90. In certain seeds, the endosperm becomes mucilaginous which is due to the presence of galactomannan and glycomannans
91. In seeds of coffee, perisperm stores food reserve materials
92. Lipids are glycosides of fatty acids
93. Starch and hemicelluloses are the two forms of protein
94. Mitochondria is the respiratory site of cell
95. The hypocotyl becomes the lower stem of the plant
96. Vegetative propagules are also seeds
97. The storage structure of dicot is scutellum
98. Sunflower is an achene type of fruit

♦ Answer Keys ♦

Q.No.	Ans.	Q.No.	Ans.	Q.No.	Ans.	Q.No.	Ans.	Q.No.	Ans.	Q.No.	Ans.
1	True	2	True	3	False	4	False	5	False	6	True
7	True	8	True	9	True	10	True	11	True	12	True
13	False	14	True	15	True	16	False	17	True	18	True
19	True	20	False	21	False	22	True	23	True	24	True
25	True	26	True	27	False	28	True	29	True	30	False
31	True	32	True	33	True	34	False	35	False	36	False
37	False	38	True	39	True	40	False	41	True	42	False
43	False	44	True	45	True	46	True	47	True	48	True
49	True	50	True	51	True	52	True	53	False	54	True
55	True	56	True	57	True	58	True	59	True	60	True
61	True	62	True	63	False	64	True	65	False	66	False
67	False	68	True	69	True	70	False	71	True	72	False
73	True	74	True	75	False	76	True	77	False	78	True

Q.No.	Ans.	Q.No.	Ans.	Q.No.	Ans.	Q.No.	Ans.	Q.No.	Ans.	Q.No.	Ans.
79	True	80	True	81	True	82	True	83	True	84	False
85	True	86	True	87	False	88	True	89	True	90	True
91	True	92	True	93	False	94	True	95	True	96	True
97	False	98	True								

1.6. CHEMICAL COMPOSITION OF SEED

1.6.1. Choose the correct answer

1. Globulin is mainly present in
   (a) Endosperm (b) Embryo and aleurone layer
   (c) ER (d) Ribosome
2. The most widely used oils that are in monounsaturated fats are
   (a) Soybean (b) Peanut oil
   (c) Palm oil (d) Corn oil
3. If rice contains more of mealy endosperm, this type is amenable for
   (a) No breakage in milling (b) Less breakage in milling
   (c) Non-sticky in milling (d) High breakage in milling
4. Which seed contains starchy endosperm?
   (a) Rice (b) Maize
   (c) Sorghum (d) Pearl millet
5. The respiratory quotient values in the deteriorated seed is
   (a) High (b) Low
   (c) Medium (d) Absent
6. The protein content of wheat is
   (a) 10-12% (b) 10-14%
   (c) 10-16% (d) 10-18%
7. The oil content of castor is
   (a) 60-62% (b) 60-64%
   (c) 60-66% (d) 60-68%
8. The carbohydrate content of maize is
   (a) 80-82% (b) 80-84%
   (c) 80-86% (d) 80-88%
9. The amylose content of starch is
   (a) 15-20% (b) 20-25%
   (c) 25-30% (d) 30-35%
10. What is the percentage of amylopectin present in starch?
    (a) 50-75% (b) 60-75%
    (c) 70-85% (d) 80-95%
11. Albumin is soluble in
    (a) Water (b) Saline
    (c) Alchol (d) Dilute alkali

12. Prolamines are soluble in
    (a) Water (b) Saline
    (c) Alcohol (d) Dilute alkali
13. Which one of the following crop seed has high oil?
    (a) Wheat (b) Redgram
    (c) Blackgram (d) Groundnut
14. What is the calorific value of carbohydrate seed?
    (a) 4000 (b) 5000
    (c) 5700 (d) 9500
15. What is the calorific value of proteinaceous seed?
    (a) 4000 (b) 5000
    (c) 5700 (d) 9500
16. Which one of the following oil seed contains high saturated fatty acids?
    (a) Castor oil (b) Brassica oil
    (c) Coconut oil (d) Sunflower oil
17. Which one of the following oil seed contains high unsaturated fatty acids?
    (a) Castor oil (b) Brassica oil
    (a) Coconut oil (d) Sunflower oil

♦ Answer Keys ♦

Q.No.	Ans.	Q.No.	Ans.	Q.No.	Ans.	Q.No.	Ans.	Q.No.	Ans.	Q.No.	Ans.
1	b	2	b	3	d	4	c	5	a	6	b
7	c	8	a	9	b	10	a	11	a	12	c
13	d	14	a	15	c	16	c	17	c		

1.6.2. Fill in the blanks

1. In cereals, the major food reserve is ....
2. Most of the legumes are rich in ....
3. In general, the oil content of oil seed crop ranges from ....
4. In peas, the food materials are stored in ....
5. In castor, the food materials are stored in ....
6. Starch is the polymers of ....
7. .... are linked by polypeptide chain to form protein.

♦ Answer Keys ♦

Q.No.	Ans.	Q.No.	Ans.	Q.No.	Ans.	Q.No.	Ans.
1	Carbohydrate	2	Protein	3	40-65%	4	Cotyledon
5	Endosperm	6	Glucose	7	Amino acids		

1.7. SEED DORMANCY

1.7.1. Choose the correct answer

1. Abscisic acid first time isolated from young cotton fruit by
   (a) Letham (b) Lang
   (c) Addicott (d) Yabuta and Sumuki
2. Decline in growth rate of the seed accompanies the buildup of
   (a) Auxin (b) Cytokinin
   (c) Gibberellin (d) Abscisic acid
3. Dormancy induced by darkness is called as
   (a) Photo dormancy (b) Skoto dormancy
   (c) Thermo dormancy (d) Physico dormancy
4. Moist chilling is otherwise called as
   (a) After ripening (b) Stratification
   (c) Photoblastism (d) Skotoblastism
5. Seeds with hard impermeable seed coat are called as
   (a) Synthetic seed (b) Hard seed
   (c) Coated seed (d) Dead seed
6. Hard seeds are mainly occurring in the family of
   (a) Malvaceae (b) Solanaceae
   (c) Rootaceae (d) Umbelliferae
7. The hard seed coat can be removed by
   (a) Acid scarification (b) After ripening
   (c) Drying (d) Stratification
8. After imbibition the release of hormone from embryo signals the seed to break dormancy
   (a) Auxin (b) Cytokinin
   (c) Enzyme (d) Gibberellins
9. .... regulates plants response to light throughout its life
   (a) Phytochrome (b) Photosynthesis
   (c) Chloroplast (d) Golgi complex
10. Triggers the breaking of seed dormancy
    (a) Pfr (b) Pr
    (c) Cytochrome (d) Chloroplast
11. Phytochrome receptor present in
    (a) Mitochandria (b) Nucleus
    (c) Cytoplasm (d) Plasma membrane
12. The type of seed dormancy in *Acacia nilotica* is
    (a) Physical (b) Physiological
    (c) Morphological (d) Morphophysiological

13. Storing of seeds in moist dark conditions for 4 weeks at 1-4^{0}C (stratification) can break the dormancy of
(a) *Tectona grandis* (b) *Eucalyptus tereticornis*
(c) *Azadirachta indica* (d) *Santalum album*

14. Inhibitor coumarin is present in
(a) Paddy (b) Coriander
(c) Groundnut (d) Cotton

15. Stratification is given to seed for
(a) Creating moisture stress resistance (b) Dormancy breaking
(c) Disease resistance (d) Increasing vigour

16. Skoto dormancy is caused due to
(a) Light (b) Temperature
(c) Chemicals (d) Dark

17. Inhibitor present in cabbage seed is
(a) ABA (b) Coumarin
(c) Hydroxy urea (d) Methyl urea

18. The hormone used for breaking dormancy is
(a) Auxin (b) CKs
(c) GA (d) ABA

19. Suspension of germination by unfavourable environmental conditions is called
(a) Dormancy (b) Embryo dormancy
(c) Quiescence (d) Resting period

20. The wavelength of 700 to 800 nm carries
(a) Plant pfr (b) Fungal pfr
(c) Algal pfr (d) Bacterial pfr

21. The seed which will not germinate under normal environmental conditions is called as
(a) Live seed (b) Dead seed
(c) Dormant seed (d) Rejected seed

22. In paddy, endosperm is separated by a layer known as
(a) Silicon layer (b) Endothelium
(c) Pericarp (d) Epithelium

23. Possible reasons for seed dormancy is
(a) Presence of pathogen (b) Immature embryo
(c) Cracking of hulls (d) Green distortion

24. Dormancy caused due to dark is called
(a) Photo dormancy (b) Thermo dormancy
(c) Skoto dormancy (d) Innate dormancy

25. Physiological dormancy is caused due to
(a) Hard seed coat (b) Under developed embryo
(c) Hard endocarp (d) Thick pericarp

26. Leaching of seeds is the common practice in
(a) Redgram (b) Coriander
(c) Tomato (d) Nutmeg

27. Scientist associated with seed dormancy
(a) Nikolaeva (b) Bewley
(c) Khan (d) Roberts

28. The natural inhibitor found in the seed is
(a) 2,4,D (b) MH
(c) Kinetin (d) ABA

29. The plant product which is used to break the dormancy
(a) Strigol (b) Kharanjia
(c) Pongamol (d) Nimbicidin

30. A chemical which can be used to induce dormancy in seeds during the process of seed development
(a) IAA (b) IBA
(c) ABA (d) GA

31. Embryo dormancy of seeds can be overcome by exposure to
(a) Light temperature (b) Low temperature
(c) Scarification treatment (d) Light

32. A suitable hard seed breaking treatment is
(a) Halogenation (b) Scarification
(c) Hardening (d) Leaching

♦ Answer Keys ♦

Q.No.	Ans.	Q.No.	Ans.	Q.No.	Ans.	Q.No.	Ans.	Q.No.	Ans.	Q.No.	Ans.
1	c	2	d	3	b	4	b	5	b	6	a
7	a	8	d	9	a	10	a	11	d	12	a
13	b	14	b	15	b	16	d	17	c	18	c
19	c	20	a	21	c	22	d	23	b	24	c
25	b	26	b	27	a	28	d	29	a	30	c
31	b	32	b								

1.7.2. Fill in the blanks

1. Hard seed development is common in .... crops
2. Embryo dormancy can be broken by ....
3. The chemical used for breaking dormancy in groundnut is ....
4. Dormancy is due to accumulation of .... substance
5. Method used to break hard seededness in blackgram is ....
6. .... family producing dormant seeds having physical barrier to water
7. .... form of the phytochrome is physiologically active in overcoming seed dormancy

8. Physiomorphic dormancy is due to ....
9. Hard seeds are formed due to .... soil conditions
10. Tephrosia establishment in field occurs due to .... phenomenon
11. An inexpensive chemical used for breaking dormancy is ....
12. Incubation of seeds in moist cold condition is called ....
13. Dormancy in sunflower is caused by presence of .... in pericarp
14. Hard seeds can be softened by .... treatment
15. Give an example for dormancy breaking chemical ....
16. Occurrence of hard seed is due to ....
17. Dormancy in *Arabidopsis* can be overcome by promoting factors as ....
18. Seed dormancy is must to prevent germination in temperate species until ....
19. Light promotes .... synthesis
20. Seed absorb ethylene in large amount from the soil and cause ....
21. Changes occurring in dormant seeds that permits germination is ....
22. The inability to germinate even in the presence of favourable environmental conditions is called as ....
23. The seeds do not germinate when they have ....
24. .... is the widely used chemical to break the dormancy in groundnut
25. Exposure of seeds to low temperature is called ....
26. Tobacco seeds have .... dormancy
27. Presence of .... acid is essential for inducing seed dormancy
28. Inhibitor present in coriander seed which causes dormancy is ....
29. Rubbing of seed against hard surface to remove dormancy is ....
30. Chemical used to break the seed dormancy in paddy is ....
31. A solution of KNO_3 (0.2%) is prepared by dissolving .... of KNO_3 in 100 ml of water.
32. To break dormancy of groundnut, preheat the seeds at .... C.
33. Wheat seeds can be prechilled at .... C for breaking dormancy.
34. To break the dormancy of grass seed, the light required is .... lux.
35. To prepare 0.05% GA_3, 500g GA_3 is dissolved in .... litre of water.
36. The best example for prewashing of seed to break dormancy is ....

♦ Answer Keys ♦

Q.No.	Ans.	Q.No.	Ans.	Q.No.	Ans.
1	Leguminous	2	Stratification	3	Ethrel
4	Toxic	5	Scarification	6	leguminaceae
7	Pfr	8	Under developed embryo	9	Dry
10	Hard seed	11	KNO_3	12	Stratification
13	Chlorogenic acid	14	Scarification	15	GA_3
16	High temperature	17	High temperature	18	Spring season
19	GA3	20	Ecological suicide	21	After-ripening
22	Quiescence	23	Dormancy	24	Ethrel
25	Vernalization	26	Thermo	27	Abscisic
28	Coumarin	29	Scarification	30	KNO_3
31	2 g	32	50°	33	5-10°
34	750–1250 lux	35	1 litre	36	Beta/Xanthium/Fraxinus

1.7.3. State True or False

1. Coumarin is an alkaloid compound present in pericarp and testa of coriander seed
2. Catechol affects seed coat permeability in peas
3. In maize, kernels sprayed with fluridone prevent viviparous germination
4. *Hordeum* spp shows only embryo dormancy
5. Endospermal membrane cause dormancy in *Pyrus malus*
6. Dormancy is not linked with temperature
7. *Lactuca sativa* shows both photo and skoto dormancy
8. Coumarin alkaloid is present in *Eleaguns angustifolia*
9. Coumarin at a very low concentration is an inhibitor
10. Dormancy is common in cultivated species
11. Presence of hard seed coat is a reason for physical dormancy
12. Breaking of dormancy is retest of response
13. In darkness the Pr is converted in Pfr
14. Potato's response to light is an example of cell signal process
15. Phytochrome is non photoreceptor
16. Dormancy in rice seed is caused by hull and pericarp
17. Scarification is the treatment to improve germination
18. Inability of a viable seed to germinate under favorable condition is called dormancy
19. In cucurbits, the seed coat is not permeable to gases
20. Hard seed results in dormancy
21. Dormancy is clearly revealed by TZ test
22. Incubation of seeds in moist cold condition is stratification

23. Dormancy in sunflower is caused by the content of chlorogenic acid in the pericarp
24. In dark/light the Pfr is slowly converted into Pr
25. In kolingi seeds, premature embryo is the cause for seed dormancy
26. Ethrel is used to break dormancy in groundnut
27. KNO_3 soaking breaks dormancy of seeds in marigold
28. β-tubulin is used as indication of germination progress in non dormant seeds for testing purpose
29. Seed lipases are generally specific and cannot hydrolyse a wide variety of triglycerides
30. Citric acid cycle takes place in cytoplasm (mitochondria)
31. The phytochrome Pfr is inducing dormancy
32. Gibberellic acid is the substitute for light to break the dormancy
33. Hard seeds are otherwise known as hardened seeds
34. Potato seeds exhibit dormancy
35. The commonly used chemical to break the seed dormancy of groundnut is 2,4,D
36. Dormancy breaking method for coriander is hot water soaking

♦ Answer Keys ♦

Q.No.	Ans.	Q.No.	Ans.	Q.No.	Ans.	Q.No.	Ans.	Q.No.	Ans.	Q.No.	Ans.
1	True	2	True	3	False	4	False	5	False	6	False
7	True	8	True	9	False	10	False	11	True	12	True
13	False	14	True	15	False	16	True	17	True	18	True
19	True	20	True	21	False	22	True	23	True	24	True
25	False	26	True	27	True	28	True	29	False	30	False
31	False	32	True	33	False	34	True	35	False	36	False

1.8. SEED GERMINATION

1.8.1. Choose the correct answer

1. Net ATP production during seed germination
   (a) 36 (b) 4
   (c) 2 (d) 32
2. Which stage of cotyledon development provides nutrients to the embryo
   (a) Globular stage (b) Heart stage
   (c) Mature embryo stage (d) Torpedo stage
3. Seed germination is stimulated by the hormone
   (a) Cytokinin (b) Auxin
   (c) Gibberrellin (d) Kinetin

4. Cytokinin induces
   (a) Cell elongation (b) Apical dominance
   (c) Cell differentiation (d) Cell division
5. A naturally occurring sesquiterpene compound which inhibits seed germination is
   (a) IAA (b) Cytokinin
   (c) Traumatic acid (d) ABA
6. Seed physiologist defines germination as
   (a) Protrusion of primary root (b) Protrusion of root and shoot
   (c) Production of normal seedlings (d) Production of vigourous seedlings
7. During germination, the embryo resumes
   (a) Growth
   (b) Metabolic activity
   (c) Growth and metabolic activity
   (d) Growth, metabolic activity and photosynthesis
8. During germination, oxygen is required for
   (a) Enzyme activity (b) Hydration of cells
   (c) Metabolic needs (d) Growth
9. Tomato seeds give 90% germination at
   (a) 30 DAP (b) 40 DAP
   (c) 60 DAP (d) 35 DAP
10. Precocious germination in seeds of fleshy fruits is prevented by
   (a) More moisture (b) ABA content
   (c) Osmotic stress (d) Desiccation stress
11. How the rate of gas exchange is measured?
   (a) Nanometer (b) Manometer
   (c) Warburgs respirometer (d) PCR
12. Which one of the growth regulator is a newly developed one?
   (a) Auxin (b) GA_3
   (c) Brassinosteriod (d) Cytokinin
13. Which one of the following chemical is a light substitute?
   (a) Urea (b) KCl
   (c) Thiourea (d) $KMnO_4$
14. Which is the living tissue of seed
   (a) Endosperm (b) Seed coat
   (c) Aleurone layer (d) Micropyle
15. Critical moisture content for soybean seed germination
   (a) 30% (b) 40%
   (c) 50% (d) 60%
16. The enzyme occur in seed only after imbibition
   (a) α-amylase (b) β-amylase
   (c) Dehydrogenase (d) Carboxylase

17. Optimum temperature for brussels sprouts germination
    (a) 10-30^{0}C (b) 20-40^{0}C
    (c) 30-50^{0}C (d) 40-60^{0}C
18. GA induces synthesis of which enzyme?
    (a) Amylase (b) Catalase
    (c) Peroxidase (d) Phosphatase
19. Light promotes
    (a) IBA accumulation (b) GA synthesis
    (c) ABA synthesis (d) IAA synthesis
20. The process of initial water uptake by seeds prior to germination is
    (a) Dehydration (b) Imbibition
    (c) Desiccation (d) Impermeation
21. The emergence of seedling from seed is called as
    (a) Germination (b) Dormancy
    (c) Senescence (d) Vigour
22. The type of germination in onion is
    (a) Hypogeal (b) Epigeal
    (c) Semi hypogeal (d) Semi epigeal
23. Low seed germination in papaya is due to
    (a) Embryolessness (b) Sarcotesta
    (c) Inhibitors (d) Hard seed coat
24. The hard endocarp inhibits germination of
    (a) Ber (b) Neem
    (c) Banana (d) Jamun
25. Epigeal germination occurs in
    (a) Paddy (b) Groundnut
    (c) Sorghum (d) Maize
26. Light is required for germination in
    (a) Paddy (b) Amaranthus
    (c) Onion (d) Sunflower
27. Imbibition is a
    (a) Chemical process (b) Physical process
    (c) Physiochemical process (d) Biochemical process
28. In enzyme activation, the respiration process begin in
    (a) Metabolic process (b) Phase II
    (c) Phase I (d) Both phase I & II
29. The first event in seed germination is
    (a) Imbibition (b) Enzyme activity
    (c) Respiration (d) Hydrolysis

30. Amount of energy released in anaerobic respiration is
(a) 2 ATP (b) 40 ATP
(c) 38 ATP (d) 6 ATP

31. Mitochondria also participate in photo respiration in
(a) C 4 plants (b) C 3 plants
(c) CAM plants (d) C 5 plants

32. Site of respiration is
(a) Cytoplasm (b) Protoplasm
(c) ETC (Election Transport Chain) (d) Mitochondria

33. Respiration is essential for
(a) AMP production (b) ADP production
(c) NADPH production (d) ATP production

34. Direct enhancing effect of light on the rate of respiration has also been reported by
(a) Emerson and Lewis (1943) (b) Seikh *et al.* (2001)
(c) Honson *et al.* (1959) (d) Hopkins (1927)

35. Occurs in green plants only when light is available
(a) Germination (b) Photosynthesis
(c) Anaerobic respiration (d) Fermentation

36. Which hormone comes under shock hormone
(a) Auxin (b) Gibberellin
(c) ABA (d) Cytokinin

37. Which hormone is used for growth
(a) Dormin (b) GA
(c) ABA (d) Florigen

38. GA synthesis and signaling dominate the transition of
(a) Dormancy (b) Sensitivity
(c) Germination (d) Receptivity

39. Brassinosteroid comes under which hormone group
(a) Growth hormone (b) Stress hormone
(c) Shock hormone (d) Sensitivity hormone

40. The first developed hormone is
(a) ABA (b) CKs
(c) Auxin (d) GA_3

41. Precursor for ethylene synthesis is
(a) Methionine and B-alanine (b) Tryptophan
(c) Phytin (d) Glutaline

42. Which organelle is the power house of cell
(a) Mitochondria (b) Glyoxysome
(c) Golgi bodies (d) Ribosome

43. The major 'auxin' in developing seed is
    (a) IAA (b) ABA
    (c) BOAA (d) BOA

44. The phase after imbibition and before radicle protrusion during germination is called as
    (a) Interphase (b) Lag phase
    (c) Prophase (d) Log phase

45. Gibberellic acid was first isolated from
    (a) Fungus (b) Minerals
    (c) Animals (d) Bacteria

46. Uptake of water from the atmosphere by cells or tissue in the seed coat is called
    (a) Wetting of the seed (b) Adsorption
    (c) Adhesion (d) Absorption

47. Drying from lack of oxygen, e.g. respiring seeds stored in an oxygen free atmosphere is termed as
    (a) Noxious condition (b) Oxidated condition
    (c) Fermented condition (d) Anoxia

48. Impaired viability caused by exposure to low temperature is called as
    (a) Drought injury (b) Chilling injury
    (c) Mechanical injury (d) Chilling tolerance

49. Group of enzymes catalyzing reactions involving transfer of hydrogen from a substrate to a hydrogen acceptor is named as
    (a) Glutamase (b) Hydrogenase
    (c) Methionase (d) Dehydrogenase

50. Which hormone is involved in flowering induction
    (a) GA (b) Auxin
    (c) Florigen (d) NAA

51. Seeds germinating while still attached to the parent plant, e.g. *Rhizophora* spp. and extremely difficult to store are known as
    (a) Vigour (b) Tumbling
    (c) Viviparous (d) Imbibition

52. At a time, thermo gradient plate is used to maintain different levels of
    (a) RH (b) Temperature
    (c) Light (d) CO_2

53. Seeds which require light are called
    (a) Skotoblastic (b) Photo blastic
    (c) Hydrophilic (d) Hydrophobic

54. Pre-sowing seed treatment recommended for rainfed/drought condition
    (a) Seed pelleting (b) Seed priming
    (c) Seed hardening (d) Seed coating

55. One of the requirement for seed germination is
    (a) Soil (b) Water
    (c) Fertilizer (d) Seed

56. The respiration stimulant in germination of seed is
    (a) KNO_3 (b) GA_3
    (c) H_2O_2 (d) Thiourea

♦ Answer Keys ♦

Q.No.	Ans.	Q.No.	Ans.	Q.No.	Ans.	Q.No.	Ans.	Q.No.	Ans.	Q.No.	Ans.
1	a	2	b	3	c	4	d	5	d	6	a
7	c	8	c	9	b	10	b	11	c	12	c
13	c	14	c	15	c	16	a	17	a	18	a
19	b	20	b	21	a	22	b	23	a	24	a
25	b	26	b	27	b	28	c	29	a	30	a
31	b	32	d	33	d	34	a	35	b	36	b
37	a	38	c	39	c	40	c	41	b	42	a
43	a	44	b	45	a	46	d	47	d	48	b
49	d	50	c	51	c	52	b	53	b	54	c
55	b	56	c								

1.8.2. Fill in the blanks

1. .... enables the rice seed to germinate under water
2. .... is the first biochemical event in the process of germination
3. Non-germination of seed due to more production of ethylene is ....
4. A seed which require pre chilling for germination is ....
5. .... of the embryo emerges first during germination
6. Respiratory Quotient RQ = ...
7. Rate of gas exchange is measured by ....
8. In ...., carbohydrates are used as substrate
9. .... is the exothermic process
10. In epigeal germination, the cotyledons remain .... the soil
11. The dark loving seeds for its germination are known as ....
12. Light substituting chemical is ....

♦ Answer Keys ♦

Q.No.	Ans.	Q.No.	Ans.	Q.No.	Ans.
1	Anerobic respiration	2	Hydrolysis (GA3 release → α amylase)	3	Ecological suicide
4	Lettuce	5	Radicle	6	$\frac{QCO_2}{QO_2}$
7	Warburg's Respirometer	8	Respiration	9	Respiration
10	Above	11	Skotoblastic seeds	12	Thiourea

1.8.3. State True or False

1. The use of energy to construct components of cells is called anabolism
2. Amino acid codons and 13 termination codons stop the process of translation
3. Corn seeds require 50% moisture for germination
4. α-amylase is present in dry state of seeds
5. Seeds with lower moisture content will absorb more moisture than seeds with more moisture content
6. Temperate seeds require lower temperature than tropical seeds
7. Wild species require more temperature than domestic
8. *Brassica oleracea* requires 4 and 42°C as minimum and maximum temperatures respectively for germination
9. *Lycopersicon esculentum* require 12 and 36°C as minimum and maximum temperatures, respectively for germination
10. Nitrogen affects germination
11. Alcohol dehydrogenase activity paralleled decline in seed ethanol levels
12. Pr is useful for seed germination
13. Blue light increases sensitivity to ABA
14. Photo conversion of Pr to Pfr takes place when seed is hydrated above 8%
15. Red light affects germination and far red promotes germination
16. Seed science and technology is a multi-disciplinary science
17. Hydroxyl urea present in cabbage seed will affect radicle protrusion
18. CCAAT box binding protein complex (nuclear factor) and LECI mediating both ABA inhibition of germination and slight induction of light harvesting complex/chlorophyll binding protein
19. The type of germination in paddy is epigeal
20. Hot water treatment stimulates crop seed germination, but kills weed seed because of high thermal sensitivity of the later
21. Cell division occurs during pre-germination stage in the process of germination

22. The first process that occurs during germination is respiration
23. Swelling force on the cell wall is turgor pressure
24. In mitochondria, the ATP molecules are formed during anaerobic respiration
25. The organelles of CO_2 release during respiration increase the pollution
26. The rate of respiration depends on the quality of protoplasm present in the cell
27. The optimum temperature of respiration is about 30°C
28. Oxygen is a participant in the reactions of aerobic respiration
29. In heterotrophic germination, respiration is excepted to be high
30. Salicylates (SAs) is grouped under stress hormone
31. Brassinosteroids is a newly developed growth hormone
32. Ethylene helps in fruit ripening
33. BAs inhibits root growth and leaf abscission
34. ABA inhibits endosperm rupture, but not testa rupture
35. Auxins are synthesized in leaf tissue
36. CK promotes chlorophyll production and photosynthesis
37. ABA stimulates seed germination and inhibit seed dormancy
38. Prolamins are soluble in water
39. DNA repair mechanism seems to be unaffected by presence of ABA during imbibition
40. Light is an essential factor for seed germination
41. The survival of seedlings is decided by presence of safe sites than number of seedlings
42. Seeds which are not requiring light for germination are called photoblastic
43. Seeds, other than hard seeds which remain firm and apparently viable at the end of germination test period is called dead seed
44. Gibberellins are growth stimulating substances
45. In hypogeal germination plants the cotyledons remain below the surface of the soil
46. Potassium nitrate is the most widely used chemical for promoting seed germination

♦ Answer Keys ♦

Q.No.	Ans.	Q.No.	Ans.	Q.No.	Ans.	Q.No.	Ans.	Q.No.	Ans.	Q.No.	Ans.
1	True	2	False	3	False	4	False	5	True	6	True
7	False	8	True	9	True	10	False	11	True	12	False
13	True	14	True	15	True	16	True	17	False	18	True
19	False	20	True	21	True	22	False	23	True	24	False
25	True	26	True	27	True	28	True	29	True	30	False
31	True	32	True	33	True	34	False	35	False	36	True

Q.No.	Ans.	Q.No.	Ans.	Q.No.	Ans.	Q.No.	Ans.	Q.No.	Ans.	Q.No.	Ans.
37	False	38	False	39	True	40	True	41	True	42	False
43	False	44	True	45	True	46	True				

1.9. SEED DISPERSAL

1.9.1. Choose the correct answer

1. Which one of the following plants self disperses its seeds?
   (a) Dandelion (b) Apple
   (c) Gorse (d) Strawberry
2. For sexual reproduction in a plant, which one of the following conditions must be met?
   (a) There must be nectar or scent present in the flower.
   (b) The male and female gametes must meet.
   (c) The pollen must be transferred by insects
   (d) The stamens must be inside the petals
3. Spreading of seed as far away as possible from the parent plant is known as
   (a) Germination (b) Fertilization
   (c) Asexual reproduction (d) Dispersal
4. How are dandelion seeds dispersed?
   (a) Animal (b) Self
   (c) Water (d) Wind
5. Seed dispersal does not reduce competition between the seedlings and the parent plant for
   (a) Minerals (b) Light
   (c) Oxygen (d) Water
6. Coconut is an example of fruit being dispersed by
   (a) Water (b) Animal
   (c) Wind (d) Self
7. Which one of the following is not a method of seed dispersal ?
   (a) Animal (b) Germination
   (c) Wind (d) Self

♦ Answer Keys ♦

Q.No.	Ans.	Q.No.	Ans.	Q.No.	Ans.	Q.No.	Ans.
1	c	2	b	3	d	4	d
5	c	6	a	7	b		

1.10. MATCH THE FOLLOWING

1. Question

1.	ABA	A.	Triggering agent of germination
2.	Heart stage	B.	Apple
3.	Lag phase	C.	Dormancy induction
4.	Heavy uptake of water	D.	Sorghum
5.	GA	E.	Sunflower
6.	α – amylase	F.	Embryo development
7.	Stratification	G.	Bhendi
8.	Dunken layer formation	H.	Phase II of imbibition
9.	Achene	I.	Phase I of imbibition
10.	Hair line crack	J.	Starch degradation

♦ Answer Keys ♦

Q.No.	Ans.	Q.No.	Ans.	Q.No.	Ans.	Q.No.	Ans.	Q.No.	Ans.
1	C	2	F	3	H	4	I	5	A
6	J	7	B	8	D	9	E	10	G

2. Question

1.	*Citrus* spp.	A.	Combined dormancy
2.	*Triticum vulgare*	B.	Andromonoecious
3.	*Tectona grandis*	C.	Tassels
4.	*Malus domestica*	D.	Recalcitrant seed
5.	*Beta vulgaris*	E.	+ photoplastic seed
6.	*Vigna radiata*	F.	Orthodox seed
7.	*Ricinus communis*	G.	Under developed embryo
8.	*Zea mays*	H.	Intermediate seed
9.	*Lactuca sativa*	I.	Hard seed
10.	*Azadirachta indica*	J.	Multigerm seeds

♦ Answer Keys ♦

Q.No.	Ans.	Q.No.	Ans.	Q.No.	Ans.	Q.No.	Ans.	Q.No.	Ans.
1	D	2	F	3	A	4	G	5	J
6	I	7	B	8	C	9	E	10	H

3. Question

1.	Paddy	A.	Often cross pollinated cop
2.	Maize	B.	Protogyny
3.	Cotton	C.	Hard seed
4.	Sunflower	D.	Twisted embryo

5.	Bajra	E.	Self pollinated crop
6.	Blackgram	F.	Bees pollination
7.	Groundnut	G.	Rich in oil and protein
8.	Soybean	H.	Cross pollinated crop
9.	Tomato	I.	Monoecious
10.	Cauliflower	J.	Epigeal germination

♦ Answer Keys ♦

Q.No.	Ans.	Q.No.	Ans.	Q.No.	Ans.	Q.No.	Ans.	Q.No.	Ans.
1	E	2	I	3	A	4	F	5	B
6	C	7	J	8	G	9	D	10	H

4. Question

1.	Gymnosperm seed	A.	Anthurium
2.	Protoandry	B.	Mango
3.	Spadix	C.	Paddy
4.	Captiulam	D.	Grapes
5.	Druple	E.	Maize
6.	Panicle	F.	Beet root
7.	Berry	G.	Sunflower
8.	Siliqua	H.	Pollination by water
9.	Aggregate fruit	I.	Pine
10.	Hydrophily	J.	Sesamum

♦ Answer Keys ♦

Q.No.	Ans.	Q.No.	Ans.	Q.No.	Ans.	Q.No.	Ans.	Q.No.	Ans.
1	I	2	E	3	A	4	G	5	B
6	C	7	D	8	J	9	F	10	H

5. Question

1.	Fusion of male and female gamete	A.	Turning of fruit colour to yellow
2.	Fusion of male gamete with female gamete and polar nuclei	B.	Fruit colour from green to red or orange
3.	Soybean	C.	Cracking of bolls
4.	Brinjal	D.	Warm stratification
5.	Bittergourd	E.	Fertilization
6.	Ribbedgourd	F.	Acid scarification

7.	Jatropha	G.	Pods and seeds changed from green to yellow
8.	Cotton	H.	Double fertilization
9.	Exposing the seeds at 5 to 10°C	I.	Browning of dried fruits
10.	Lucerne	J.	Capsules turned to yellowish

♦ Answer Keys ♦

Q.No.	Ans.	Q.No.	Ans.	Q.No.	Ans.	Q.No.	Ans.	Q.No.	Ans.
1	E	2	H	3	G	4	A	5	B
6	I	7	J	8	C	9	D	10	F

1.11. FIND OUT INCORRECT STATEMENT

1. a. Flower pistils mature before stamen is called as protandry
b. Maize is an example for protandry
c. Protandry is one of the reasons for cross pollination

2. a. Hard seed development is common in leguminous crops
b. Dormant seeds of legumes is having a physical barrier to water
c. When legume seeds are directly soaked in water, soaking injury does not occur

3. a. In a monocot, food reserves are stored in endosperm
b. In a dicot seed, the food reserves are stored in cotyledon
c. In castor seed, the food reserves are stored in cotyledon

4. a. Self incompatibility in crops can be overcome by cross pollination
b. Cross pollination is called as Autogamy
c. Complete self pollination is called as cleistogamy

5. a. Dormancy due to under developed embryo is physiomorphic dormancy
b. Occurrence of hard seed is due to high temperature
c. ABA stimulates seed germination and inhibits seed dormancy

6. a. GA triggers germination
b. GA triggers hydrolase production
c. ABA promotes germination release

7. a. Maximum dry matter accumulation and high moisture
b. Maximum germination and vigour
c. Maximum germination and low moisture

8. a. Mechanical dormancy is broken by sand scarification
b. Mechanical dormancy is broken by acid scarification
c. Mechanical dormancy is broken by stratification

9. a. In phase I of germination, more uptake of water
b. In phase II of germination, more uptake of water
c. In phase III of germination, more uptake of water

10. a. Cell division follows radicle protrusion
b. Radicle protrusion follows cell division
c. Radicle protrusion follows cell cycle division

11. a. Recalcitrant seed is susceptible to freezing injury
b. Orthodox seed is not susceptible to chilling injury
c. Orthodox seed is susceptible to freezing injury

12. a. Indeterminate flowering is continuous
b. Determinate flowering is continuous
c. Determinate flowering is not continuous

♦ Answer Keys ♦

Q.No.	Ans.	Q.No.	Ans.	Q.No.	Ans.	Q.No.	Ans.	Q.No.	Ans.	Q.No.	Ans.
1	a	2	c	3	c	4	b	5	c	6	c
7	a	8	c	9	b	10	b	11	c	12	b

1.12. ARRANGE IN ORDER

1. Stages of embryo development

Tarpedo stage – Heart shaped – Proembryo – Globular

Answer: Proembryo – Globular – Tarpedo stage – Heart shaped

2. Processes of seed development

Harvestable maturity – Physiological maturity – Accumulation of dry weight – Accumulation of fresh weight

Answer: Accumulation of fresh weight – Accumulation of dry weight – Physiological maturity – Harvestable maturity

3. Stages of seed dispersal

Abscission of seeds – Abscission of fruits – Physiological maturity – Seed dispersal

Answer: Physiological maturity – Abscission of fruits – Abscission of seeds – Seed dispersal

4. Pollination – Anthesis – Fertilization – Seed set – Fruit set – Flowering

Answer: Flowering – Anthesis – Pollination – Fertilization – Fruit set – Seed set

5. Sucrose - Glucose - Imbibition - GA synthesis - Break down of starch - Hydrolytic enzyme

Answer: Imbibition – GA synthesis – Hydrolytic enzyme – Breakdown of starch – Sucrose – Glucose

6. Imbibition – Embryo growth – Break down of storage material – Cell elongation – Cell division – Cell differentiation

Answer: Imbibition – Breakdown of storage material – Embryo growth – Cell division – Cell elongation – Cell differentiation

7. GA – Dormancy induction – ABA – Dormancy release – germination

Answer: ABA – Dormancy induction – GA – Dormancy release – Germination

8. Moisture loss – Endosperm development – Embryo development – Fertilization – Dry matter accumulation – Germination

Answer: Fertilization – Endosperm development – Embryo development – Dry matter accumulation – Germination – Moisture loss

9. Double fertilization – Seed – Zygote – Fruit – Pollination – Endosperm development – Embryo development

Answer: Pollination – Double fertilization – Zygote – Endosperm development – Embryo development – Fruit – Seed

10. Megagametogenesis – Male gamete – Microsporogenesis – Megasporogenesis – Microgametogenesis – Female gamete

Answer: Megasporogenesis – Megagametogenesis – Microsporo-genesis – Microgametogenesis – Female gamete

1.13. DIFFERENTIATE THE FOLLOWING

1. Differentiate between drupe and berry with examples

Drupe	Berry
Fleshy fruit with single seed	Fleshy fruit with more than one seed
It contains hard and stony endocarp	It contains soft pericarp
Eg. Apple, Pear	Eg. Grapes, Papaya, Sapota

2. Differentiate between dehiscent fruit and indehiscent fruit with examples

Dehiscent fruit	Indehiscent fruit
Dry fruits that split at maturity	Dry fruits that do not split at maturity
Occurrence of shattering loss	No shattering loss
Seed extraction is easy	Seed extraction is difficult
Eg. Grams, Bhendi, Sesamum	Eg. Sunflower, Wheat, Maize

3. Differentiate between determinate inflorescence and indeterminate inflorescence with examples

Determinate inflorescence	Indeterminate inflorescence
Determinate simple inflorescence generally called as cymose	Indeterminate simple inflorescence generally called racemose
Main kind of cymose inflorescence is the cyme	Main kind of racemose inflorescence is the raceme
Flowering period stops at period of time	Continuous flowering
Once over harvesting	Multiple harvesting
Mechanical harvesting is possible	Mechanical harvesting is impossible
Does not show seed polymorphism	Exhibits seed polymorphism
Eg. Paddy, Wheat, Sorghum	Eg. Greengram, Tomato

4. Differentiate between raceme and umbel with examples

Raceme	Umbel
Unbranched indeterminate inflorescence	Branched indeterminate inflorescence
Short floret stalks and florets along the axis	Multiple florets pedicels of equal length arise from a common point
Eg. Castor	Eg. Onion, Carrot

5. Differentiate between compound raceme and compound umbel with examples

Compound raceme	Compound umbel
The inflorescence may or may not end with a final raceme	Single flowers are replaced by smaller umbels called umbels
Compound raceme is called as panicle	Compound umbel is called as umbellates
Eg. Paddy	Eg. Coriander

6. Differentiate between monoecious and dioecious with examples

Monoecious	Dioecious
Species having functional male and female organs on the separate places on the same plant	Male and female functional organs present in separate plant
Eg. Maize, Ashgourd	Eg. Papaya, Palmyrah

7. Differentiate between perfect flowers and imperfect flowers with examples

Perfect flowers	Imperfect flowers
Have both male (androecium) and female (gynoecium) reproductive structures in the same flower	Reproductive structure that is either functionally male or female
Also called as bisexual flowers	Also called as incomplete flowers
Aids self pollination	Aids cross pollination
Eg. Lily, rose, Paddy, Groundnut	Eg. Maize, Bajra, Cucurbits

8. Differentiate between gynomonoecious and andromonoecious with examples

Gynomonoecious	Andromonoecious
Plant species having each individual bearing both female and hermaphrodite flowers on the same plant	Plant species producing both bisexual and male flowers on the same plant
Eg. Chenopodium	Eg. Cucumis melo

9. Differentiate between homogamy and dichogamy with examples

Homogamy	Dichogamy
Male and female sex organ mature at same time	Maturation of anthers and stigma at different times
Helps in the occurrence of self pollination	Prevents the self pollination, but cross pollination occurs
Eg. Rice, Wheat, Barley, Oat	Eg. Maize, Bajra

10. Differentiate between hydrophily and anemophily with examples

Hydrophily	Anemophily
Pollination brought about by means of water	Pollination brought about by means of wind
Aids in cross pollination	Aids in cross pollination
Eg. Aquatic plants water hyacinth, Vallisneria	Eg. Coconut, Maize, Sugarcane, Date palm

11. Differentiate between entomophily and ornithophily with examples

Entomophily	Ornithophily
Cross pollination brought about by insects	Cross pollination brought about by birds
Insects attract the flowers by their special adaptations include colour, nectar and scent	
Eg. Orchids, Sunflower, Alfalfa	Eg. Erythrina, Bombax, Bignonia

12. Differentiate between saturated fatty acids and unsaturated fatty acids with examples

Saturated fatty acids	Unsaturated fatty acids
Hydrocarbon chain without double bond	Possess one to more double bonds in their hydrocarbon chain
Solid at room temperature	Liquid at room temperature
Relatively higher melting point	Relatively lower melting point
Occurrence in most animal fats	Occurs in most plant cells
These are long lasting and do not get spoiled quickly	These get spoiled quickly
Eg. Butter, Peanut, Vegetable oil	Eg. Avocado, Soybean, Coconut oil, Olive oil

13. Differentiate between simple fruit and aggregate fruit with examples

Simple fruit	Aggregate fruit
Developed from a single pistil of a single flower	Developed from a single flower that has many separate carpels

Simple fruit	Aggregate fruit
The pistil may be composed of several fused carpels	Each carpel will develop into a fleshy globule that contains one or many seeds
Eg. Tomato, Peach, Mango	Eg. Black berries, Annona squamosa

14. Differentiate between cross pollination and self pollination with examples

Cross pollination	Self pollination
Transfer of pollen grains from the anther of a flower to the stigma of a flower of a different plant of the same species	Transfer of pollen grains from the anther to the stigma of the same flower or of another flower of the same plant
Pollinating agent is required Wind, Insects, Water, Birds etc.,	No pollinating agent is required
Eg. Apple, Grape, Cumbu, Maize	Eg. Paddy, Groundnut

15. Differentiate between epigeal germination and hypogeal germination with examples

Epigeal germination	Hypogeal germination
Type of germination in which cotyledons and shoots are carried above the soil level	In which, cotyledon or comparable structure remain in the soil
Hypocotyl is elongated	Epicotyl is elongated
Eg. Maize	Eg. Blackgram, Greengram

16. Differentiate between physiological maturity and harvestable maturity with examples

Physiological maturity	Harvestable maturity
It is the stage when a fruit and seed are capable of further development or ripening when it is harvested	It is the stage followed by the physiological maturity
Physiological maturity attains when the development is over	Harvestable maturity attains 5-6 days after physiological maturity
Seed possess maximum dry weight, vigour and germination	Dry weight, vigour index and germination started to decline
Low moisture content	Very low moisture content
Correct stage for harvesting seed crop	Harvestable maturity leads to loss of seed quality
Eg : 25-30% in orthodox species 35-50% in recalcitrant species	Eg : 15-20%

17. Differentiate between exalbuminous seed and albuminous seed with examples

Exalbuminous seed	Albuminous seed
Lack endosperm	Endosperm is present
Cotyledons are thick and fleshy as they store food	Cotyledons are thin and membranous
Eg. Gram, Pea, Mango, Mustard, Orchids	Eg. Custard apple, Poppy seeds

18. Differentiate between dicot seeds and monocot seeds with examples

Dicot seeds	Monocot seeds
Have two halves	Do not divide into two halves
There are two cotyledons	There is only one cotyledon called scutellum
Cotyledons are thick and fleshy	Cotyledons are thin and narrow
Endosperm is absent	Endosperm is present
Embryo is large	Embryo is small
Plumule large and plumule leaves folded	Plumule very small and plumule leaves rolled
Aleurone layer is not present	Aleurone layer is present
Eg : Redgram, Blackgram, Groundnut etc.	Eg : Rice, Maize, Sorghum, etc.,

19. Differentiate between osmosis and plasmolysis

Osmosis	Plasmolysis
Osmosis is a physical process	Plasmolysis is a vital process, concerned with living cells
Water molecules move through semi-permeable membrane from region of higher diffusion pressure to lower diffusion pressure i.e. towards concentrated solution	Water diffuses out of the cell as a result of exosmosis. This causes protoplasm to contract away from the cell wall

20. Differentiate between phloem and xylem with examples

Phloem	Xylem
Transportation of food and nutrients such as sugar and amino acids from leaves to storage organs and growing part of plant	Water and mineral transport from roots to aerial parts of the plant
Bidirectional (moves up or down the plant stem from `source to sink')	Unidirectional (move upwards only)
Phloem occur on outer side of the vascular bundle	Xylem occupy the center of the vascular bundle

Phloem	Xylem
Living tissue with little cytoplasm, but no nucleus/tonoplast	Dead tissue at maturity, so it is hollow with no cell contents

21. Differentiate between protandry and protogyny with examples

Protandry	Protogyny
Anthers mature and release pollen before the stigmas are receptive	Stigmas become receptive before pollen release
Aids in cross pollination	Aids in cross pollination
Eg : Maize	Eg : Bajra

22. Differentiate between dent corn and flint corn

Dent corn	Flint corn
Dent kernals contain both hard and soft starch that become indented at maturity	Hard outer shell and kernels with a range of colors from white to red
Presence of horny endosperm at the sides and back of the kernels while the central core is a soft, floury endosperm extending to the crown of the endosperm	Kernels are smooth and rounded and the ear long and slender with a comparatively small number of rows or kernels

23. Differentiate between sweet corn and flour corn

Sweet corn	Flour corn
Primarily eaten as cob and seldom used for feed or flour	Flour corn is used in baked goods, because it has a soft, starch, filled, kernel that is easy to grind
Sweet corn is extra sweet, because it contains more natural sugars (10% sugar).	Entire endosperm is floury
Starch in the kernels is genetically controlled	Flour corns are easy for processing and high recovery of fine flour is desirable characteristics

24. Differentiate between photo dormancy and thermo dormancy

Photo dormancy	Thermo dormancy
Light plays a major role in development or releases from dormancy	Temperature plays a role in the development or release of dormancy
Dormancy is released by exposure of light	Dormancy is released by stratification
Eg. *Cecropia obtusifolia, Ficus* spp. *Eucalyptus* spp, tobacco	Eg. *Pinus* spp, Abies spp, *Quercus* spp. etc.,

25. Differentiate between aerobic respiration and anaerobic respiration

Aerobic respiration	Anaerobic respiration
Aerobic respiration occurs in presence of oxygen	Anaerobic respiration occurs in absence of oxygen
CO_2 and water is produced as a result	Lactic acid or alcohol is produced as a result
Lot of energy is liberated (38 ATP)	Relatively small energy is liberated (2 ATP)
Occurs in cytoplasm and mitochondria	Occurs in cytoplasm
Occurs during the stages of glycolysis, Kreb Cycle	Occurs during the stages of glycolysis and fermentation

26. Differentiate between absorption and adsorption

Absorption	Adsorption
Intake of water molecules in to a seed from the ambient air	Accumulation and adhesion of a thin layer of water or gases on the surface or from another surface substance
It is a bulk phenomenon	It is surface based phenomenon
Absorption will be uniform throughout the bulk of the material	In this phenomenon, the concentration is same throughout the material
It's rate remains same throughout the process	It's rate is high in the beginning and then decreases till equilibrium is attained

27. Differentiate between mitosis and meiosis

Mitosis	Meiosis
It occurs in vegetative or somatic cells	It occurs in reproductive cells
Two daughter cells are produced after mitosis	Four daughter cells are produced after meiosis
Daughter cell possess the same chromosome number as that of parent cell	Chromosome number of daughter cells is reduced to half
It is of short duration	It is of long duration
Daughter cells are similar to parent cell genetically	Daughter cells have genetic differences from the parent cell
There is no crossing over in mitosis	Crossing over takes place between the nonsister chromatids
Homologous chromosomes are not arranged in pairs in the equatorial plate	Homologous chromosomes are arranged in pairs in the equatorial plate

Mitosis	Meiosis
DNA replication is completed in the interphase	DNA replication is not completed in the interphase
In metaphase, the centromeres are lined up on the equatorial plane and the arms extend into the cytoplasm	In metaphase, the centromeres of the homologous chromosomes lie towards the two opposite poles and their arms extend towards the equator.

28. Differentiate between animal cell and plant cell

Animal cell	Plant cell
Animal cell is comparatively smaller in size	Plant cells are larger than animal cells
Cell wall is absent	Plant cell has rigid wall on the outside
Plastids are usually absent	Plastids are found in plant cell
Chlorophyll is absent	Plant cells exposed to sunlight posses chlorophyll containing plastids called chloroplast
Nucleus usually lies in the centre	Nucleus lies on one side in the peripheral cytoplasm
Centrioles are present in animal cell	Centrioles are usually absent except in lower plants
Cytoskeleton contain intermediate fibres	Cytoskeleton does not contain intermediate fibres
Mitochondria are generally more numerous	Mitochondria are comparatively fewer
Glyoxysomes are absent	Glyoxysomes may be present
Cytokiensis occurs by constriction	Cytokiensis occurs by cell plate method
Lysosomes occur in animal cells	Lysosomes are rare. Their activity is performed by specialized vacuoles

29. Differentiate between cell and tissue

Cell	Tissue
Cells are smallest unit of life	Group of like minded cells or similar cells
Developed by the process of mitosis and meiosis	Tissue repairs through regeneration and fibrosis
Cell performs the functions of growth, metabolism, creation and protein synthesis	Group of similar cells come together to perform a similar function and to form organs

30. Differentiate between physical dormancy and physiological dormancy

Physical dormancy	Physiological dormancy
Caused by water impermeable seed and fruit coat such as hard seed coat, thick pericarp, fibry mesocarp etc.	Caused by a physiologically inhibiting mechanism of the embryo that pre-vents emergence of the radicle (fully development, but dormant embryo)
The break of physical dormancy is related to formation of an opening in the seed or otherwise, a mechanical, chemical or physical corrosion of the seed coat so that its impermeability is elevated and the seed can get imbibed with water	Environmental factors responsible for the physiological dormancy and also involved in the stimulation of germination
Overcome by mechanical scarification and acid scarification using commercial grade sulphuric acid Eg. Malvaceae, Anacardiaceae, Cucurbitaceae	Seed soaking with growth regulators (GA, ethrel and chemical solution - KNO_3, thiourea) Eg. Fraxinus, Polygonum spp, Apple

31. Differentiate between cell division and cell elongation

Cell division	Cell elongation
Cell division is the post germination phenomena concerned with axis growth and establishment of seedling	Cell can enlarge and elongate as fast as the growth of their cell wall allows
Radicle elongation is a result of cell division, but not radicle protrusion	Cell elongation plays an important role in root and shoot growth
Cell division is due to mitosis	Cell elongation is due to vacuole enlargement and cell wall loosening

32. Differentiate between mechanical scarification and acid scarification

Mechanical scarification	Acid scarification
Involves breaking or weakening the seed coat and can be carried out using sand paper or mechanical scarifier to abrade the seed coat	Soak the seeds with commercial grade sulphuric acid
Applicable only to hard seed coated seeds of leguminous group of species	It is applicable to many species not only leguminous and malvaceous seeds
Treatment to bulk seed is not possible	Most effective method of bulk treatment of very hard coated seeds

Mechanical scarification	Acid scarification
Tools like sand paper, hammer, soil cutter, knife, mechanical scarifier and seed gun are needed to practice this method	Requires no special equipment
Seeds should be used immediately after scarification	Seeds can be stored for a period after treatment
It is of relatively low cost when compared to acid treatment	It is expensive
Risk of over treatment is small, provided that the radicle region is avoided	Seeds are in risk of being damaged by over treatment
Repeated washing and neutralization with lime solution is not needed	Repeated washings and lime solution neutralization is must to make the seed free of acid
Free from pollution and environmental hazard	Causing environmental hazard and injurious to human being

33. Differentiate between quiescent seed and dormant seed

Quiescent seed	Dormant seed
Quiescence is a state of suspended growth of the embryo or a resting condition of the seed	Dormancy is a state that requires a special event or `trigger' before the embryo can resume growth
Seeds cannot germinate in unfavourable condition	Seeds cannot germinate under favourable condition

34. Differentiate between embryo dormancy and endosperm dormancy

Embryo dormancy	Endosperm dormancy
A block that inhibits extension growth and therefore, excised embryos do not grow	The combination of an embryo with low growth potential and mechanical constraint from the seed covering layers like endosperm can result in dormancy
A dormant embryo is characterized by high ABA/GA ratio. Embryo dormancy release involves remodeling of hormone biosynthesis and degra-dation towards a low ABA/GA ratio. A decrease in ABA sensitivity and increase in GA sensitivity	GA by increasing the embryo growth potential and by promoting endosperm weakening which is achieved through ABA through ABA independent and ABA inhibited mechanism or by chilling/after ripening
It requires a specific period of cold (or heat) with available moisture and oxygen	It requires afterripening for endosperm weakening
Eg. *Nicotiana* sp, *Helianthus annuus*	Eg. *A. thaliana* and Solanaceous species

35. Differentiate between prechilling and afterripening

Prechilling	Afterripening
Prechilling of seeds is given to break internal dormancy, which stimulates cold winter conditions	A period of usually several months of dry storage at room temperature of freshly harvested, mature seeds is a common method to release dormancy and to promote germination
At cold temperature, more oxygen is soluble in water, so the oxygen requirements of the embryos are better satisfied Eg. Camellia, Maple, Oak	Afterripening can be characterized by a widening of temperature range for germination and decrease in ABA level and an increase in GA level Eg. Tobacco

36. Differentiate between vernalization and stratification with example

Vernalization	Stratification
Artificial exposure of plants or seeds to low temperature of 0°C	Process of subjecting seeds to both cold and moist conditions. Typically, temperature between 1°C and 5°C
A longer cold treatment (several weeks) of seedlings or young plants which promote flowering	A brief cold treatment (few days) in moisture medium to break seed dorm-ancy, results in an even germination
Example. Umbelliferae, Cruciferae, Alliaceae	Example. Annonaceae, Winteraceae

37. Differentiate between warm stratification and cold stratification

Warm stratification	Cold stratification
Exposing the presoaked or premoistened seeds to cold temperature of 3-5°C for 48h and then to warm temperature of 20-25°C. Constant temperature and 20°C and 30°C alternate temperature in moist substratum of sand or peat	Exposing the presoaked or premoistened seeds at low temperature 3-5°C over a moisture substratum like sand or peat
It requires the exposure time of two weeks to 16 weeks depending upon species	It requires 2-3 days to several months
Can be stored at refrigerator for cold stratification and then at constant and alternate temperature	Can be stored in the refrigerator
Both for breaking physical and physiological dormancy and increase endosperm GA level	Triggers biochemical activity and reduces microorganism activity

Warm stratification	Cold stratification
Example. *Areca palms, Fraxinus* spp.	Example. Cherry, Oilpalm, Mustard seeds

38. Differentiate between physical scarification and bioscarification

Physical scarification	Bioscarification
This is the manual treatment that access the seed coat permeability with out damaging the embryo	Seed coats are broken down as they pass through the digestive system track of animals and birds guts, because of the action of the strong digestive chemicals
Cut, drill or file a small hole in the seed coat and abrasion by sand paper Eg. Legumes	Eg. Birds and animals; Acacias and neem Eg. *Tectona grandis*; Eating or damaging the pericarp and fibrous mesocarp by termites
Seed coat is softened which helps in the movement of water and oxygen	Reduced the mass of seed coat or seed coat is removed which aids in entry of water and oxygen

39. Differentiate between phase-I germination and phase–II germination

Phase-I germination	Phase–II germination
Imbibition is a purely physical process which occur whether the seed is dor-mant or non dormant (except physical dormancy), viable or non viable	During this phase, the seed mobilizes the stored food reserves such as protein and starch and metabolic enzymes become active
Rate of flow is very high	Rate of water uptake is very slow

40. Differentiate between phase-I and phase–III of germination process

Phase-I germination	Phase –III germination
It is the active water uptake and imbibition phase	It is a germination, post germination and seedling growth phase
In Phase-I, water uptake (or imbibi-tion) is high and increase fast with time of germination	Water uptake is also high
DNA and mitochondria repairing, respiration and synthesis of protein	Food reserves converted to simple food which are mobilized
Seed/cell is preparing for initiation of metabolic activity	Appearance of epicotyl, hypocotyl, and cotyledons, and growth of seedling
New protein, RNA and mitochondria synthesis	Cell division is post germination phenomena by cell cycle activity

41. Differentiate between one step germination and two step germination with examples

One step germination	Two step germination
Usually occurs in seed without endo-sperm (non endospermic). Testa rup-ture and initial radicle elongation result in the completion of germination	Rupture of testa (seed coat) and endo-sperm are two separate events. This is wide spread in the plant kingdom
Radicle does not rupture endosperm since it does not exist. ABA does not inhibit imbibition and testa rupture, but inhibits phase III water uptake and the transition from germination to post germination growth	ABA appears to inhibit endosperm rupture, but not testa rupture and appears to be counteracted by GA
Eg. Brassica napur and peas	Eg. Lepedium sativum, Arabidopsis thaliana, Nicotiana, Petunia, Chenopodium

42. Differentiate between phase-II germination and phase –III germination

Phase-II germination	Phase –III germination
Following imbibition, a lag phase (phase II) is characterized by little or no water uptake	It is a post germination and seedling growth phase
Mitochondria become active and respiration and ATP synthesis increases	Water uptake is also high
Protein synthesis increases and new proteins are synthesized from new mRNAs	Food reserves converted to simple food which are mobilized
Metabolism of storage reserves begins	Appearance of epicotyls, hypocotyls, and cotyledons, and growth of seedling
Specific enzymes are produced such as hydrolases (• amylase) which is triggered by GA synthesis	Cell division is post germination phenomena by cell cycle activity

43. Differentiate between seed development and seed maturation

Seed development	Seed maturation
After fertilization, development of fertilized ovule (zygote) into a seed involving different stages	Accumulation of reserve food material and loss of moisture content after seed formation
Sequence of events controlled by the genotypes	A set of successive stages in prepar-ation for successful seed germination

Unit 2

SEED PRODUCTION

2.1. BREEDING METHODS

2.1.1. Choose the correct answer

1. The phenomenon of heterosis of the progeny over better parent is

(a) Heterosis (b) Heterobeltiosis

(c) Relative heterosis (d) Standard heterosis

2. Heterosis over mid parent value

(a) Heterosis (b) Heterobeltiosis

(c) Relative heterosis (d) Standard heterosis

3. Heterosis over the check parent

(a) Heterosis (b) Heterobeltiosis

(c) Relative heterosis (d) Standard heterosis

4. Which one of the following formula is used to calculate relative heterosis

(a) $\frac{F - BP \times 100}{BP}$ (b) $\frac{F - MP \times 100}{MP}$

(c) $\frac{F - CV \times 100}{CV}$ (d) $\frac{F - CP \times 100}{CP}$

5. Which one of the following formula is used to calculate heterobeltiois

(a) $\frac{F - BP \times 100}{BP}$ (b) $\frac{F1 - MP \times 100}{MP}$

(c) $\frac{F - CV \times 100}{CV}$ (d) $\frac{F1 - CP \times 100}{CP}$

6. Which one of the following formula is used to calculate standard heterosis

(a) $\frac{F - BP \times 100}{BP}$ (b) $\frac{F1 - MP \times 100}{MP}$

(c) $\frac{F - CV \times 100}{CV}$ (d) $\frac{F1 - CP \times 100}{CP}$

7. The plant tissue or organ excised and used for *in vitro* culture is known as
   - (a) Transgenic plant
   - (b) Bulb
   - (c) Explant
   - (d) Callus
8. Triticale has been developed by crossing wheat with
   - (a) Barley
   - (b) Phalaris
   - (c) Rye
   - (d) Bajra
9. Rice gene bank is situated at
   - (a) India
   - (b) Philippines
   - (c) China
   - (d) Japan
10. Different types of hybrids available in sorghum
    - (a) Mono cross
    - (b) Double cross
    - (c) Triple cross
    - (d) Single cross
11. A × B crossing in pearl millet is done at the stage of
    - (a) Certified seed
    - (b) Breeder seed
    - (c) Foundation seed
    - (d) Registered seed
12. Top cross is otherwise known as
    - (a) Inbred variety cross
    - (b) Single cross
    - (c) Double cross
    - (d) Triple cross
13. In three way cross hybrid seed production, the pedigree involved in
    - (a) A × B
    - (b) (A × B) × C
    - (c) (A × B) × open pollinated variety
    - (d) (A × B) × C (C × D)
14. The first generation resulting from crossing of two inbreds is
    - (a) Hybrid
    - (b) Synthetics
    - (c) Composites
    - (d) Variety
15. The phenomenon of heterosis of the progeny over better parent is
    - (a) Heterosis
    - (b) Heterobeltiosis
    - (c) Relative heterosis
    - (d) Standard heterosis
16. The first generation resulting from the controlled crossing of an approved inbred line and open pollinated variety is
    - (a) Single cross
    - (b) Double cross
    - (c) Top cross
    - (d) Four way cross
17. A cross between parents differing in a single gene is
    - (a) Double hybrid
    - (b) Mono hybrid
    - (c) Single cross hybrid
    - (d) Triple cross hybrid
18. An alloautogamous oilseed is
    - (a) Safflower
    - (b) Groundnut
    - (c) Sesame
    - (d) Mustard

19. The average performance of a strain or genotype in a series of crosses or hybrid combinations and a measure of additive gene action is known as
(a) General combining ability (b) Specific combining ability
(c) Standard error (d) Directed combining ability

20. The heterosis estimated over the superior parent is known as
(a) Heterobeltiosis (b) Clonal degeneration
(c) Standard heterosis (d) Inbreeding depression

21. The superiority of an F_1 hybrid over both its parents in terms of yield and some other characters is defined as
(a) Inbreeding depression (b) Heterosis
(c) Clonal degeneration (d) Heterobeltiosis

22. The progeny obtained by hybridization between two or more genetically dissimilar individuals, strains, pure lines etc., is called as
(a) Embryo (b) Explant
(c) Cybrid (d) Hybrid

23. The homozygous line developed by continued inbreeding, usually selfing, accompanied by selection in cross pollinated crops is called as
(a) Hybrid derivative (b) Inbred line
(c) Ideotype (d) Cybrid line

24. The reduction or loss in vigour and fertility as a result of inbreeding is called as
(a) Inbreeding depression (b) Co-heritability
(c) Heterosis (d) Heritability

25. Two lines, which are identical with respect to all other genes, except for the gene under investigation are known as
(a) Multilines (b) Isogenic lines
(c) Hybrid lines (d) Transition lines

26. The appearance of a plant with respect to particular character, such as plant height, flower colour etc., is known as
(a) Genotype (b) Phenotype
(c) Pathotype (d) Biotype

27. The observable variation present in a character in a population, which includes genotypic and environmental components of variation, is known as
(a) Phenotypic variation (b) Genotypic variation
(c) Pathotype variation (d) Biotype variation

28. National Biodiversity Board in India is situated at
(a) Bangalore (b) New Delhi
(c) Mumbai (d) Chennai

29. Progeny of a single homozygous plant of a self pollinated species is called as
(a) Dihybrid (b) Pure line
(c) Inbred (d) Monohybrid

30. The variety produced by crossing in all combinations of a number of lines that combine well with each other is called as
(a) Varietal mixtures (b) Hybrid variety
(c) Composite variety (d) Synthetic variety

31. The cross between a single cross hybrid (A × B) and inbred (C) is termed as
(a) Three way cross (b) Double cross
(c) Single cross (d) Poly cross

32. The male sterile parent is crossed with which one of the following line to produce hybrid seed
(a) A line (b) B line
(c) R line (d) C line

33. Hybrid rice seed production is carried out by
(a) One line system (b) Multi line system
(c) Three line system (d) Four line system

34. The first generation progeny of a cross between two single cross hybrids
(a) Single cross hybrid (b) Double cross hybrid
(c) Back cross hybrid (d) Three way cross hybrid

35. The dominance hypothesis is first proposed by
(a) Shull (b) Davenport
(c) Bruce (d) Jones

36. A line became homozygous by repeated selfing upto
(a) 3 generations (b) 5 generations
(c) 8 generations (d) 12 generations

37. Over dominance hypothesis was independently proposed by
(a) Shull (b) Collins
(c) Jones (d) East and Shull

38. The male parent of CORH 4 rice hybrid is
(a) C 20 R (b) IR 1098-66 R
(c) CB 87 R (d) CB 174 R

39. Single cross hybrid in maize is between
(a) A × B (b) (C × D) × (C)
(c) (A × A) × (C x D) (d) (A × B) × (C × D) × (E × F)

♦ Answer Keys ♦

Q.No.	Ans.	Q.No.	Ans.	Q.No.	Ans.	Q.No.	Ans.	Q.No.	Ans.	Q.No.	Ans.
1	b	2	c	3	d	4	b	5	a	6	c
7	c	8	c	9	b	10	d	11	c	12	a
13	b	14	a	15	b	16	c	17	b	18	c
19	a	20	a	21	b	22	d	23	b	24	a
25	b	26	b	27	a	28	d	29	b	30	d
31	a	32	a	33	c	34	b	35	b	36	c
37	d	38	d	39	a						

2.1.2. Fill in the blanks

1. H = V (W-1) H-value means .... value.
2. A gene usually dominant that effectively overcomes the effect of male sterile cytoplasm of a male sterile line is called as ....
3. Ability of an allele to express itself in the heterozygote state is called as ....
4. .... are alternative forms of same gene and are located at the same point in homologous chromosomes
5. Functionally .... is the unit of inheritance
6. The genetic engineering is a tool to develop .... crops.
7. Nucleus seed production is done by adopting ....
8. The recent concept in rice hybrid is ....
9. The term heterosis was proposed by ....
10. The term mutation was proposed by ....
11. .... hybrid is a cross between two single crosses
12. A cross involving the hybrids and one of its parents is named as ....
13. Superiority of F_1 over the parent is known as ....
14. Progeny from hybridization between two inbred lines is named as ....
15. A nearly homozygous line developed by continued selfing including, usually selfing accompanied by selection is called as ....
16. A sudden heritable change in a characteristic of an organism is called as ....

♦ Answer Keys ♦

Q.No.	Ans.	Q.No.	Ans.	Q.No.	Ans.
1	Heterogeneity	2	Restorer gene	3	Dominance
4	Alleles	5	Gene	6	Genetically modified
7	Maintenance Breeding	8	Two line breeding	9	Dr GH Shull
10	De Vries	11	Double cross	12	Backcross
13	Heterosis	14	Hybrid	15	Inbred line
16	Mutation				

2.1.3. State True or False

1. In early days, cultivation of crops is brought out through introduction of indigenous varieties
2. A double cross is a first generation hybrid between two inbred lines
3. B line is called as maintainer line
4. F_1 denotes the first generation off spring resulting from mating of two parents
5. R line restores the sterility of 'A' line
6. In hybrid seed production programme, the maintenance breeding is practiced in truthful labeled stage
7. The term heterosis was first used by Dr. Singh
8. Hybrid vigour was first reported in tobacco
9. Heterosis is expressed as improved function of any biological quality
10. The heterozygotes for hooded gene in maize show higher rate of photosynthesis than two homozygous
11. Breeding for disease resistance is known as combination breeding
12. Breeding for high yield is known as combination breeding
13. Inbreeding in maize was first published by East
14. Seedless watermelon is a triploid plant
15. Backcross is a cross between F_1 hybrid with a recessive parent
16. In maize double cross hybrid, three parents are involved
17. Luxuriance is the increased vigour and size of intra specific hybrids
18. Dominance hypothesis is the result of prevention of expression of harmful recessives by their dominant alleles
19. The term heterosis was first used by Shull in 1914

♦ Answer Keys ♦

Q.No.	Ans.	Q.No.	Ans.	Q.No.	Ans.	Q.No.	Ans.	Q.No.	Ans.	Q.No.	Ans.
1	True	2	False	3	True	4	True	5	False	6	False
7	False	8	True	9	True	10	False	11	True	12	False
13	True	14	True	15	False	16	False	17	False	18	True
19	True										

2.2. VARIETY RELEASE

2.2.1. Choose the correct answer

1. Notification is done under seed act to bring the variety
   (a) For national level cultivation
   (b) For state level cultivation
   (c) For quality seed production
   (d) Under purview of seed certification

2. **The notified variety or a farmers variety that is in public domain is called**
   (a) Derived variety (b) Extant variety
   (c) Farmers variety (d) Extinct variety
3. **The variety that has been traditionally cultivated and evolved by farmers, or is a wild relative or land race in common knowledge of farmers is called**
   (a) Derived variety (b) Extant variety
   (c) Farmers variety (d) Extinct variety
4. **What is the seed renewal period for varieties?**
   (a) 1-2 years (b) 2-3 years
   (c) 3-4 years (d) 4-5 years
5. **What is the seed renewal period for hybrid?**
   (a) Every year (b) Once in 2 years
   (c) Once in 3 years (d) Once in 4 years

♦ Answer Keys ♦

Q.No.	Ans.	Q.No.	Ans.	Q.No.	Ans.	Q.No.	Ans.	Q.No.	Ans.
1	d	2	b	3	c	4	c	5	a

2.2.2. Fill in the blanks

1. The authority empowered for the release of the new crop variety is ....
2. .... the authority empowered for the release of the new crop variety at state level
3. IVT means ....
4. AVT means ....
5. IHT means ....
6. MLT means ....

♦ Answer Keys ♦

Q.No.	Ans.	Q.No.	Ans.	Q.No.	Ans.
1	State/Central variety release committee	2	State variety release committee	3	Initial Varietal Trial
4	Advanced Varietal Trial	5	Initial Hybrid Trial	6	Multilocation Trial

2.2.3. State True or False

1. State varieties are notified by state variety release committee
2. All varieties released by the breeders are subjected to certification
3. GMS system is effectively used in different crops for hybrid seed production
4. The development of new varieties usually takes 7-8 years

♦ Answer Keys ♦

Q.No.	Ans.	Q.No.	Ans.	Q.No.	Ans.	Q.No.	Ans.
1	False	2	False	3	False	4	False

2.3. DETERIORATION OF VARIETY

2.3.1. Choose the correct answer

1. A rogue is
   (a) Normal plant (b) Crop plant
   (c) Off type plant (d) A line
2. The genetic purity of a variety can be tested by
   (a) Grow out test (b) Germination test
   (c) Seed health test (d) Seed vigour test
3. In maize, the effect of foreign pollen of same generation is known as
   (a) Metaenia (b) Xenia
   (c) Synthetic effect (d) Mutation
4. The most important factor which cause varietal deterioration
   (a) Mutation
   (b) Techniques of plant breeder
   (c) Selective influence of diseases
   (d) Mechanical admixture
5. Volunteer plants are
   (a) Plants of previous season (b) Low vigour plants
   (c) Diseased plants (d) Weeds
6. The developmental variation occurs in the varieties due to
   (a) Same environment (b) Same photoperiod
   (c) Different growth phase (d) Different environment
7. Mechanical mixtures commonly occur at the time of
   (a) Sowing (b) Seedling stage
   (c) Flowering (d) Pod formation stage
8. In sexually propagated crops, the genetic contamination takes place because
   (a) Self pollination (b) Sib mating
   (c) Self depression (d) Natural crossing
9. The deterioration of varieties due to natural crossing occurs due to any one of the following
   (a) Natural crossing with pests affected plants
   (b) Natural crossing with same variety
   (c) Natural crossing with undesirable types
   (d) Natural crossing with closely related plants

10. Who suggested the method for maintenance of genetic purity?
    (a) Abdul Baki and Anderson (1973) (b) Hartman and Kestor (1968)
    (c) Baskin and Delouche (1973) (d) Woodstock and Bewley (1968)
11. What is the isolation distance for onion certified seed production?
    (a) 500 m (b) 750 m
    (c) 1000 m (d) 1250 m
12. What is the isolation distance for foundation seed production of coriander?
    (a) 400 m (b) 800 m
    (c) 1200 m (d) 1600 m

♦ Answer Keys ♦

Q.No.	Ans.	Q.No.	Ans.	Q.No.	Ans.	Q.No.	Ans.	Q.No.	Ans.	Q.No.	Ans.
1	c	2	a	3	b	4	d	5	a	6	d
7	a	8	d	9	c	10	b	11	a	12	b

2.3.2. Fill in the blanks

1. The plants which cause genetic contamination of a seed crop is called as ....
2. Visual symptom of genetic contamination in maize seed is ....
3. Resultant seed should be .... in seed production while it may or may not be in grain production
4. The genetic purity is maintained in the field by an operation called ....

♦ Answer Keys ♦

Q.No.	Ans.	Q.No.	Ans.	Q.No.	Ans.	Q.No.	Ans.
1	Rogue	2	Xenia	3	True to type	4	Roguing

2.3.3. State True or False

1. Pollen shedder causes genetic contamination in hybrid seed production
2. Morphological traits of varieties are environmentally caused
3. Seed size variation is governed by genetic factor within a genome
4. Live barrier is more suitable than dead barrier in arresting pollen contamination
5. Xenia and metaxenia are one and the same
6. The seed renewal period for hybrid is 2 years
7. Hard seed development is controlled by both genetic and environmental factors
8. Genetic contamination is very frequent in self pollinated crop
9. The presence of var. CO 26 in CO 25 seed production of sorghum is called as an off type

10. Xenia is an example for genetic contamination
11. Not only storage potential and disease resistance, but also germination and vigour are the seed production characteristics that are genetically inherited
12. Presence of B line plants in A line are called partials
13. Off-type plants cause physical contamination
14. Genetic purity deterioration is very fast in groundnut
15. Genetic contamination is more in cross pollinated crop than in self pollinated crop
16. Border rows prevents cross contamination

♦ Answer Keys ♦

Q.No.	Ans.	Q.No.	Ans.	Q.No.	Ans.	Q.No.	Ans.	Q.No.	Ans.	Q.No.	Ans.
1	True	2	True	3	True	4	False	5	False	6	False
7	True	8	False	9	True	10	True	11	True	12	True
13	False	14	False	15	True	16	True				

2.4. SEED GENERATION SYSTEMS

2.4.1. Choose the correct answer

1. The progeny of the breeder seed is
   (a) Certified seed (b) Foundation seed
   (c) Nucleus seed (d) Registered seed
2. The tag colour of foundation seed class is
   (a) Blue (b) White
   (c) Yellow (d) Red
3. Breeder seed production plot is inspected by
   (a) Monitoring team (b) Seed Certification Officers
   (c) University personnel (d) Agricultural Officer
4. The progeny of foundation seed is
   (a) Nucleus seed (b) Breeder seed
   (c) Certified seed (d) Registered seed
5. Foundation seed is the progeny of
   (a) Truthfully labeled seed (b) Foundation seed stage I
   (c) Nucleus seed (d) Breeder seed
6. Breeder seed is the progeny of
   (a) Foundation seed (b) Breeder seed
   (c) Nucleus seed (d) Certified seed
7. Certified seed is the progeny of
   (a) Breeder seed (b) Foundation seed
   (c) Nucleus seed (d) Truthfully labeled seed

8. Five generation system of seed multiplication can be followed in
   (a) Self pollinated crops (b) Often cross pollinated crops
   (c) Cross pollinated crops (d) Hybrids
9. Registered seed production is practiced in
   (a) India (b) Pakistan
   (c) Canada (d) America
10. The original seed of a variety available with the producing breeder or any other recognized breeder of the crops is called
   (a) Breeder seed (b) Certified seed
   (c) Nucleus seed (d) Foundation seed
11. In India, normally how many generation system seed are produced?
   (a) 1 (b) 3
   (c) 2 (d) 4
12. The seed which farmers generally used is called
   (a) Breeder seed (b) Nucleus seed
   (c) Foundation seed (d) Certified seed

♦ Answer Keys ♦

Q.No.	Ans.	Q.No.	Ans.	Q.No.	Ans.	Q.No.	Ans.	Q.No.	Ans.	Q.No.	Ans.
1	b	2	b	3	a	4	c	5	d	6	c
7	b	8	a	9	d	10	c	11	b	12	d

2.4.2. Fill in the blanks

1. Breeder seed is the progeny of ....
2. Stage I foundation seed is the progeny of ....
3. Certified seed is produced from ....
4. Progeny of breeder seed is ....
5. .... is used for multiplication of foundation seed
6. The tag colour prescribed for foundation seed is ....
7. Five generation model usually followed in .... crops
8. The foundation seed stage is the progeny of ....
9. Different classes of seeds used in seed production ...., BS, FS & CS
10. Different classes of seeds produced under certification are .... and .... seeds
11. The monitoring team inspects .... class of seeds
12. Seed multiplication ratio of paddy is .... while for cotton it is ....
13. The generation system of multiplication comprises of ...., ... and ....
14. The generation system of seed multiplication is .... for sorghum and .... for groundnut

15. Production and supply of .... are directly controlled by the originating or the sponsoring breeder or institution
16. Quality of breeder seed during production is monitored by ....

♦ Answer Keys ♦

Q.No.	Ans.	Q.No.	Ans.	Q.No.	Ans.
1	Nucleus seed	2	Breeder seed	3	Foundation seed
4	Foundation seed	5	Breeder seed	6	White
7	Highly self pollinated	8	Breeder seed	9	NS
10	Foundation, Certified	11	Breeder	12	1 : 80, 1 : 50
13	Breeder, foundation and certified seed	14	Three, Five	15	Nucleus and breeder seed
16	Monitoring team				

2.4.3. State True or False

1. The tag colour of breeder seed class is golden yellow
2. Generation system of seed multiplication is followed for maintaining the genetic purity of crop varieties
3. Foundation seed is the progeny of certified seed
4. To produce certified seeds, the source seeds to be used is breeder seeds
5. Breeder seed is the progeny of foundation seed
6. Breeder seed production is planned based on the indent or demand every year
7. The seed harvested from the breeder seed is the certified seed

♦ Answer Keys ♦

Q.No.	Ans.	Q.No.	Ans.	Q.No.	Ans.	Q.No.	Ans.
1	True	2	True	3	False	4	False
5	False	6	True	7	False		

2.5. SEED PRODUCTION FACTORS

2.5.1. Choose the correct answer

1. In seed production, the selection of land should be primarily free from
   (a) Off-types (b) Rogues
   (c) Volunteer plants (d) Weeds
2. The extent of varietal contamination depends on
   (a) The population of weeds
   (b) The number of diseased plants
   (c) The amount of natural cross pollination
   (d) The number of border rows

3. The genetic purity of the seed shall be assured through
   (a) Source of contamination (b) Selection of seed
   (c) Roguing (d) Field inspection
4. Effect of pollen grains on the maternal tissues or the effect of foreign pollen on the endosperm colouration is called as
   (a) Paramutation (b) Metaxenia
   (c) Penetrance (d) Regulon
5. Basic input of agriculture is
   (a) Irrigation (b) Fertilizer
   (c) Soil (d) Seed
6. The direct and immediate visible effects of the foreign pollen on the seed coat colour during the formation of seed is called as
   (a) Fertilized effect (b) Pollen impact
   (c) Xenia (d) Penetrance
7. In nurseries, soil sterilization by fumigation is done with
   (a) Methyl bromide (b) Silver nitrate
   (c) Ammonium chloride (d) Calcium chloride
8. Which one of the following chemical is used for reclamation of alkaline soil
   (a) Lime (b) Gypsum
   (c) Sodium hydride (d) Bleaching powder
9. Selfed bolls/fruits in hybrid seed production of cotton, tomato, brinjal and muskmelon are
   (a) Genetic contaminant (b) Physical contaminant
   (c) Health quality (d) Physiological quality
10. The plants that are very similar and resemble closely to the seed crop plants in habit and growth characters
    (a) Off-types (b) Inseparable other crop plants
    (c) Objectionable weeds (d) Rogues
11. Rain during maturation promoted the production of following seed in pulses
    (a) Off coloured seed (b) Mottled seed
    (c) Dimbled seed (d) Blackened seed
12. Plants exhibits variation in morphological characters refers to
    (a) Off-types (b) Objectionable weeds
    (c) Pollen shedders (d) Designated disease plants
13. A dry sunny period is highly essential for seed production during
    (a) Vegetative phase (b) Flowering phase
    (c) Post flowering phase (d) Harvest phase
14. Isolation distance for a seed crop depends on
    (a) Flowering behaviour (b) Pollination behaviour
    (c) Fertilization capacity (d) Receptivity of stigma

15. What is the pollinating agent for bajra?
    (a) Wind (b) Beetle
    (c) Wasps (d) Water
16. Precocious germination occurs due to delayed harvest in seed crop and this is due to
    (a) Heavy rain (b) Heavy wind
    (c) Too wet conditions of soil (d) Too dry conditions of soil
17. The pollen will desiccate, if the environment is
    (a) Heavy rain (b) High wind speed
    (c) High temperature (d) High RH
18. Lack of P delays
    (a) Pollination (b) Fertilization
    (c) Seed set (d) Flowering

♦ Answer Keys ♦

Q.No.	Ans.	Q.No.	Ans.	Q.No.	Ans.	Q.No.	Ans.	Q.No.	Ans.	Q.No.	Ans.
1	c	2	c	3	c	4	b	5	d	6	c
7	a	8	b	9	a	10	a	11	a	12	a
13	d	14	b	15	a	16	a	17	c	18	d

2.5.2. Fill in the blanks

1. High rainfall may cause .... during seed production
2. .... and .... are the biotic factors affecting seed production
3. Prevalence of high rainfall during flowering causes ...., ...
4. High temperature and low humidity induces hard seed production in ....
5. Among the major factors of seed production, the edaphic, environmental and .... ones assume great importance
6. .... and pest and disease incidence affect seed quality
7. Use of .... with low vigour and viability as well as inadequate nutrition will affect the quality of seeds
8. Regions with .... and .... are ideal for seed production
9. Adverse environment results in production of .... in pulses and bhendi
10. The time isolation of a crop is .... days
11. .... and .... climate condition is suitable for quality seed production
12. .... and .... are the major problems of consideration in selecting land for seed production
13. Isolation distance is fixed based on pollination behavior and .... activity

14. Isolation distance will be .... for self pollinated crop, while it will be .... for cross pollinated crop
15. High humidity may cause .... during seed production
16. Volunteer plant in the seed plot occur from ....
17. Seeds should be planted at a depth of ....
18. Volunteer plants are ....
19. Rogues and off-types should be removed before .... and ....
20. In tomato and chilli, spraying of .... will arrest flower drop
21. .... deficiency causes marsh spot in peas
22. Boron deficiency causes .... in peas
23. .... deficiency causes whiptail in crucifer
24. Iron deficiency causes ....
25. Spraying of .... increase seed set in pulses
26. In sorghum, formation of .... spot on the aleurone layer is indication of physiological maturity.

♦ Answer Keys ♦

Q.No.	Ans.	Q.No.	Ans.	Q.No.	Ans.
1	Poor seed set	2	Weeds, Pests and Diseases	3	Pest and diseases, flower shedding
4	Pulses	5	Biotic	6	Soil factors, Unfavourable Environment
7	Genetically impure seed	8	Moderate rainfall, good sunshine	9	Off coloured seed
10	30	11	Cool, dry	12	Isolation distance, volunteer plants
13	Insect	14	Less, more	15	Poor seed set
16	Previous crop	17	2-4 cm	18	Physical contaminants
19	Flowering, pollination	20	NAA	21	Mn
22	Hollow heart	23	Molybdenum	24	Sterility
25	2% DAP	26	Dunken black		

2.5.3. State True or False

1. Dry winds may have desiccating effect on seeds during seed production
2. When seed production is taken up in calcareous soils, DAP 2% spray can be given
3. Irrigation at vegetative phase of the seed crop decides the quality of the resultant seeds
4. Long day conditions + high temperature induce more of pistillate flowers in cucurbits
5. High temperature during flowering cause pollen (contabescence) abortion

6. Nitrogen application is recommended to overcome the problem of heterostyly in bhendi
7. For grain purpose, the crop is harvested at physiological maturity and for seed purpose the crop is harvested at field maturity
8. Seeds at 20% moisture resist mechanical damage
9. Volunteer plants are self sown crop plants
10. Seed should be harvested at physiological maturity stage
11. Iron chlorosis causes female sterility of crops
12. $ZnSO_4$ spray enhances the pollen viability
13. Zinc sulphate enhances the stigma receptivity
14. Zinc sulphate enhances the pollen viability
15. Trueness to variety refers to physical purity

♦ Answer Keys ♦

Q.No.	Ans.	Q.No.	Ans.	Q.No.	Ans.	Q.No.	Ans.	Q.No.	Ans.	Q.No.	Ans.
1	True	2	False	3	False	4	True	5	True	6	True
7	False	8	False	9	True	10	True	11	True	12	True
13	False	14	True	15	False						

2.6. SEED PRODUCTION PLANNING

2.6.1. Choose the correct answers

1. When the central seed certification board was formed?
   (a) 1982 (b) 1972
   (c) 1962 (d) 1952
2. When the seed control order was passed?
   (a) 1983 (b) 1973
   (c) 1963 (d) 1953
3. Monsanto belongs to which country
   (a) UK (b) USSR
   (c) USA (d) UAE
4. When NSP was implemented?
   (a) 1955-56 (b) 1965-66
   (c) 1975-76 (d) 1985-86
5 SRR = X/Y × 100. In this formula Y is
   (a) Quantity of seed produced (b) Quantity of seed sown
   (c) Quantity of seed required (d) Quantity of seed supplied
6. The quantity of quality seed that have replaced the actual need of seed
   (a) Seed multiplication ratio (b) Seed replacement rate
   (c) Seed rate (d) Seed yield

7. Seed Replacement Rate for cabbage is
   (a) 100 (b) 98
   (c) 87 (d) 200

♦ Answer Keys ♦

Q.No.	Ans.	Q.No.	Ans.	Q.No.	Ans.	Q.No.	Ans.
1	b	2	a	3	c	4	c
5	a	6	b	7	a		

2.6.2. Fill in the blanks

1. The formula for working out seed multiplication ratio is ....
2. The formula for working out seed replacement rate is ....
3. The formula for calculating the quantity of seed required is ....
4. The quantity of quality seed that have replaced the actual need of seed is ....
5. Seed multiplication ratio of tomato is ....
6. .... are the seed multiplication ratio for paddy and groundnut, respectively.

♦ Answer Keys ♦

Q.No.	Ans.	Q.No.	Ans.
1	$\frac{\text{See yield}}{\text{Seed rate}}$	2.	$\frac{\text{Quantity of actual seed sown}}{\text{Quantity of quality seed required}} \times 100$
3	Area x Seed rate	4	Seed Replacement Rate
5	1 : 400	6	1 : 100 and 1 : 8

2.7. HYBRID SEED PRODUCTION

2.7.1. Choose the correct answer

1. The planting ratio for hybrid rice seed production is
   (a) 8 : 2 (b) 10 : 2
   (c) 3 : 1 (d) 6 : 2
2. The breeding tool for cotton hybrid seed production
   (a) Emasculation and dusting (b) Cytoplasmic male sterility
   (c) Self incompatibility (d) Genetic male sterility
3. Presence of B line in A line of paddy hybrid seed production field is
   (a) Pollen shedder (b) Off-type
   (c) Rogue (d) Weed
4. Supplementary pollination is required for
   (a) Rice (b) Safflower
   (c) Sorghum (d) Cotton

5. Rope pulling is practiced for good seed set in hybrid seed production of
   (a) Sorghum (b) Rice
   (c) Sunflower (d) Castor
6. Gibberellic acid spray to hybrid rice seed production plots are given for
   (a) Delayed flowering (b) Enhanced flowering
   (c) Complete exertion of panicle (d) Enhanced maturation
7. Genetic contaminants present in hybrid rice field
   (a) Shedding tassel (b) Selfed bolls
   (c) Pollen shedder (d) Diseased plants
8. The genetic contaminant present in hybrid maize field
   (a) Off-type plants (b) Weed plants
   (c) Diseased plants (d) Shedding tassel
9. Isolation distance for hybrid rice seed production
   (a) 200 m (b) 400 m
   (c) 50 m (d) 3 m
10. The breeding tool for redgram hybrid seed production is
   (a) Cytoplasmic male sterility (b) Emasculation and dusting
   (c) Self incompatibility (d) Genetic male sterility
11. The seed replacement rate for hybrids
   (a) 40% (b) 75%
   (c) 50% (d) 100%
12. Hand rubbing is practiced for good seed set in hybrid seed production of
   (a) Sorghum (b) Rice
   (c) Sunflower (d) Castor
13. Staggered sowing is practiced for
   (a) Good field establishment (b) Tolerance of pest and diseases
   (c) Uniform flowering (d) Easy processing
14. Detasseling is the technique used for hybrid seed production in
   (a) Castor (b) Cotton
   (c) Maize (d) Tomato
15. Presence of B line in A line in sunflower hybrid seed production plot is called
   (a) Off-type (b) Pollen shedder
   (c) Rogue (d) Weed
16. Isolation could be modified based on the number of border rows in
   (a) Maize (b) Paddy
   (c) Sunflower (d) Castor

17. Gibberellic acid sprayed to hybrid rice seed production plots to
(a) Control weeds (b) Increase seed set
(c) No effect (d) Increase growth

18. The technique used for hybrid seed production of sunflower is
(a) CMS (b) GMS
(c) CGMS (d) Emasculation and dusting

19. Which lines are used for producing hybrid cucumber
(a) Gynoecious (b) Gynomonoecious
(c) Monoecious (d) Androgynoecious

20. Planting ratio recommended for hybrid seed production of tomato is
(a) 12 : 1 (b) 8 :2
(c) 5 : 2 (d) 3 :1

21. The crop in which the first Indian vegetable hybrid was released
(a) Bottlegourd (b) Ashgourd
(c) Ridgegourd (d) Bittergourd

22. Pusa Deepthi is a hybrid of
(a) Tomato (b) Chilli
(c) Brinjal (d) Bhendi

23. In double cross hybrid seed production of maize, number of inbreds involved are
(a) 2 (b) 3
(c) 4 (d) 5

24. Chemical hybridizing agent
(a) Sodium arsenate (b) $ZnSO_4$
(c) $MnSO_4$ (d) GA_3

25. 'R' line in CMS hybrid seed production
(a) Male fertile (b) Male sterile
(c) Female sterile (d) Female fertile

26. Emasculation and dusting technique is practiced in
(a) Maize (b) Paddy
(c) Sunflower (d) Cotton

27. A line is referred as
(a) Male parent (b) Female parent
(c) F_1 line (d) Fertile plant

28. Presence of 'R' line plants in 'A' line are called as
(a) Volunteer plant (b) Weeds
(c) Pollen shedder (d) Rogue

29. Chemicals used for killing the male gamete
(a) Sodium Methyl arsenate (b) Sodium chloride
(c) Sodium azide (d) Sodium acetate

30. A type of male sterility that is conditioned by the cytoplasm rather than by the nuclear genes and transmitted only through the female parent is called as
    (a) Genetic male sterility (b) Cytoplasmic genetic male fertility
    (c) Cytoplasmic male sterility (d) Self incompatibility
31. The male sterility that is dependent upon the action of genes carried in the nucleus with a particular cytoplasm is called as
    (a) Genetic male sterility (b) Cytoplasmic genetic male sterility
    (c) Cytoplasmic male sterility (d) Self incompatibility
32. The chemicals, which are used to induce male sterility in crop plants is called as
    (a) Nematicides (b) Herbicides
    (c) Fungicides (d) Gametocides
33. Normally, male sterility in plants can be used to develop
    (a) Purelines (b) Synthetics
    (c) Inbreds (d) Hybrids
34. The maintainer line in hybrid seed production denotes
    (a) A line (b) R line
    (c) B line (d) C line
35. The hybrid seed production technique adopted in rice is
    (a) CGMS (b) GMS
    (c) Block method (d) CMS
36. Cytoplasmic male sterile line are also known as
    (a) A line (b) B line
    (c) R line (d) Inbred

♦ Answer Keys ♦

Q.No.	Ans.	Q.No.	Ans.	Q.No.	Ans.	Q.No.	Ans.	Q.No.	Ans.	Q.No.	Ans.
1	a	2	a	3	a	4	a	5	b	6	c
7	c	8	d	9	a	10	d	11	d	12	c
13	c	14	c	15	b	16	a	17	b	18	a
19	b	20	a	21	a	22	b	23	c	24	a
25	a	26	d	27	b	28	d	29	c	30	c
31	b	32	d	33	d	34	c	35	a	36	a

2.7.2. Fill in the blanks

1. The method of hybrid seed production in tomato is ....
2. Foliar spray of .... improve the female flower production in cucurbits
3. Presence of B line in the A line is ....
4. Presence of B line in A line of rice seed production plot is called as ....

5. The seed replacement rate for hybrid is ....
6. Method adopted to get uniform flowering in hybrid seed production field is ....
7. .... is the latest sunflower hybrid released from TNAU
8. The first Indian Vegetable hybrid is .... in ....
9. Techniques used for okra hybrid seed production is ....
10. Rope pulling is practiced to get higher hybrid seed setting in .... crop
11. Detasseling is practiced in hybrid seed production of .... crop
12. .... technique is used in production of cotton hybrids
13. .... is followed in maize hybrid seed production.
14. .... and zinc methyl arsenate, malic hydrazide are the gametocides used in hybridization programme
15. The "B" line of sorghum hybrid is also known as .... line
16. The parental lines are multiplied in .... stages in CGMS system of hybrid seed production
17. In rice hybrid, high pollination is achieved through ....
18. In all hybrids, synchronization is achieved through ....
19. GMS system is followed in .... hybrid seed production
20. The fertilizer used for synchronization of parental line in sorghum is ....
21. Selfed bolls are identified for maintenance of genetic purity in .... crop
22. The indeterminate flowering of castor is responsible for multiplication of ....
23. .... and .... are the tools involved in hybrid seed production of maize and cotton respectively
24. Border rows at a edge of the hybrid seed field are used to ....
25. The female flower production in cucurbits can be improved by foliar application of ....
26. The female line of paddy hybrid CORH 3 is also known as ....
27. The hybrid (F_1) seeds are produced in .... seed stage
28. Jerking is the operation recommended for synchronization in .... crop
29. Staggered sowing is recommended for production of commercial hybrids for effective ....
30. In redgram, the system of hybrid seed production is known as ....
31. The .... of tomato hybrids can be assessed through the presence of selfed fruits
32. The temperature required for hybrid seed setting is ....

33. Self incompatibility system is used for hybrid seed production of .... crops.
34. Sterility system adopted in hybrid rice is ....
35. Sterility system adopted in pigeon pea hybrid is ....
36. Method of hybrid seed production in cotton ....
37. Chemical used to induce male sterility is ....
38. The gametocide used to induce male sterility in rice is ....
39. Genetic male sterility is used for hybrid seed production in ....
40. Different types of hybrids available in maize are double, triple and ....
41. The method used in the hybrid seed production of sorghum or bajra is ....
42. The method adopted for the production of hybrid seeds in maize is .... and ....
43. In cotton hybrid seed production, ... to be opened next day are to be emasculated.
44. Chemicals used to bring about death of male gamete is named as ....
45. Line used for maintaining a cytoplasmic male sterile line is called as ....
46. Absence of functional male gametes/pollen grain in plant is called as ....
47. Ability to produce viable progeny is called ....
48. In hybrid seed production .... line is to be harvested first
49. In maize crossing block, the term shedding tassel is always refers to the tassels found in the ....
50. Seed replacement rate for double cross hybrid is ....
51. Two line breeding is practiced in production of .... hybrid
52. Pistillate lines are used for hybrid seed production in ....
53. Nicking is done for .... of flowering in hybrid seed production
54. The "B" line of sorghum hybrid is also known as ....
55. The parental lines are multiplied in .... stage in CGMS system of hybrid seed production
56. In rice hybrids, isolation is permitted by growing live fences like ....
57. In sorghum hybrid seed production withholding .... is to promote flowering of late flowering parent.
58. In cotton hybrid seed production, pollination of emasculted flowers is to be done in the .... hours.
59. Hybrids in cotton are produced by .... technique
60. The onion hybrid seeds are produced commercially by use of .... system

61. The function of `R' line is ....
62. Adjusting the planting dates to coincide the flowering in seed plots is called as ....
63. For hybrid seed production of rice, the planting ratio adopted is ....
64. The chemicals which are used to induce male sterility are called as ....
65. Removal of male flowers from the female parent in maize is known as ....
66. The hybrid seed production in sorghum involves the .... sterility system.
67. The method suggested for emasculation in cotton is ....
68. The seed multiplication ratio for paddy hybrids is ....
69. The male sterility system used in hybrid seed production for rice is ....
70. For hybrid seed production in rice, the isolation distance maintained for foundation seed is ....
71. To obtain synchronized flowering during hybrid seed production in rice, ... type of sowing will be beneficial.
72. In rice hybrid seed production, single line breeding system is based on .... and ....
73. PGMS (Photosensitive Genic Male Sterility) and TGMS (Thermo-sensitive Genic Male Sterility) systems have been developed successfully in ....
74. The father of hybrid rice is ....
75. Examples of an ideal gametocides are .... and ....
76. In rice, CORH-1 is also called as ....
77. Spraying of .... enhances complete panicle exertion in hybrid seed production of rice.
78. The technique involved to supplement the pollination mechanism in rice hybrid seed production is ....
79. Presence of pollen grains in the `A' line of hybrid seed production is called as ....
80. The seed multiplication ratio for maize hybrids is ....
81. The parentage for COH 1 maize is ....
82. The cross between a single cross hybrid (A × B) × inbred (C) is termed as ....
83. Examples for double cross hybrid are .... and ....
84. Isolation distance for maize single cross hybrid is ....
85. In hybrid seed production of maize, the planting ratio for single cross hybrid is ....

86. The planting ratio for double and three way cross hybrid maize is ....
87. In hybrid seed production for sorghum, at foundation seed stage the planting ratio should be ....
88. In sorghum hybrid seed production, at certified seed stage the planting ratio should be ....
89. The lines involved in CGMS for hybrid seed production in cumbu are .... and ....
90. Isolation distance for foundation seed in bajra hybrids is ....
91. In cotton, ... is used as a source of CGMS.
92. Isolation distance for foundation seed stage in cotton hybrids ....
93. The maximum permitted percentage of off types for cotton hybrids is ....
94. The male sterility employed for hybrid seed production is ....
95. Isolation distance adopted for sunflower hybrids during foundation seed stage is ....
96. For the production of sunflower hybrid seed, the planting ratio adopted is ....
97. Isolation distance for hybrid seed production in redgram is ....
98. The gametocide used in sorghum is ....
99. Spraying of gametocides induces ....
100. Top cross is otherwise known as ....
101. Female sterility in sorghum is influenced by ....
102. Male sterility in mustard is controlled by .... gene
103. Nuclear male sterility is seen in .... crops

♦ Answer Keys ♦

Q.No.	Ans.	Q.No.	Ans.	Q.No.	Ans.
1	Emasculation and dusting	2	Ethrel	3	Pollen shedder
4	Pollen shedder	5	100%	6	Staggered sowing
7	CO (SFH) 2	8	Pusa, Bottlegourd	9	Emasculation and dusting
10	Rice	11	Maize	12	Emasculation and dusting
13	Detasseling	14	Sodium methyl arsenate	15	Maintainer
16	2	17	Rope pulling/rod driving	18	Staggered sowing
19	Redgram	20	Urea	21	Cotton
22	A line	23	Detasseling emasculation	24	Prevent genetic contamination from foreign pollen

Q.No.	Ans.	Q.No.	Ans.	Q.No.	Ans.
25	Ethrel 200 ppm	26	Male sterile line	27	Certified
28	Cumbu	29	Seed set	30	Genetic male sterility
31	Genetic purity	32	30 - 32°C	33	Sunflower
34	CGMS	35	GMS	36	Emasculation and dusting
37	Gametocides	38	Sodium Methyl Arsenate	39	Redgram
40	Single cross	41	CGMS	42	Detasselling, controlled pollination
43	Flower buds	44	Chemical emasculator/ gametocide	45	Maintainer line
46	Male sterility	47	Fertility	48	Male
49	Female line	50	100%	51	Rice
52	Castor	53	Synchronization	54	Maintainer line
55	Breeder seed/ Foundation seed	56	Barrier	57	Irrigation
58	Morning	59	Emasculation and dusting	60	CMS
61	Pollen restorer	62	Staggered sowing	63	8:2
64	Gametocides	65	Detasselling	66	CGMS
67	Doak's method	68	1 : 100	69	CGMS
70	200m	71	Staggered	72	Apomixis tissue culture
73	China	74	Yuan Long Ping	75	Maleic hydrazide ethrel
76	MGR-1	77	GA3	78	Rope pulling/Rod driving
79	Partials	80	1 : 100	81	UMI 29 × UMI 51
82	Three way cross hybrid	83	Deccan hybrid COH 3	84	400m
85	4:2	86	6 : 2	87	4:2 ratio
88	5:2	89	A, B, R lines	90	1000 m
91	*G harkenssii*	92	50 m	93	0.05%
94	CGMS	95	600m	96	4:1
97	200m	98	FW 450	99	Female flower production
100	Inbred variety cross	101	Micro nutrients	102	Nuclear
103	Mustard				

2.7.3. State True or False

1. Cytoplasmic male sterility system is followed in bhendi hybrid seed production
2. Emasculation and dusting is the hybrid seed production technique in tomato
3. The seed replacement rate for hybrids is 100%
4. Downgrading is not applicable for hybrids
5. Presence of B line in A line of hybrid seed production field is a pollen shedder
6. Staggered sowing is an effective method of synchronization in hybrids
7. Emasculation and dusting is the method of hybrid seed production in sunflower
8. The seed replacement rate for hybrid is higher than the variety
9. Emasculation means removal of staminate flowers in monoecious crop
10. Emasculation and wind pollination technique is followed in sorghum hybrid seed production
11. B-line is a non pollen restorer line
12. Breeding technique for commercial hybrid seed production of sorghum is cytoplasmic genetic male sterility
13. Embryo rescue technique is more suitable in case of interspecific crossing
14. CGMS is the main tool employed for hybrid seed production in rice
15. GMS system is effectively used in different crops for hybrid seed production
16. PGMS lines can be effectively used under short day conditions
17. Gynomonoecious lines are used in hybrid seed production of water melon
18. Hybrid seeds of tomato are produced by adopting CGMS as a tool
19. In sorghum, cytoplasmic genetic male sterility is used for hybrid seed production
20. In double cross maize hybrid, 2 inbreds are involved
21. In cotton, block method of planting is followed for hybrid seed production
22. A type of male sterility condition by the cytoplasm rather than by nuclear genes and transmitted only through the female parent is genetic male sterility
23. Hand emasculation techniques are common method of hybrid seed production in cotton
24. XYZ system is used for producing wheat hybrid
25. Cytogenetic male sterile lines are also known as B line
26. EGMS lines are used for hybrid seed production in castor
27. In sorghum, hybrids are developed by using GMS

28. Spraying of silver thiousulphate at 2-3 leaf stage is encouraged for production of more pistillate flower
29. If both seed and pollen parent is incompatible, then the planting ratio should be 2:1

♦ Answer Keys ♦

Q.No.	Ans.	Q.No.	Ans.	Q.No.	Ans.	Q.No.	Ans.	Q.No.	Ans.	Q.No.	Ans.
1	False	2	True	3	True	4	True	5	True	6	True
7	False	8	True	9	False	10	False	11	False	12	True
13	False	14	True	15	False	16	True	17	True	18	False
19	True	20	False	21	True	22	False	23	True	24	True
25	False	26	False	27	False	28	False	29	False		

2.8. SEED PRODUCTION IN CEREALS

2.8.1. Choose the correct answer

1. In paddy seed production, the time isolation given from MS line to other varieties
   (a) 20 days (b) 1 month
   (c) 3 days (d) 2 months
2. Split husk commonly occurs in
   (a) Barley (b) Wheat
   (c) Oats (d) Paddy
3. The recommended number of border rows for paddy hybrid seed production is
   (a) 4 (b) 12
   (c) 10 (d) 8
4. Fodder sorghum variety is
   (a) CO 24 (b) CO 26
   (c) CO 27 (d) CO 28
5. An example for once over harvest is
   (a) Redgram (b) Greengram
   (c) Paddy (d) Bhendi
6. CSH 5 is a hybrid of
   (a) Sorghum (b) Cumbu
   (c) Paddy (d) Maize
7. For commercial hybrid seed production of sorghum, the isolation distance of foundation seed on the presence of Johnson grass is
   (a) 100 m (b) 200 m
   (c) 300 m (d) 400 m
8. Cob sorting in maize is done to identify
   (a) Diseased cob (b) Matured cob
   (c) Big sized cob (d) Off type cob

9. Detaselling is employed in
(a) Paddy (b) Cotton
(c) Maize (d) Cumbu

10. Once over harvest is in
(a) Pulses (b) Paddy
(c) Gourds (d) Tomato

11. Time isolation is followed in
(a) Maize (b) Paddy
(c) Groundnut (d) Redgram

12. Gametocide used as chemical emasculant during seed production
(a) Ethrel (b) H_2SO_4
(c) $MnSO_4$ (d) $ZnSO_4$

13. Spacing adopted for sorghum seed production is
(a) 30 × 10 cm (b) 45 × 10 cm
(c) 30 × 15 cm (d) 45 × 15 cm

14. What is the concentration of urea to be sprayed as foliar spray to the lagging parent for synchronization of flowering in sorghum hybrid seed production ?
(a) 0.25% (b) 0.50%
(c) 0.75% (d) 1.00%

15. At what stage, the urea foliar spray should be undertaken for synchronization of flowering in sorghum hybrid seed production
(a) Primordial initiation stage (b) Late primordial stage
(c) Early flowering stage (d) Late flowering stage

16. How many irrigation to be withhold to the late parent to make early flowering in sorghum hybrid seed production?
(a) One irrigation (b) Two irrigations
(c) Three irrigations (d) Four irrigations

17. What chemical is to be sprayed to the advancing parent to delay flowering in hybrid sorghum seed production ?
(a) GA (b) ABA
(c) MH (d) IAA

18. In rice hybrid seed production, GA is sprayed for exertion of panicle at the dosage of
(a) 25 g/ha (b) 50 g/ha
(c) 75 g/ha (d) 100 g/ha

19. How many splits of GA are to be sprayed for rice hybrid seed production ?
(a) One split (b) Two splits
(c) Three splits (d) Four splits

20. Planting ratio for sorghum hybrid seed production is
(a) 4 : 2 (b) 6 :2
(c) 5 :2 (d) 8 :2

21. Planting ratio for bajra hybrid seed production is
    (a) 6 : 2 (b) 4 :2
    (c) 4 :1 (d) 5 :1
22. The pollen parent of CORH 2 rice hybrid is
    (a) IR 66 R (b) C 20 R
    (c) 10198-66-2-R (d) KMR 3
23. The isolation distance recommended in metres to avoid contamination with Johnson grass in hybrid sorghum is
    (a) 100 m (b) 200 m
    (c) 300 m (d) 400 m
24. Female sterility in sorghum is deficiency symptom of
    (a) Fe (b) Zn
    (c) Boron (d) Mn
25. The name of the chemical, which on spraying will enhance complete panicle exertion in hybrid seed production of rice is
    (a) Glutamic acid (b) Ethanol
    (c) Lactic acid (d) Gibberellic acid
26. The ratio of economic yield to biological yield is described as
    (a) Heterosis (b) Inbreeding effect
    (c) Harvest index (d) Yield index
27. The techniques in which early formed ear heads of the first tiller are pulled so as to enhance the flowering of all the tillers in bajra is called as
    (a) Emasculation (b) Detasseling
    (c) Desilking (d) Jerking
28. The parentage of COH 1 maize is
    (a) CM 137 × CM 138 (b) UMI 31 × UMI 56
    (c) CM 511 × CM 617 (d) UMI 29 × UMI 51
29. The isolation distance normally followed for foundation seed production of paddy varieties is
    (a) 3 m (b) 50 m
    (c) 10 m (d) 1000 m
30. The isolation distance normally followed for foundation seed production of paddy hybrids is
    (a) 3 m (b) 50 m
    (c) 10 m (d) 200 m
31. The isolation distance normally followed for foundation seed production of bajra hybrids is
    (a) 30 m (b) 50 m
    (c) 100 m (d) 1000 m

32. The seed multiplication ratio for rice varieties is normally
(a) 1:80 (b) 1:40
(c) 1:20 (d) 1:150

33. The seed multiplication ratio for maize hybrids is normally
(a) 1:50 (b) 1:40
(c) 1:20 (d) 1:100

34. In hybrid seed production of maize, the planting ratio for single cross hybrids is
(a) 4:2 (b) 8:2
(c) 6:3 (d) 12:3

35. In hybrid seed production of maize, the planting ratio for double and three way cross hybrids is
(a) 1:1 (b) 10:1
(c) 6:2 (d) 12:3

36. Isolation distance for certified seed in sorghum varieties is
(a) 3 m (b) 10 m
(c) 200 m (d) 100 m

37. Isolation distance for certified seed in forage sorghum varieties is
(a) 3 m (b) 10 m
(c) 200 m (d) 100 m

38. Detasselling is practiced in
(a) Paddy (b) Sorghum
(c) Maize (d) Cotton

39. Isolation distance for paddy variety seed production in certified class is
(a) 3 m (b) 30 m
(c) 100 m (d) 200 m

40. Designated objectionable weed in wheat seed crop
(a) *Chicorum intybus* (b) *Cucurbita* sp
(c) *Convolvulus arvensis* (d) *Chicorum* sp

41. For hybrid paddy seed production, the recommended number of border rows is
(a) 2 (b) 3
(c) 4 (d) 6

42. The minimum number of border rows in hybrid seed production of sorghum is
(a) 1 (b) 2
(c) 4 (d) 6

43. Ganga 5 is a .... hybrid
(a) Sorghum (b) Maize
(c) Rice (d) Bajra

44. The detasseling operation in seed parent rows must be completed before the emergence of receptive silks that reaches a level of
    (a) 3% (b) 5%
    (c) 7% (d) 9%
45. Rope pulling is practiced for effective seed set in hybrid seed production of
    (a) Pearl millet (b) Sorghum
    (c) Rice (d) Redgram
46. Staggered sowing is a common practice for effective synchronization in
    (a) Sorghum (b) Redgram
    (c) Castor (d) Cotton
47. WCC 75 is a composite variety of
    (a) Pearl millet (b) Maize
    (c) Samai (d) Sorghum
48. Debearders are used in processing of
    (a) Barley (b) Cowpea
    (c) Sunflower (d) Rice
49. The presence of B line parent in A line in sorghum hybrid seed production is called
    (a) Pollen shedder (b) Partials
    (c) Rogue (d) Objectionable weed
50. Seed production plots are isolated based on
    (a) Distance (b) Sowing date
    (c) Barrier (d) Pollination behavior
51. Isolation distance for certified seed production of paddy hybrid is
    (a) 3 m (b) 100 m
    (c) 400 m (d) 200 m
52. Isolation distance could be reduced with border rows in
    (a) Cumbu (b) Rice
    (c) Maize (d) Pulses
53. The isolation distance for foundation seed is the maximum in
    (a) Cumbu (b) Pulses
    (c) Cotton (d) Sunflower
54. In rice, the hybrid development is by using
    (a) Milo cytoplasm (b) Tift 23 A cytoplasm
    (c) Wild abortive cytoplasm (d) V20 A
55. Seed multiplication for ragi is
    (a) 400 (b) 420
    (c) 430 (d) 410

56. Barrier isolation can be permitted in
    (a) Sorghum (b) Cumbu
    (c) Paddy (d) Maize

57. Harvestable maturity indices for maize seed crop is turning of the sheath colour from
    (a) Green to yellow (b) Green to white
    (c) Green to grey (d) Dark green to light green

58. Ragi attains physiological maturity at
    (a) 36 DAA (b) 48 DAA
    (c) 50 DAA (d) 55 DAA

♦ Answer Keys ♦

Q.No.	Ans.	Q.No.	Ans.	Q.No.	Ans.	Q.No.	Ans.	Q.No.	Ans.	Q.No.	Ans.
1	a	2	d	3	a	4	d	5	c	6	a
7	d	8	d	9	c	10	b	11	b	12	a
13	b	14	d	15	a	16	a	17	b	18	b
19	b	20	c	21	a	22	b	23	d	24	a
25	d	26	c	27	d	28	d	29	a	30	d
31	d	32	a	33	d	34	a	35	c	36	d
37	c	38	c	39	a	40	c	41	c	42	c
43	b	44	b	45	c	46	a	47	a	48	a
49	a	50	d	51	b	52	c	53	a	54	c
55	b	56	c	57	a	58	b				

2.8.2. Fill in the blanks

1. Sorghum is an .... pollinated crop
2. .... and .... are important post harvest operation to be followed in maize and are also specific to this crop
3. In production of organic seed, the botanical recommended for seed hardening is ....
4. A special post harvest operation followed for maize with respect to seed certification is ....
5. An example for once over harvest is ....
6. An example for mechanical harvester is ....
7. Nicking can be induced by .... in hybrid rice
8. Seeds with high .... are preferred for malting in barley
9. Base sterility is common in ....
10. Objectionable weed plant in rice is ....
11. Different methods of isolation adopted in seed production are ...., and ....
12. Isolation distance of a crop depends on .... and ....
13. Rogues and off-types should be removed before .... and ....

14. The type of injury caused to the seeds when the moisture content is high during threshing is ....
15. Father of green revolution in India is ....
16. Give an example for paddy hybrid ....
17. Panicle exertion in female parent of paddy is enhanced by ....
18. Planting ratio adopted for hybrid seed production of maize is ....
19. Volunteer plants in the seed plot occur during ....
20. Maize is an example for ....
21. Quality seed alone ensures the higher seed yield up to ....
22. Staggered sowing is followed for ....
23. Objectionable weed in sorghum is ....
24. Decreasing isolation distance in a maize hybrid seed plot is possible by increasing number of ....
25. A method of providing isolation by raising a tall plants around the seed crop is called as .... isolation
26. The emergence of panicles from the boot leaf is called as ....
27. Jerking is practiced in .... hybrid seed production
28. Time isolation is accepted in .... for seed certification
29. The chemical used for synchronization of parental line in sorghum is ....
30. .... is the effect of less pollination in same season due to foreign pollen
31. In wheat, the isolation distance is .... and in case of loose smut infection the ID is ....
32. The number of border rows for sorghum is ....
33. In IR 20 paddy seed production crop, the presence of IR 50 is called ....
34. Ecological male sterility is common in .... crop
35. In hybrid rice seed production, GA_3 is used for ....
36. .... chemical is associated with ecological suicide
37. The Seed Replacement Rate period for hybrid rice is ....
38. Seed renewal period of paddy is ....
39. Iron deficiency in sorghum causes ....
40. Seed renewal period for hybrid is .... while for the varieties it is ....
41. The area of production and productivity of paddy in Tamil Nadu is .... and .... respectively
42. Seasons practiced in Tamil Nadu for rice cultivation based on the availability of water are .... and ....
43. In pearl millet, CO 7, CO (Cu) 9 is a grain type while .... is fodder variety

44. To boost up the yield of sorghum, ... is applied as a foliar spray
45. In calcareous soil, foliar spray with .... is recommended for better seed set
46. .... crop is responsible for occurrence of volunteer plants in the seed production plot.
47. The .... method is used for maize hybrid seed production
48. Differential blooming technique of isolation is not permitted for the crops with .... types of flowering
49. An example for multiple harvest is ....
50. Female parent of CO (R) H 3 rice hybrid is ....
51. Minimum isolation distance for foundation seed programme in sorghum is ....

♦ Answer Keys ♦

Q.No.	Ans.	Q.No.	Ans.	Q.No.	Ans.
1.	Often cross	2.	Cob sorting, Shelling	3.	Pungam leaf extract
4.	Cob sorting	5.	Paddy	6.	Combined harvester
7.	Staggered sowing	8.	Germination	9.	Rice
10.	Wild rice	11.	Space, Time, Barrier isolation	12.	Pollen load, Pollen weight
13.	Flowering, Pollination	14.	Bruising	15.	Dr. M.S. Swaminathan
16.	CORH 4	17.	GA spray	18.	6 : 2
19.	Vegetative phase	20.	Protandrous	21.	15 - 20%
22.	Hybrid sorghum and rice	23.	Cuscuta	24.	Border rows
25.	Barrier	26.	Exertion	27.	Pearl millet
28.	Paddy	29.	Maleic hydrazide	30.	Metaxenia
31.	3 m, 150 m	32.	4	33.	Rogue
34.	Sorghum	35.	Panicle exertion	36.	Ethylene
37.	One year	38.	4 years	39.	Female sterility
40.	Once in a year, 3 years	41.	17 lakhs ha, 3.4 ton/ha	42.	Samba, Kuruvai
43.	CO 8	44.	2% DAP	45.	Ferrous sulphate
46.	Previous season	47.	Detasseling	48.	Indeterminate
49.	Bajra	50.	TNAUCMS2A	51.	400 m.

2.8.3. State True or False

1. Tip sterility is common in maize crop
2. Isolation distance is fixed based on area of seed production
3. The colour of the tag accompanying the foundation seed lots shall be cream white
4. The isolation distance requirement for paddy is 50 m

5. Differential blooming dates for modifying the isolation distance is not permitted
6. Harvest loss is more by the use of combined harvester when compared to manual harvesting
7. Partials and pollen shedders are one and same
8. Seed polymorphism is more frequent in determinate plant types rather than indeterminate ones
9. Seeds from a crop of double cross hybrid maize should not be used for sowing
10. Physiological isolation is provided by altitude difference
11. Protandry is the mechanism that aid in cross pollination in bajra
12. The maize silk remains receptive for 7 days
13. Tip sterility is a common feature in sorghum.
14. Base sterility in rice is due to improper panicle exertion
15. Roguing is done to remove the off-type plants from the seed field
16. Base sterility is common in paddy hybrid
17. Detasseling is continued for a duration of 7 days
18. The isolation distance can be modified in the case of maize hybrid
19. Ragi is the crop with highest multiplication ratio
20. Red rice is an objectionable weed plant in rice
21. In rice hybrid seed production, draining of water in `R' line alone can delay its flowering by 2-3 days for synchronization of flowering
22. In CORH 3 hybrid seed production, irrigating the `R' line can delay its flowering for synchronization of flowering
23. To obtain synchronization of flowering in rice hybrid seed production, apply urea @ 35 kg/ha, to the delayed flowering parent
24. An isolation distance of atleast three metres is necessary for seed production of paddy
25. Detasseling is done for sorghum hybrid seed production
26. Planting ratio of female to male is essential for hybrid seed production
27. Time isolation is allowed in determinate crops
28. Roguing is followed in hybrid seed production only
29. Isolation distance is more in self pollinated crops than in cross pollinated crops
30. The seed crop must not be rogued carefully at pre-flowering, flowering and maturity stages
31. Tripping mechanism is associated with paddy crop
32. Leaf clipping is practiced for synchronization of flowering in rice
33. Leaf clipping is practiced for synchronization of flowering in sorghum
34. In sorghum, the isolation distance required for grassy panicle is 400 m
35. The isolation distance is increased by 100m, if the kernel colour is different in maize

36. The contamination in maize hybrid seed production plot is shedding tassel
37. Insect pollination is adopted for hybrid seed production in sorghum
38. FDA test is used to assess stigma receptivity
39. The isolation distance is increased by 100m, if the kernel colour is different
40. Zinc sulphate enhance the stigma receptivity
41. Zinc sulphate enhance the pollen viability
42. Harvestable maturity indices for paddy is turning of 90% of seed to straw or golden yellow.

♦ Answer Keys ♦

Q.No.	Ans.	Q.No.	Ans.	Q.No.	Ans.	Q.No.	Ans.	Q.No.	Ans.	Q.No.	Ans.
1	False	2	False	3	True	4	False	5	True	6	False
7	False	8	False	9	True	10	True	11	False	12	False
13	False	14	True	15	True	16	True	17	False	18	True
19	True	20	False	21	True	22	False	23	False	24	True
25	False	26	True	27	True	28	False	29	False	30	False
31	False	32	True	33	False	34	True	35	True	36	True
37	False	38	False	39	True	40	False	41	True	42	True

2.9. SEED PRODUCTION IN PULSES

2.9.1. Choose the correct answer

1. Isolation distance for varietal foundation seed production in redgram
   (a) 50 m (b) 100 m
   (c) 400 m (d) 200 m
2. An example for pre-harvest sanitation spray chemical is
   (a) Endosulphan (b) Urea
   (c) DAP (d) Bavistin
3. Planting ratio for redgram hybrid seed production is
   (a) 6 : 2 (b) 4 :2
   (c) 4 :1 (d) 3 :1
4. The first hybrid in pigeon pea in the world is
   (a) ICPH 12 (b) PH 102
   (c) IIPH 4 (d) ICPH 8
5. Example for often cross pollinated crop
   (a) Greengram (b) Cumbu
   (c) Paddy (d) Redgram
6. Tendril clipping is a common practice in
   (a) Redgram (b) Peas
   (c) Cowpea (d) Tomato

7. Pulses seed set could be improved by foliar spray during flowering with
   (a) DAP (b) $ZnSO_4$
   (c) Borax (d) Urea
8. The chemical recommended for pre-harvest sanitation spray to avoid storage pest infestation in pulses is
   (a) Endosulphan (b) Bavistin
   (c) Rogar (d) Dithane
9. Greengram seeds can be harvested when seed turn
   (a) Dark green and hard (b) Light green and hard
   (c) Light green (d) Dark green
10. Peas attains physiological maturity at
   (a) 40 DAA (b) 35 DAA
   (c) 30 DAA (d) 45 DAA
11. The largest source of vegetable protein is
   (a) Paddy (b) Soybean
   (c) Blackgram (d) Sunflower
12. Which of the following crop is highly sensitive to mechanical damage?
   (a) Sorghum (b) Cumbu
   (c) Blackgram (d) Soybean
13. The seed rate of cowpea is
   (a) 1- 6 Kg (b) 2- 6 Kg
   (c) 3- 6 Kg (d) 4 – 6 Kg
14. Which one of the following crop meets the protein requirement of vegetarian population
   (a) Sorghum (b) Paddy
   (c) Redgram (d) Sunflower
15. Isolation distance required for redgram seed production
   (a) 25 m (b) 50 m
   (c) 75m (d) 100 m
16. In pulses, commonly used biofertilizers for presowing seed treatment is
   (a) *Azospirillum* (b) *Azotobacter*
   (c) Rhizobium (d) Phosphobacteria

♦ Answer Keys ♦

Q.No.	Ans.	Q.No.	Ans.	Q.No.	Ans.	Q.No.	Ans.	Q.No.	Ans.	Q.No.	Ans.
1	d	2	a	3	d	4	d	5	d	6	c
7	a	8	a	9	a	10	b	11	b	12	d
13	b	14	c	15	d	16	c				

2.9.2. Fill in the blanks

1. Direct soaking of soybean seeds in water leads to ....
2. Method used to moistening the soybean seeds is ....
3. Chilling damage is noticed in ...., ... and ....
4. Fish mouth in soybean is caused by ....
5. Foliar spray of .... is recommended to improve the seed set in soybean crop
6. Rain during maturation removes malignant pigments and causes off coloured seeds in ....
7. .... deficiency cause marsh spot in peas
8. Dimple seeds are common feature of ....
9. The isolation distance recommended for the certified seed production of lablab is ....
10. Seed rate for lablab is ....
11. Soybean is also known as ....
12. The seed rate of soybean is ....
13. Maturity symptom of soybean is 70% of pods turn .... and few at the base turn .... colours .
14. Minimum pure seeds per cent for soybean is .... for both foundation and certified classes seeds.
15. Cowpea otherwise called as ....
16. Shape of the cowpea seeds are ....
17. Maturity symptom of cowpea is pods turn ....
18. .... isolation distance is required for greengram seed production.
19. Hard seeds are commonly occur in ....
20. To break hard seed nature in greengram, ... is required.
21. Minimum seed germination per cent required for greengram is ....
22. Redgram seeds hardened with .... for ....
23. .... is used as a pre-harvest sanitation spray to control bruchids in redgram.
24. Isolation distance required for blackgram seed production is ....
25. During seed hardening of blackgram, ... is important to avoid soaking injury

♦ Answer Keys ♦

Q.No.	Ans.	Q.No.	Ans.	Q.No.	Ans.
1	Soaking injury	2	Solid matric priming	3	Lima bean, Peas
4	Alternate wetting and drying	5	NAA	6	Redgram

7	Manganese	8	Redgram	9	5m
10	8 to 10 kg/acre	11	Golden bean	12	16 kg/acre
13	Yellow brown	14	98%	15	Black-eyed pea
16	Kidney shape	17	Straw colour	18	5m
19	Green gram	20	100 ml H_2SO_4/kg of seed	21	75%
22	05% $ZnSO_4$ 3h	23	Endosulphan	24	5m
25	Pre-conditioning				

2.9.3. State True or False

1. Pigeonpea is an often cross pollinated crop
2. Fish mouth is associated with soybean seed
3. Horse gram seed crop should be harvested when the pods turn yellowish brown in colour.
4. Lablab is self pollinated crop.
5. In soybean, spraying of DAP 2% during flowering stage will enhance the seed yield.
6. Cowpea is highly cross pollinated crop.
7. Redgram is an example for cross pollination.
8. Delay in harvest will lead to shattering of pods in redgram
9. Yellow mosaic disease is common in pulses

♦ Answer Keys ♦

Q.No.	Ans.	Q.No.	Ans.	Q.No.	Ans.	Q.No.	Ans.	Q.No.	Ans.	Q.No.	Ans.
1	True	2	True	3	True	4	True	5	True	6	False
7	False	8	True	9	True						

2.10. SEED PRODUCTION IN OILSEEDS

2.10.1. Choose the correct answer

1. Fish mouth symptoms are seen in
   (a) Groundnut (b) Soybean
   (c) Peas (d) Pulses
2. Important nutrient for groundnut seed production is
   (a) Copper (b) Magnesium
   (c) Manganese (d) Calcium
3. Seed multiplication ratio is low in
   (a) Sunflower (b) Castor
   (c) Gingelly (d) Groundnut
4. The causes for ill-filled seeds in sunflower is due to
   (a) Development of late flowers (b) Micronutrients deficiency
   (c) Poor seed set (d) Poor insect activity

5. In groundnut, the process of removing the pods either mechanically/ manually is known as ....
   (a) Jerking (b) Stripping
   (c) Topping (d) Clipping
6. Seed multiplication ratio for groundnut is
   (a) 1:80 (b) 1:60
   (c) 1:8 (d) 1:10
7. White seeded sesame variety
   (a) VRI 1 (b) TMV 2
   (c) SVPR 1 (d) VRI 2
8. For which crop, the specific gravity separator separated seeds are in positive association with vigour
   (a) Cotton (b) Sunflower
   (c) Casuarina (d) Cashewnut
9. Removal of seed from pod in groundnut is called
   (a) Stripping (b) Decortication
   (c) Extraction (d) Threshing
10. Synchronization of flowering of parental lines is achieved in sunflower seed production by
    (a) Use of quality seed (b) Staggered sowing
    (c) Simultaneous planting (d) Seasonal sowing
11. Hand pollination practiced to increase seed set per cent in
    (a) Maize (b) Paddy
    (c) Sunflower (d) Sesamum
12. Planting ratio for sunflower hybrid seed production is
    (a) 4 : 1 (b) 5 :1
    (c) 6 :1 (d) 7 :1
13. Planting ratio for castor hybrid seed production is
    (a) 5 : 1 (b) 4 :1
    (c) 3 :1 (d) 2 :1
14. In sunflower, pollination is brought about by
    (a) Honey bees (b) Bungle bees
    (c) Wind (d) Butterflies
15. Temperature dependent pistillate lines are used for hybrid seed production in
    (a) Redgram (b) Maize
    (c) Sorghum (d) Castor
16. Isolation distance for foundation seed in hybrid sunflower is
    (a) 50 m (b) 100 m
    (c) 1000 m (d) 600 m

17. The isolation distance normally practiced during seed production for groundnut varieties is
    (a) 3 m (b) 50 m
    (c) 10 m (d) 200 m
18. Inhibitor present in the seed which causes dormancy in sunflower is
    (a) Hydrocyanic acid (b) Coumarin
    (c) Zeatin (d) Phytine
19. The isolation distance for castor foundation seed production is
    (a) 100 m (b) 200 m
    (c) 300 m (d) 400 m
20. Hollow heart is the common symptom in
    (a) Groundnut (b) Peas
    (c) Beans (d) Oats
21. Gypsum is applied during earthing up in
    (a) Groundnut (b) Soybean
    (c) Gingelly (d) Blackgram
22. Darkened plumule of groundnut occurs due to the deficiency of
    (a) Boron (b) Calcium
    (c) Iron (d) Zinc
23. Season plays an important role in hybrid seed production of
    (a) Castor (b) Cotton
    (c) Bhendi (d) Sunflower
24. Five tier system of seed multiplication is adopted in
    (a) Groundnut (b) Paddy
    (c) Tomato (d) Pulses

♦ Answer Keys ♦

Q.No.	Ans.	Q.No.	Ans.	Q.No.	Ans.	Q.No.	Ans.	Q.No.	Ans.	Q.No.	Ans.
1	b	2	d	3	d	4	b	5	b	6	c
7	c	8	b	9	b	10	b	11	c	12	a
13	c	14	a	15	d	16	d	17	a	18	a
19	c	20	a	21	a	22	b	23	a	24	a

2.10.2. Fill in the blanks

1. In sunflower, to enhance the seed set, the pollination is supplemented by ....
2. Removal of .... bud will improve the seed set in sunflower
3. .... and .... deficiency leads to production of single seeded pods and darkened plumule type of seeds in groundnut
4. Sunflower flower always face sun and this is due to ....
5. Improving seed set in sunflower is by ....

6. Does a sorption isotherm vary with the composition of seed? Which kind of seed has low value? ....
7. Castor seed is extracted by .... method
8. Physiological maturity symptom of castor is ....
9. Sex reversion is noticed in castor during ....
10. Dark plumule in groundnut is an indication of .... deficiency
11. Boron application in groundnut seed production ....
12. Boron application in groundnut reduces ....
13. Sunflower is predominantly a .... crop
14. Nipping is practiced in .... hybrid seed production
15. Seeds of the groundnut variety .... has the mottled seed coat
16. Rubbing with muslin cloth is practiced for enhancing seed set in ....
17. The indeterminate flowering of castor is responsible for multiplication of ....
18. Removal of auxillary bud will improve the .... in sunflower
19. .... plumules in groundnut is caused by calcium deficiency
20. To arrest apical dominance in sunflower, ... is used
21. Ca deficiency in groundnut causes ....
22. The important commercial value of sunflower other than edible oil is ....
23. Sunflower belongs to the family .... and the seed is known as ....
24. "Hollow Heart" in groundnut seed is due to .... deficiency
25. Physiological maturity index for sunflower is ....

♦ Answer Keys ♦

Q.No.	Ans.	Q.No.	Ans.	Q.No.	Ans.
1	Rubbing of heads	2	Auxillary	3	Boron, Calcium
4	Phototropisim	5	Supplementary pollination	6	Oilseed
7	Dry extraction	8	Capsule cracking	9	Hot temperature
10	Calcium	11	Improves seed filling	12	Single seed pods
13	Cross pollinated	14	Sunflower	15	TMV 10
16	Sunflower	17	Female line	18	Seed set
19	Darkened	20	Malic hydrazide	21	Darkened plumple
22	Live stock feed	23	Compositae, Achene	24	Boron
25	Thalamus turn to lemon yellow				

2.10.3. State True or False

1. Rope pulling is the method of supplementary pollination in sunflower

2. Groundnut is the crop with least multiplication ratio
3. Boron deficiency causes lack of seed filling in groundnut
4. Calcium deficiency causes dark plumule in groundnut
5. Supplementary pollination is important to improve the seed oil content/ seed filling in sunflower
6. Calcium deficiency causes production of more of single compartment pods in groundnut
7. Isolation distance for groundnut is 25 m
8. In sunflower, pollination is mainly brought about by wind
9. The flower racemes of castor are usually dioecious
10. Five generation model is being followed in groundnut seed production
11. VP 1 is the pistillate line of castor
12. Block system for production of hybrid seed is adopted in castor

♦ Answer Keys ♦

Q.No.	Ans.	Q.No.	Ans.	Q.No.	Ans.	Q.No.	Ans.	Q.No.	Ans.	Q.No.	Ans.
1	False	2	True	3	True	4	True	5	False	6	False
7	False	8	False	9	True	10	True	11	True	12	False

2.11. SEED PRODUCTION IN FIBRE CROPS

2.11.1. Choose the correct answer

1. The acid used for cotton delinting is
   (a) Citric acid (b) Nitric acid
   (c) Sulphuric acid (d) Acetic acid
2. The chemical used for delinting in cotton is
   (a) H_2SO_4 (b) HCl
   (c) KNO_3 (d) KCl
3. Removal of fuzz in cotton is known as
   (a) Delinting (b) Defuzzing
   (c) Dehusking (d) Debearding
4. Which one of the following micronutrient can be sprayed for increasing flower number in cotton
   (a) $MnSO_4$ (b) KH_2PO_4
   (c) $ZnSO_4$ (d) $MgSO_4$
5. Planting ratio for cotton hybrid seed production is
   (a) 4 : 1 (b) 5 :1
   (c) 6 :1 (d) 7 :1
6. The prescribed isolation distance for a certified crop of cotton is
   (a) 10 m (b) 30 m
   (c) 50 m (d) 75 m

7. In cotton cultivar MCU 5, MCU stands for
   (a) Madras Combodian Uganda (b) Madurai Cotton Uganda
   (c) Madras Cotton Uganda (d) Madurai Combodian Uganda
8. Uppam cotton is
   (a) *G.arboreum* (b) *G.herbaceum*
   (c) G.barbadense (d) G.hirsutum
9. In cotton, tirak disorder refers to
   (a) Zn deficiency (b) Mg deficiency
   (c) Bad opening of bolls (d) N_2 deficiency
10. Cotton hybrid was first released in the year
    (a) 1991 (b) 1970
    (c) 2000 (d) 1996
11. Indian cotton is
    (a) *G. herbaceum* (b) *G. hirsutum*
    (c) *G. arboreum* (d) *G. barbadense*
12. The isolation distance normally followed for foundation seed production of cotton hybrids will be
    (a) 30 m (b) 50 m
    (c) 100 m (d) 1000 m
13. The maximum percentage of pollen shedders, that is permitted in casc of cotton hybrid is
    (a) 0.05% (b) 1.0%
    (c) 5.0% (d) 10%
14. "Savitha" is the hybrid of
    (a) Cotton (b) Maize
    (c) Bhendi (d) Bittergourd
15. DCH 32 is the hybrid of
    (a) Sorghum (b) Maize
    (c) Paddy (d) Cotton
16. Genetic purity standards for foundation stage of hybrids cotton
    (a) 95.0% (b) 85.0%
    (c) 90.0% (d) 80.0%
17. Block system is adopted for hybrid seed production in
    (a) Cotton (b) Safflower
    (c) Bhendi (d) Maize
18. The ploid condition of *Gossypium hirsutum* is
    (a) Diploid (b) Tetraploid
    (c) Triploid (d) Monoploid
19. Nipping is practiced in male parent to induce side branches to get additional pollen in hybrid seed production plot in
    (a) Bajra (b) Rice
    (c) Redgram (d) Cotton

20. In gas delinting of cotton, the gas used is
    (a) Dry sulphuric acid gas (b) Dry hydrochloride acid gas
    (c) Nitrogen peroxide gas (d) Nitrate nitrite gas
21. In jute seed production, the first rogueing is to be carried out when the plants are
    (a) 30 days old (b) 25 – 35 days old
    (c) 30 – 40 days old (d) 35 – 45 days old

♦ Answer Keys ♦

Q.No.	Ans.	Q.No.	Ans.	Q.No.	Ans.	Q.No.	Ans.	Q.No.	Ans.	Q.No.	Ans.
1	c	2	a	3	a	4	c	5	a	6	b
7	a	8	b	9	c	10	b	11	c	12	b
13	a	14	a	15	d	16	c	17	a	18	b
19	d	20	b	21	c						

2.11.2. Fill in the blanks

1. The processing equipment used for seed grading based on seed density is ....
2. The method used for delinting cotton is ....
3. During ginning, care should be taken to avoid .... in cotton
4. During kapas sorting .... and .... are to be removed as its seeds are poorer in quality
5. Specific gravity separator grades the seeds according to .... of seeds
6. An eco-friendly and efficient method of delinting cotton seed is ....
7. Specific gravity grading is useful for improving the marginal seed lots of .... crop
8. The isolation distance between male and female parent in cotton is ....
9. Spraying .... improves pollen viability in cotton
10. The most practical and widely adopted type of isolation for cotton is ....
11. The seed renewal period for hybrid cotton is ....
12. Boll shedding in cotton is controlled by spraying ....
13. Selfed bolls are identified for maintenance of genetic purity in .... crop
14. The maturity index for cotton is ....
15. .... crop is considered to be the Golden fibre

♦ Answer Keys ♦

Q.No.	Ans.	Q.No.	Ans.	Q.No.	Ans.
1	Specific gravity separator	2	Acid scarification	3	Mechanical injury

Q.No.	Ans.	Q.No.	Ans.	Q.No.	Ans.
4	Hard locks, skewed bolls	5	Density	6	Gas delinting
7	Cotton	8	5m	9	$ZnSO_4$
10	Spatial	11	One year	12	NAA
13	Cotton	14	Cracking of bolls	15	Jute

2.11.3. State True or False

1. Suguna is a CMS based hybrid in cotton
2. Acid scarification is followed for delinting in cotton
3. Acid used for cotton seed delinting is sulphuric acid
4. Hybrid seed production in cotton is performed through hand emasculation technique
5. Hand emasculation and dusting is practiced in cotton hybrid seed production

♦ Answer Keys ♦

Q.No.	Ans.	Q.No.	Ans.	Q.No.	Ans.	Q.No.	Ans.	Q.No.	Ans.
1	False	2	True	3	True	4	True	5	True

2.12. SEED PRODUCTION IN VEGETABLES

2.12.1. Choose the correct answers

1. The chemical used for seed extraction in tomato is
   (a) HCl (b) HNO_3
   (c) H_2SO_4 (d) $CaCl_2$
2. The isolation distance recommended for foundation and certified classes of cucurbits is
   (a) 800 : 400 (b) 50 : 25
   (c) 400 : 200 (d) 1600 : 1000
3. The best method of seed extraction in tomato is
   (a) Acid method (b) Fermentation method
   (c) Alkali (d) Mechanical method
4. Closer spacing in cucurbits leads to
   (a) Blight (b) Downy mildew
   (c) Flower drop (d) Seed set
5. In brinjal, anthesis takes place during
   (a) 5-7 am (b) 6-8 am
   (c) 7-9 am (d) 8-10 am
6. In cauliflower, flowers are emasculated to remove stamens using
   (a) Needles (b) Scissors
   (c) Knites (d) Forceps

7. Planting ratio for tomato hybrid seed production is
   (a) 5 : 1 (b) 4 :1
   (c) 3 :1 (d) 2 :1
8. Planting ratio for chilli hybrid seed production is
   (a) 2 : 1 (b) 3 :1
   (c) 4 :1 (d) 5 :1
9. Planting ratio for brinjal hybrid seed production is
   (a) 10 :1 (b) 8 :1
   (c) 6 :1 (d) 4 :1
10. Planting ratio for cucurbit hybrid seed production is
    (a) 5 :1 (b) 4 :1
    (c) 3 :1 (d) 2 :1
11. Formation of an elongated stem or seed stalk and occurs generally during the second season of growth in case of biennial plants is called as
    (a) Pome (b) Achene
    (c) Bolt (d) Caryopsis
12. The harvestable maturity indices for beetroot seed crop when middle portion of the inflorescence turns
    (a) Brown (b) Black
    (c) Brownish black (d) Light black
13. Seed rate for tomato seed production is
    (a) 1 kg/ha (b) 100g/ha
    (c) 500g/ha (d) 350g/ha
14. In tomato seed extraction, maximum seed recovery per cent
    (a) Alkali method (b) Mechanical method
    (c) Fermentation method (d) Acid method
15. What is the final count for tomato seed germination test?
    (a) 8 days (b) 14 days
    (c) 10 days (d) 21 days
16. Seed recovery per cent of brinjal is
    (a) 1-2% (b) 5-6%
    (c) 30-35% (d) 10-15%
17. Seed rate for chilli seed production is
    (a) 1 kg/ha (b) 300g/ha
    (c) 500g/ha (d) 2kg/ha
18. Plant harmone used to control flower dropping in chilli is
    (a) NAA (b) GA
    (c) IAA (d) IBA
19. Chilli seeds attain physiological maturity at
    (a) 45 DAA (b) 35 DAA
    (c) 50 DAA (d) 90 DAA

20. Chilli seed is extracted by
    (a) Acid method (b) Fermentation method
    (c) Alkali method (d) Dry seed extraction
21. Germination standards for foundation stage of chilli is
    (a) 70% (b) 80%
    (c) 50% (d) 60%
22. The seed rate for bhendi seed production is
    (a) 1-2kg/ha (b) 350-500g/ha
    (c) 3-4 kg/ha (d) 10-12 kg/ha
23. Techniques used to remove insect damaged and ill-filled seeds in bhendi is
    (a) Size grading (b) Water floatation
    (c) Egg floatation (d) Shape grading
24. Incubation, drying and separation technique (IDS) is used to sepatrate viable and dead seed in
    (a) Bhendi (b) Chilli
    (c) Tomato (d) Brinjal
25. Germination standards for foundation and certified seed production of bhendi is
    (a) 80% (b) 70%
    (c) 75% (d) 65%
26. Seed rate for bitter gourd seed production is
    (a) 5-7kg/ha (b) 3-4 kg/ha
    (c) 8-10kg/ha (d) 1-2 kg/ha
27. Isolation distance for bitter gourd foundation seed production is
    (a) 100m (b) 300 m
    (c) 400m (d) 1500m
28. Germination standard for bitter gourd is
    (a) 70% (b) 60%
    (c) 50% (d) 80%
29. Isolation distance for snake gourd foundation seed production is
    (a) 100m (b) 1000m
    (c) 1500m (d) 500m
30. Isolation distance for ridge gourd certified seed production is
    (a) 1000m (b) 500m
    (c) 100m (d) 200m
31. Which cucurbitaceous vegetables having a large volume of fibrous portion when fully ripe.
    (a) Snake gourd (b) Bitter gourd
    (c) Sponge gourd (d) Bottle gourd

32. Isolation distance required for sponge gourd certified seed production is
    (a) 400m (b) 800m
    (c) 200m (d) 1000m
33. Carrot belongs to family
    (a) Cruciferae (b) Brassicaceae
    (c) Chenopodiaceae (d) Umbelliferae
34. In which method, the nucleus seed production of radish is taken up?
    (a) Seed to seed method (b) Root to seed method
    (c) Stump to seed method (d) Bud to seed method
35. How the carrot certified seed production is done?
    (a) Seed to seed method (b) Root to seed method
    (c) Stump to seed method (d) Bud to seed method

♦ Answer Keys ♦

Q.No.	Ans.	Q.No.	Ans.	Q.No.	Ans.	Q.No.	Ans.	Q.No.	Ans.	Q.No.	Ans.
1	a	2	d	3	a	4	c	5	b	6	d
7	a	8	d	9	a	10	b	11	c	12	a
13	d	14	d	15	b	16	b	17	a	18	a
19	b	20	d	21	d	22	d	23	b	24	a
25	d	26	a	27	d	28	b	29	b	30	b
31	c	32	a	33	d	34	b	35	b		

2.12.2. Fill in the blanks

1. Designated weed plant of bhendi seed crop is ....
2. Hybrid seed production technology for bhendi is ....
3. The hybrid seed production system followed in onion is ....
4. .... technique is used to upgrade the seeds of bhendi
5. .... method of seed extraction is the best suited method for extraction of tomato seeds
6. .... chemical is used in alkali method of seed extraction
7. The breeding tool utilized for tomato hybrid seed production is ....
8. .... spray alters sex expression in cucurbits
9. Spraying .... induces bolting in cabbage
10. Flower induction in cole crops is by ....
11. Free crossing of varieties and their wild parents is frequent in ....
12. For the positive sex expression in ashgourd, ... can be sprayed
13. Flower drop in chilli is controlled by spraying .... at flowering stage

14. The isolation distance recommended for the foundation seed production of bhendi is ....
15. The number of female flowers in bittergourd can be increased by spraying of ....
16. Wet seed extraction method is followed in ....
17. Purple long is the variety of ....
18. Bulb is the main method of propagation of ....
19. Arka anamika is the variety of ....
20. Indicate the isolation distance to be maintained for certified seed production of cabbage ....
21. NaOH chemical is used in .... method of seed extraction
22. Isolation distance for certified class of carrot seed production is ....
23. Seed recovery percentage for tomato is ....
24. Seed rate for snake gourd seed production is ....
25. Spraying of .... are increasing the female flowers and seed yield in snake gourd.
26. Recovery per cent of snake gourd is ....
27. A seed rate of .... required for ridge gourd seed production.
28. Spraying of .... is increasing the seed yield in caluliflower and radish.
29. Visual maturity symptom of turnip is plants turn to ....
30. In cabbage production, at head formation stage .... cut is made for seed stalk emergence.
31. In cauliflower seed production, to induce seed stalk formation .... cut is made on the curd.

♦ Answer Keys ♦

Q.No.	Ans.	Q.No.	Ans.	Q.No.	Ans.
1	Wild okra	2	Emasculation and dusting	3	CMS
4	Water floatation	5	Acid	6	NaOH
7	Emasculation and Dusting	8	Ethrel	9	GA3
10	Over wintering	11	Cucurbits	12	Ethrel
13	NAA	14	100 m	15	Ethrel
16	Tomato	17	Brinjal	18	Onion
19	Bhendi	20	1000 m	21	Alkali
22	800 m	23	1%	24	15kg/ha
25	Maleic hydrazide and ethrel	26	45 per cent	27	3.5 to 5 kg/ha
28	DAP @2%	29	Parchment colour	30	Cross
31	Horizontal				

2.12.3. State True or False

1. Flower induction in cole crops is called steckling
2. Seedlings of bhendi are transplanted at the age of 30-35 DAS
3. Isolation distance requirement for bhendi is more than that for bittergourd
4. 1000 seed weight of bittergourd is approximately 500 g
5. Changing of yellow colour capsules is the maturity symptom of onion seeds
6. Functional male sterility has been exploited in hybrid seed production of capsicum
7. In cabbage and cauliflower, the isolation distance required for maintenance of parental lines is 1000 m
8. All types of crosses hybrid is available in potato
9. The harvestable maturity indices for onion seed crop is when seed become black on ripening in silvery capsules.
10. Fruits from first 8 to 10 pickings are used for seed purpose in tomato.
11. Visual maturity symptom of tomato is fruit turning red
12. Maximum germination per cent required for tomato is 70 per cent
13. In brinjal, long and medium styled flowers only produced fruits
14. In brinjal, pseudo short and short styled flowers also sets fruits.
15. In capsicum, foliar application of 2,4-D at fortnightly during flowering will increases the flowering.
16. Bhendi can be classified as often cross pollinated crop.
17. Hard seededness is a major cause of erractic germination in bhendi
18. Pure seed per cent for bhendi is 99%
19. Bitter gourd is popularly known as a plant insulin
20. Snake gourd seeds are extracted by manually
21. Dry seed extraction method is employed for sponge gourd seed extraction

♦ Answer Keys ♦

Q.No.	Ans.	Q.No.	Ans.	Q.No.	Ans.	Q.No.	Ans.	Q.No.	Ans.	Q.No.	Ans.
1	False	2	False	3	False	4	False	5	False	6	False
7	False	8	True	9	True	10	True	11	True	12	True
13	True	14	False	15	True	16	True	17	True	18	True
19	True	20	True	21	True						

2.13. SEED PRODUCTION IN FORAGE LEGUMES AND GRASSES

2.13.1. Choose the correct answer

1. Inseparable other crop seed of lucerne
   (a) Glarn (b) Sesbania
   (c) Cuscuta (d) Niger

2. The seed rate for lucerne seed production is
(a) 4 kg/ha (b) 3 kg/ha
(c) 2 kg/ha (d) 4.5 kg/ha

3. In alfalfa, flowers are usually tripped for efficient pollination by
(a) Wind (b) Rain
(c) Bees (d) Honey bees

4. HCN is present in fodder sorghum at
(a) 40-50 days (b) 90-100 days
(c) 80-90 days (d) 20-30 days

5. Daincha seed crop attain physiological maturity at
(a) 40 DAA (b) 45 DAA
(c) 50 DAA (d) 60 DAA

6. What is the isolation distance for foundation seed production of napier grass?
(a) 10 m (b) 15 m
(c) 20 m (d) 25 m

7. What is the isolation distance for certified seed production of forage sorghum?
(a) 50 m (b) 100 m
(c) 150 m (d) 200 m

8. Common name of hedge lucerne is
(a) Velimasal (b) Kudiraivali
(c) Samai (d) Muyal masal

9. Seed rate of velimasal is
(a) 30-35 kg/ha (b) 20-25 kg/ha
(c) 50-60 kg/ha (d) 10-20 kg/ha

10. Stylo seeds posses
(a) Hard seed coat (b) Immature embryo
(c) Inhibitors (d) Appendages

11. *Clitoria* ternatea is
(a) Indeterminate vine (b) Determinate vine
(c) Hedge (d) Bush

12. Alfalfa of the tropics is
(a) Stylo (b) Cumbu napier
(c) Desmodium (d) Sorghum

13. Mimosine is a toxic amino acid found in
(a) Subabul (b) Lucerne
(c) Sorghum (d) Fodder maize

14. Isolation distance for certified seed production of berseem is
(a) 100 m (b) 200 m
(c) 300 m (d) 400 m

15. The berseem seed crop should be harvested when pods turned
    (a) Brown (b) Yellow
    (c) Red (d) Black
16. The isolation distance for lucerne crop is
    (a) 50 m (b) 100 m
    (c) 150 m (d) 200 m
17. The final count for dinanath grass is
    (a) 30 days (b) 15 days
    (c) 7 days (d) 28 days
18. Minimum germination per cent requirement for berseem is
    (a) 80% (b) 75%
    (c) 65% (d) 85%

♦ Answer Keys ♦

Q.No.	Ans.	Q.No.	Ans.	Q.No.	Ans.	Q.No.	Ans.	Q.No.	Ans.	Q.No.	Ans.
1	c	2	b	3	d	4	a	5	a	6	c
7	b	8	a	9	b	10	a	11	a	12	c
13	a	14	a	15	a	16	b	17	d	18	a

2.13.2. Fill in the blanks

1. Glue spray is resorted in .... crop
2. Rainfall induces .... in lucerne
3. .... is a variety of multicut fodder sorghum
4. .... is a valuable leguminous forage and hay crop.
5. Botanical name of berseem is ....
6. Irrigation requirement for berseem is ....
7. Cumbu Napier hybrid grass derived from inter specific cross between ....
8. The suitable area for berseem seed production in India is ....
9. Alfalfa is a .... legume
10. The pollinators required for alfalfa seed production is ....
11. The world's first hybrid alfalfa is developed by ....
12. Botanical name for bird's foot trefoil (perennial forage crop) is ....
13. Clovers are most efficiently pollinated by ....
14. The botanical name for subabul is ....
15. CO(CN)4 is a variety of ....
16. Queen of forage crop is ....

♦ Answer Keys ♦

Q.No.	Ans.	Q.No.	Ans.	Q.No.	Ans.
1	Forage	2	Vegetative growth	3	CO FS 29
4	Lucerne	5	*Trifolium alexandrium*	6	710 mm
7	*Pennisetum glaucum and P purpureum*	8	Northern plain	9	Perennial forage
10	Western honey bees	11	Teweles seed company	12	*Lotus corniculatus*
13	Bumble bees	14	*Leucaena leucocephala*	15	Bajra Napier hybrid
16	Alfalfa				

2.13.3. State True or False

1. Tripping is associated with lucerne crop
2. African tall maize is a fodder crop.
3. CSH 13 sorghum variety is used for both grain and forage production
4. The cumbu napier hybrid is a triploid
5. Forage crops are shy seed yielders.
6. Alfalfa is propagated by seeds

♦ Answer Keys ♦

Q.No.	Ans.	Q.No.	Ans.	Q.No.	Ans.	Q.No.	Ans.	Q.No.	Ans.	Q.No.	Ans.
1	True	2	True	3	True	4	True	5	True	6	True

2.14. SEED PRODUCTION IN PLANTATION CROPS

2.14.1. Choose the correct answer

1. Coffee plants begin yielding from
   (a) 4th year (b) 8th year
   (c) 12th year (d) 2nd year
2. National yield of rubber is
   (a) 2.6 t/ha (b) 2.0 t/ha
   (c) 1.6 t/ha (d) 1.0 t/ha
3. Highest production of cocoa is in state of
   (a) Karnataka (b) Kerala
   (c) Tamil Nadu (d) Andhra Pradesh
4. Sundaram is a very high yielding clone of
   (a) Tea (b) Coffee
   (c) Rubber (d) Coconut
5. Viviparous seed present in
   (a) Cocoa (b) Nutmeg
   (c) Clove (d) Cinnamon

6. Cashew nut varieties in Tamil Nadu
   (a) VRI – 1 to 3 (b) Vengurla – 1 to 7
   (c) BPP (d) Jhargram – 1
7. The seeds of rubber are viable for
   (a) 7 days (b) 70 days
   (c) 7 weeks (d) 7 months
8. Rubber seeds should be harvested when
   (a) Capsules begin to turn yellow colour
   (b) Capsules begin to turn brown colour
   (c) Capsules begin to turn from green to greenish yellow colour.
   (d) Capsule begin to turn brown and open
9. The fruit type of cardamom is
   (a) Dry berry (b) Capsule
   (c) Pod (d) Dry pod
10. Large seeds of cardamom remain dormant upto
    (a) 7-8 hours (b) 7-8 days
    (c) 7-8 weeks (d) 7-8 months
11. Propagation methods of coffee is
    (a) Seeds and cuttings (b) Seeds and layerings
    (c) Seeds and buddings (d) Seeds and grafting
12. The process of picking of left berries irrespective of ripening in coffee is called as
    (a) Fly picking (b) Main picking
    (c) Stripping (d) Cleaning
13. The process of collection of fruit dropped during harvesting of coffee is called as
    (a) Stripping (b) Main picking
    (c) Fly picking (d) Cleaning
14. Flower buds of coffee are formed only in the
    (a) Oligotrophic shoots (b) Plaiotrophic shoots
    (c) Geotrophic shoots (d) Main shoots
15. Coffee flowers are
    (a) Monoecious (b) Dioecious
    (c) Hermaphrodite (d) Bisexual
16. Number of stages in fruit development of coffee is about
    (a) One (b) Two
    (c) Three (d) Four
17. Seed moisture content of coffee seeds at the time of berry stage is about
    (a) 25% (b) 50%
    (c) 75% (d) 100%

18. Techniques used for harvesting by hand in coffee is called as
(a) Stripping (b) Finger tip
(c) Fly picking (d) Main picking

19. Coffee belongs to the family
(a) Rubiaceae (d) Rosaceae
(c) Teliacease (d) Rutaceae

20. Coffee seeds shows the reduced seed germination due to the presence of the
(a) Abscisic acid (b) Stearic acid
(c) Oleic acid (d) Palmitic acid

21. The first step in processing tea is
(a) Withering (b) Cutting
(c) Tearing (d) Fermentation

22. Tea is pollinated by
(a) Insects (b) Beetles
(c) Wind (d) Air

23. Regular watering for tea is essential for
(a) To establish roots (b) To produce flowers
(c) To develop seeds (d) To harvest

24. Rubber seeds complete germination in
(a) 5 to 10 days (b) 10-15 days
(c) 15-20 days (d) 20-25 days

25. Flower type of rubber is
(a) Monoecious (b) Dioecious
(c) Bisexual (d) Unisexual

26. The moisture content of the fresh weight basis of the rubber seed is about
(a) 25% (b) 35%
(c) 45% (d) 55%

27. Rubber seeds can be stored in
(a) Plastic bags (b) Jute bags
(c) Tins (d) Card boards

28. Rubber belongs to the family of
(a) Euphorbiaceae (b) Asteraceae
(c) Rutaceae (d) Teliaceae

29. Fruits of rubber are
(a) Two lobed (b) Three lobed
(c) Four lobed (d) Five lobed

30. Rubber seeds ripen during the month of
(a) July- Sep (b) Oct-Dec
(c) Jan-Mar (d) Apr-June

31. Fruit of cocoa is
(a) Pod (b) Capsule
(c) Berry (d) Drupe
32. Viability of cardamom seed is only about
(a) 7-10 days (b) 10-12 days
(c) 12-15 days (d) 15-20 days
33. Problem in cocoa when stored beyond one week
(a) Viviparous germination (b) Looe taste
(c) Colour discolaration (d) Become hard
34. The serious pest in pepper cultivation is
(a) Pollu beetle (b) Thrips
(c) Aphids (d) Whitefly
35. Pepper starts bearing after
(a) 2 years of planting (b) 3 years of planting
(c) 4 years of planting (d) 5 years of planting
36. Harvesting season of pepper berries is from
(a) Nov- Mar (b) Apr – June
(c) July- Sep (d) Oct – Dec
37. Rubber seed is
(a) Orthodox seed (b) Recalcitrant seed
(c) Intermediate seed (d) Facultative seed
38. Fruiting branches of black pepper is called as
(a) Plagiotropic (b) Orthotropic
(c) Geotropic (d) Runner shoot
39. Inflorecsence type of balck pepper is
(a) Raceme (b) Catkin
(c) Capitulum (d) Spadix
40. Number of fruits produced per spike in black pepper is
(a) 50-60 fruits (b) 60-70 fruits
(c) 80-90 fruits (d) 90-100 fruits
41. Which part of the fruit turns green to red on ripening and black on drying in black pepper?
(a) Carp (b) Mesocarp
(c) Endocarp (d) Exocarp
42. Commercial propagation of black pepper is
(a) Root cuttings (b) Stem cuttings
(c) Leaf cuttings (d) Tip cuttings
43. Method of propagation in cardamom prevents the spread of khatte disease is
(a) Seeds (b) Cuttings
(c) Budding (d) Layerings

44. Optimum average yield of dry capsule of cardamom is about
    (a) 20-30kg per ha (b) 30-40 kg per ha
    (c) 40-50 kg per ha (d) 50-70 kg per ha
45. Pepper is used as
    (a) Root spice (b) Rhizome spice
    (c) Berry spice (d) Seedy spice
46. Pulp processing to make paper can be very polluting, if they use
    (a) Mechanical grinding (b) Sulfite processes
    (c) Enzymes (d) Proteins
47. Which of the following foods is a modified stem?
    (a) Cassava (b) Sweet potato
    (c) Carrot (d) Onion
48. Cocoa is derived from the plant's
    (a) Leaves (b) Stems
    (c) Hulls (d) Nibs (cotyledons)

♦ Answer Keys ♦

Q.No.	Ans.	Q.No.	Ans.	Q.No.	Ans.	Q.No.	Ans.	Q.No.	Ans.	Q.No.	Ans.
1	a	2	c	3	b	4	a	5	a	6	a
7	c	8	b	9	d	10	d	11	a	12	c
13	d	14	b	15	c	16	d	17	b	18	b
19	a	20	a	21	a	22	a	23	a	24	a
25	b	26	b	27	a	28	a	29	b	30	a
31	a	32	a	33	a	34	a	35	b	36	a
37	b	38	a	39	b	40	d	41	d	42	b
43	a	44	d	45	c	46	b	47	d	48	d

2.14. 2. Fill in the blanks

1. The spacing adopted for robusta coffee is ....
2. Recommended spacing for arabica coffee is ....
3. .... is the seed of the coffee plant.
4. Coffee seed possess .... germination.
5. Seedling emergence of coffee may extend up to .... days.
6. Emergence of coffee seeds at 40 days after sowing is called as .... seedlings.
7. Emergence of coffee seeds at 60 days after sowing is called as .... seedlings.
8. Flowers of coffee are arranged in .... type inflorescence.
9. *Coffea arabica* seeds mature .... days after ovule fertilization.
10. *Coffea canephora* seeds take .... days to attain maturity.
11. Seed germination in coffee is observed at the ....
12. The best harvesting stage of coffee can be identified by the presence of atleast .... % mature berry fruits.

13. Germination of tea may extend up to .... weeks .
14. Flowering in rubber is negatively affected by .... deficit.
15. Opening of dry fruits in rubber occurs during ....
16. The recommended hole size for sowing rubber seeds is ....
17. Normal seed yields of rubber vary from .... kg/ha/year.
18. The storage behavior of rubber seed is ....
19. Organism which affects the seed quality during seed storage is ....
20. Rubber seed quality can be assessed by conducting ....
21. Method of drying the cocoa beans is ....
22. Propagation of cocoa is by ....
23. Bean yield of cocoa is about .... kg/ha.
24. Cocoa pods take .... days from pollination to attain the harvest stage.
25. Harvesting of cocoa pods should be done with a ....
26. .... attack will be more, if cacoa pods when allowed to ripe on tree.
27. Planting time for the cocoa seedling is ....
28. Pepper is grown mainly as a ....
29. Spacing adopted for larger cardamom is ....
30. Spacing adopted for smaller cardamom is ....
31. Pepper vine attain full bearing stage in the .... year after planting.
32. Recovery percentage of white pepper is about .... of ripe berries.
33. The temperature maintained for cardamom processing in fuel kiln is about .... over night.
34. Process during which caffeine is removed from coffee beans is called as ....
35. Moisture content of wet parchment coffee is about ....
36. Anticoagulants used to prevent coagulation in latex is ....
37. Preservatives used to preserve the latex is ....
38. Field latex with a suitable preservative is termed as ....
39. Oldest method used to extract latex is ....
40. Direct sowing in coffee is not recommended because a there is a lack of ....

♦ Answer Keys ♦

Q.No.	Ans.	Q.No.	Ans.	Q.No.	Ans.
1	3 × 3m	2	25 × 25m	3	Coffee bean
4	Slow and asynchronous	5	120	6	Match seedling
7	Jaguar ear	8	Glomerule	9	210-250

Q.No.	Ans.	Q.No.	Ans.	Q.No.	Ans.
10	300-350	11	Green fruit stage	12	95
13	8	14	Water	15	Late summer
16	$1 \times 1 \times 1\ m^3$	17	70 to 500	18	Recalcitrant
19	Aspergillus and Penicillium	20	Tetrazolium test	21	Sun drying
22	Seeds and vegetative	23	3000	24	150-170
25	Sharp knife	26	Rodents	27	September – October
28	Rainfed crop	29	25x20 m	30	20 × 15m
31	7th or 8th	32	25%	33	50-60ºC
34	Decaffeinatral	35	50-55%	36	Sodium sulphate
37	Ammonia	38	Preserved field latex	39	Sheet method
40	Unifiormity in germination				

2.15. SEED PRODUCTION OF VEGETATIVELY PROPAGATED CROPS

2.15.1. Choose the correct answer

1. Haulm cutting in potato is adopted to avoid
   (a) Viral disease (b) Bacterial disease
   (c) Genetic contamination (d) Physical contamination
2. Seed production using polyploids is common in
   (a) Sugarcane (b) Sorghum
   (c) Beet root (d) Turmeric
3. Seed plot technique is used for
   (a) Hybrid rice seed production (b) Potato seed production
   (c) Cotton seed production (d) Maize seed production
4. What is the multiplication ratio for propagation of tapioca through seed and vegetative means?
   (a) 1:100 and 1:10 (b) 1:10 and 1:100
   (c) 1:50 and 1:10 (d) 1:10 and 1:50
5. For the production of seed potatoes, the multiplication ratio ranges from
   (a) 1:6 to 1:10 (b) 1:8 to 1:15
   (c) 1:6 to 1:8 (d) 1:8 to 1:10
6. The desiccation of pollen and poor seed set in potato are caused by
   (a) Excessive temperature (b) Extreme temperature
   (c) Very hot temperature (d) Hot temperature
7. Potato tubers are modified
   (a) Stem (b) Root
   (c) Leaves (d) Flowers

8. New stems from each eye of potato are called as
   (a) Sprouts (b) Germination
   (c) Roots (d) Shoots
9. TPS are developed from the
   (a) Open pollinated flowers (b) Cross pollinated flowers
   (c) Often cross pollinated flowers (d) Self pollinated flowers
10. Plants obtained from the TPS seeds are genetically
   (a) Homogenous (b) Heterogenous
   (c) Homozygous (d) Heterozygous
11. Number of tissue culture plants required for planting an acre of sugarcane is
   (a) 2000 (b) 3000
   (c) 4000 (d) 5000
12. The tissue culture sugarcane can be harvested from the
   (a) 7th month (b) 8th month
   (c) 9th month (d) 10th month
13. The percentage of yield increase by the tissue culture sugarcane over the conventional method is
   (a) 20% (b) 30%
   (c) 40% (d) 50%
14. How many years would be needed for the production of variety involving hybridization followed by clonal selection in a vegetatively propagated crop?
   (a) Two (b) Four
   (c) Six (d) Eight
15. Which of the following species of *Solanum* was utilized for the improvement of potato
   (a) *Solanum andigena* (b) *S.nigrum*
   (c) *S.khasianum* (d) *S.melongena*
16. At how many days after haulm cutting, the potato seed tubers are harvested?
   (a) 5-10 days (b) 10-15 days
   (c) 15-20 days (d) 20-25 days
17. Which one of the following chemical is used to break dormancy of seed tubers of potato?
   (a) 1% GA_3 (b) 1% IAA
   (c) 1% IBA (d) 10% CCC
18. What is the seed certification standards for seed size of hill potato?
   (a) 20 – 30 mm small size (b) 30 – 40 mm small size
   (c) 30 – 50 mm small size (d) 30 – 60 mm small size

19. What is the seed certification standards for weight of large size seed of hill potato?
    (a) > 100 g (b) > 130 g
    (c) > 150 g (d) > 160 g
20. What is the seed rate for true potato seed?
    (a) 50 g (b) 100 g
    (c) 150 g (d) 200 g
21. At what time, the potato berries are harvested?
    (a) 25-40 days (b) 35-50 days
    (c) 45 -50 days (d) 55 -60 days
22. What is the concentration of HCl used for seed extraction in potato?
    (a) 5% (b) 15%
    (c) 15% (d) 20%
23. The fruit type of potato is
    (a) Berry (b) Drupe
    (c) Pome (d) Schizocarp

♦ Answer Keys ♦

Q.No.	Ans.	Q.No.	Ans.	Q.No.	Ans.	Q.No.	Ans.	Q.No.	Ans.	Q.No.	Ans.
1	d	2	a	3	b	4	a	5	b	6	b
7	a	8	a	9	a	10	d	11	a	12	a
13	c	14	d	15	d	16	b	17	a	18	d
19	c	20	b	21	c	22	b	23	a		

2.15.2. Fill in the blanks

1. Potato sprouting can be overcome by spraying of ....
2. Spraying of methyl bromide in potato .... sprouting
3. The use of clonal propagation is for .... of plant genotypes
4. .... has the potential for propagation of thousands plantlets from a single genetic stock.
5. Clonal propagation is not suitable for .... crops.
6. Primary commercial propagation of potato is ....
7. Clonal progeny are highly .... in all characters
8. In potato, the apical bud and lateral buds are called as ....
9. .... method is used as a tool for eradication of potato initial material from virus disease.
10. .... technique is followed when the plants in which the viruses cannot be eradicated by meristem culture.
11. The technique used to eradicate virus in plants by heating is called as ....
12. .... involves the culture of immature embryos of plants in a special medium to prevent abortion of young embryo.

13. Disadvantage of the TPS is that they are not .... and exhibit heterogeneity.
14. The size of the meristem used for culture is about ....
15. Nodal segment culture in which the axillary and apical buds grow into new plants is predominantly used for initial shoot multiplication of ....
16. .... are miniatures developed *in vitro* under tuber inducing conditions.
17. Potato plant requires..................... conditions for flowering

♦ Answer Keys ♦

Q.No.	Ans.	Q.No.	Ans.	Q.No.	Ans.
1	Methyl bromide	2	Arrest	3	Rapid regeneration
4	Clonal propagation	5	Seeding cereal	6	Clonal propagation
7	Uniform	8	Eyes	9	Meristem
10	Heat treatment	11	Thermotherapy	12	Embryo rescue
13	Genetically pure	14	0.1-0.3mm	15	Mericlones
16	Microtubers	17	Long photoperiod		

2.16. SEED PRODUCTION OF FRUIT CROPS

2.16.1. Choose the correct answer

1. Ethylene is known to stimulate
   (a) Seed deterioration (b) Bud dormancy
   (c) Fruit ripening (d) Seed dormancy
2. Amarapali, a hybrid mango is a cross between
   (a) Dashehari x Neelam (b) Neelam x Dashehari
   (c) Neelam x Alphonso (d) Alphonso x Neelam
3. Types of apomixis in apple is
   (a) Facultative (b) Obligatory
   (c) Non-recurrent (d) Recurrent
4. Akbar is a hybrid of
   (a) Grape (b) Mango
   (c) Apple (d) Papaya
5. Queen is the variety of
   (a) Grape (b) Guava
   (c) Pineapple (d) Mango
6. Commercial method of propagation of bael is
   (a) Grafting (b) Budding
   (c) Cutting (d) Seed
7. Mango hybridization work was started in year
   (a) 1911 (b) 1901
   (c) 1921 (d) 1931

8. Average yield of mango is
   (a) 5 t/ha (b) 6 t/ha
   (c) 7 t/ha (d) 8 t/ha
9. Cultivated strawberry is
   (a) Diploid (b) Triploid
   (c) Tetraploid (d) Octaploid
10. India's share in global production of fruits is
   (a) 5% (b) 10%
   (c) 15% (d) 20%
11. The chemical by which seedless grapes can be produced is called
   (a) Gibberellins (b) Ethanol
   (c) Auxins (d) Catalyses
12. Which citrus species is mono embryonic unlike other species which are poly embryonic?
   (a) *Citrus limon* (b) *Citrus reticulata*
   (c) *Citrus grandis* (d) *Citrus aurantifolia*
13. In citrus, which growth regulator acts as a good fruit thinner
   (a) GA (b) NAA
   (c) IBA (d) ABA
14. Guava belongs to which family
   (a) Myrtaceae (b) Musaceae
   (c) Caricaceae (d) Anacardiaceae
15. What is the moisture content to maintain seed viability for jack seed.
   (a) 39 per cent (b) 10 per cent
   (c) 13 per cent (d) 8 per cent
16. Mango seed germination is improved by
   (a) Seed husking (b) Soaking in GA_3
   (c) Soaking in IAA (d) Seed husking and soaking in GA_3
17. Papaya belongs to which family
   (a) Myrtaceae (b) Musaceae
   (c) Caricaceae (d) Anacardiaceae
18. In which papaya variety vivipary occurrence is more
   (a) CO 1 (b) CO 2
   (c) CO 6 (d) CO 7
19. Family name for custard apple is
   (a) Malavaceae (b) Annonaceae
   (c) Moraceae (d) Roseaceae
20. Fruit type of mongosteen is
   (a) Drupe (b) Berry
   (c) Capsule (d) Synconium

21. Fruit type of litchi is
(a) Berry (b) Capsule
(c) Nut (d) Drupe
22. Litchi is propagated through
(a) Seed (b) Vegetative propagules
(c) Both seed and vegetative propagules (d) Tissue culture
23. Fig is propagated through
(a) Seed (b) Vegetative propagules
(c) Both seed and vegetative propagules (d) Tissue culture
24. Mangosteen is propagated through
(a) Seed (b) Cuttings
(c) Both seed and cuttings (d) Layers
25. Minimum seed moisture content for storage of mangosteen is
(a) 53 – 54% (b) 10 – 15%
(c) 7 – 8% (d) 8 – 10%
26. Apple comes under the family of
(a) Anacardiaceae (b) Juglandaceae
(c) Rosaceae (d) Actinidiaceae

♦ Answer Keys ♦

Q.No.	Ans.	Q.No.	Ans.	Q.No.	Ans.	Q.No.	Ans.	Q.No.	Ans.	Q.No.	Ans.
1	c	2	a	3	d	4	c	5	c	6	b
7	a	8	d	9	d	10	b	11	a	12	c
13	b	14	a	15	a	16	d	17	c	18	b
19	b	20	b	21	c	22	b	23	c	24	a
25	a	26	c								

2.16.2. Fill in the blanks

1. .... technique is used for rapid multiplication of banana suckers.
2. Areas prone to banana stem weevil, the suckers should be treated with .... solution.
3. Type of inflorescence in citrus species is ....
4. In citrus, cuttings are treated with growth regulators solution like .... for better rooting.
5. .... is the most successful treatment for dormancy release of grape seed.
6. Guava is commonly propagated through both ....
7. Jack fruit is popularly known as ....
8. Jack seeds are an example for ....
9. Seed rate for papaya is ....
10. The botanical name of sapota is ....

11. In approach grafting of sapota, ... is used as a rootsock in India.
12. Botanical name for pineapple is ....
13. Botanical name of pomegranate is ....
14. Inflorescence type of passion fruit is ....
15. An example for corymbos inflorescence is ....
16. .... is major pollinating agent for avocado
17. Custard apple is an example for ....
18. Rambutan attains physiological maturity ....
19. West Indian cherry attains physiological maturity at ....
20. Maturity symptom for seed collection in passion fruit is ....
21. Type of dormancy present in the pine apple seed is ....
22. Origin of kiwi fruit is ....
23. Fruit type of apple is ....
24. Apple is .... pollinated crop

♦ Answer Keys ♦

Q.No.	Ans.	Q.No.	Ans.	Q.No.	Ans.
1	Tissue culture	2	0.1% metasystax	3	Cymose
4	IBA or IAA	5	Cold stratification	6	Seeds and airlayers
7	Poor man's fruit	8	Recalcitrant seed	9	500g/ha
10	*Achras zapota*	11	*Manilkhara hexandra*	12	*Ananas comosus*
13	*Punica granatum*	14	Solitary	15	Durian
16	Insects	17	Protogynous	18	15-18 weeks after anthesis
19	24-26 days after anthesis	20	Shriveled fruits	21	Seed coat
22	Southern China	23	Pome	24	Cross

2.16.3. State True or False

1. Jack seed germination is higher at constant temperature range of 25 to 30°C and in the presence of light.
2. Papaya is highly cross pollinated crop.
3. Date palm is an example for dioecious
4. Durian is an example for recalcitrant seeds
5. Custard apple and date palm are orthodox seeds
6. Walnut is an example for monoecious

♦ Answer Keys ♦

Q.No.	Ans.	Q.No.	Ans.	Q.No.	Ans.	Q.No.	Ans.	Q.No.	Ans.	Q.No.	Ans.
1	True	2	True	3	True	4	True	5	True	6	True

2.17. SEED PRODUCTION IN FLOWER CROPS

2.17.1. Choose the correct answer

1. Inflorescence of anthurium is
   (a) Spathe (b) Spadix
   (c) Raceme (d) Solitary head
2. An example for protogyny flower is
   (a) Anthurium (b) Orchid
   (c) China aster (d) Chrysanthemum
3. Anthesis time of carnation is
   (a) 6-7 am (b) 7-8 am
   (c) 8-10 am (d) 10-12 am
4. Type of fruit found in anthurium is
   (a) Berry (b) Capsule
   (c) Achene (d) Drupe
5. Fruit type of chrysanthemum is
   (a) Berry (b) Capsule
   (c) Achene (d) Drupe
6. Example for the protandrous type of cut flower is
   (a) Gerbera (b) Orchid
   (c) Gladiolus (d) Rose
7. Pollination in orchid is carried out by
   (a) Insects (b) Beetles
   (c) Wind (d) Water
8. Fruit of rose is called as
   (a) Pod (b) Hip
   (c) Berry (d) Achene
9. Carnation belongs to the family of
   (a) Araceae (b) Caryophyllaceae
   (c) Asteraceae (d) Alstromeraceae
10. Anthurium seed attains physiological maturity at
    (a) 30 DAA (b) 40 DAA
    (c) 50 DAA (d) 60 DAA
11. Petunia can be propagated by means of
    (a) Cutting (b) Budding
    (c) Layering (d) Seeds
12. Gladiolus attains the seed maturity at
    (a) 30-40 DAA (b) 40-50 DAA
    (c) 50-60 DAA (d) 60-70 DAA
13. Carnation attains the seed maturation at
    (a) 30-35 DAA (b) 40-45 DAA
    (c) 50-55 DAA (d) 60-65 DAA

14. Branched cluster type of inflorescence can be seen in
(a) Anthurium (b) Carnation
(c) China aster (d) Gladiolus

15. Dormancy in orchid can extend up to
(a) One month (b) 2 months
(c) 3 months (d) 4 months

16. Exogenous and endogenous dormancy is exhibited in which of the following crop
(a) Rose (b) Orchid
(c) Dahlia (d) Petunia

17. The commercial propagation of gladiolus is
(a) Corms (b) Crowns
(c) Slips (d) Suckers

18. Which of the following crop seeds has a tendency of sweating ?
(a) Rose (b) Orchid
(c) Dahlia (d) Petunia

19. Celosia belongs to the family of
(a) Oleaceae (b) Amaranthaceae
(c) Apocynaceae (d) Ericaceae

20. The seed colour of bougainvillea is
(a) Red (b) White
(c) Black (d) Yellow

21. Number of seeds produced by crossandra per blossom is about
(a) One (b) Two
(c) Three (d) Four

22. The inflorescence type of cosmos is
(a) Solitary (b) Spike
(c) Raceme (d) Spadix

23. Fruit type of cactus and succulents is
(a) Achenes (b) Capsule
(c) Pod (d) Drupe

24. Number of seeds present per capsule in *Azalea* is about
(a) 200 (b) 300
(c) 400 (d) 500

25. Drooping panicle type inflorescence can be observed in
(a) *Heliconia* (b) *Hemerocallis*
(c) Ghomphrena (d) Lotus

26. Zinnia attains seed maturity at
(a) 20 DAA (b) 30 DAA
(c) 40 DAA (d) 50 DAA

27. Celosia can be pollinated by
    (a) Self (b) Insects
    (c) Bees (d) Wind
28. The pollinating agent for lotus is
    (a) Birds (b) Beetles
    (c) Insects (d) Wind
29. Marigold seeds mature when they attain
    (a) 20-30 DAA (b) 30-40 DAA
    (c) 40-50 DAA (d) 50-60 DAA
30. Rose seeds attain the maturity at
    (a) 100 DAA (b) 125 DAA
    (c) 150 DAA (d) 175 DAA
31. Chrysanthamum is a
    (a) Self pollinated crop (b) Often cross pollinated crop
    (c) Cross pollinated crop (d) Self-incompatible crop
32. Basic chromosome number of jasmine is
    (a) 11 (b) 12
    (c) 13 (d) 15
33. Seed of which annual germinate only in dark
    (a) Zinnia (b) Nigella
    (c) Marigold (d) Aster
34. RRL-85-1 is a important high yielding varieties of
    (a) Celery (b) Plum
    (c) Pear (d) Chicory

♦ Answer Keys ♦

Q.No.	Ans.	Q.No.	Ans.	Q.No.	Ans.	Q.No.	Ans.	Q.No.	Ans.	Q.No.	Ans.
1	a	2	a	3	d	4	a	5	c	6	c
7	a	8	b	9	b	10	d	11	d	12	c
13	c	14	b	15	c	16	a	17	a	18	b
19	b	20	c	21	a	22	a	23	a	24	c
25	a	26	c	27	c	28	a	29	b	30	c
31	c	32	c	33	b	34	a				

2.17.2. Fill in the blanks

1. Flowering time of chrysanthemum is from ....
2. Type of florets found in marigold are the ....
3. Ray floret in marigold is ....
4. Disc florets in marigold are .... in nature
5. Chrysanthemum exhibits the .... dormancy
6. Anthurium seed is an example of ....
7. Fruit of marigold is called as ....
8. Fruit of orchid is called as ....

9. Spathe of anthurium is called as ....
10. The fruit branch of anthurium is called as ....
11. Anthurium seeds embedded in a ....
12. Colour of China aster seed is ....
13. The storage behaviour of the chrysanthemum seeds are ....
14. Seeds of *Gomphrena* are ....
15. .... can be seen in *Gomphrena*
16. Type of fruit found in *Heliconia* is ....
17. Cactus and succulents favours ....
18. Flowering season of Celosia is ....
19. Storage behaviour of lotus seed is ....
20. Number of seeds present in a seed head of lotus is about ....

♦ Answer Keys ♦

Q.No.	Ans.	Q.No.	Ans.	Q.No.	Ans.
1	April-May	2	Ray florets and disc florets	3	Male flowers
4	Bisexual	5	Seed coat and endogenous	6	Orthodox seed
7	Achene	8	Pod	9	Candle
10	Infructescence	11	Gelatinous sap	12	Bright brown
13	Orthodox type	14	Yellowish- brown colour	15	Upright spikes
16	Drupe	17	Self and cross pollination	18	March to August
19	Orthodox seed	20	20-24 seeds		

2.17.3. State True or False

1. 1000 flowers are required for production of 30 g of hybrid seed in petunia
2. *Gaillardia* seeds exhibits the storage behaviour of orthodox seeds
3. Jasmine can be propagated by means of both sexual and asexual
4. Lotus can be propagated by means of sexual and asexual
5. *Gomphrena* belongs to the family Asteraceae
6. Number of seeds present in a rose hip is about 15
7. Flowering season of orchid is during June- September
8. Marigold can be propagated by vegetative method
9. Stage of collection of seeds for extraction in chrysanthemum is dried heads
10. Method of extraction of lilium seed is by rubbing
11. Marigold seeds can be extracted by manually

♦ Answer Keys ♦

Q.No.	Ans.	Q.No.	Ans.	Q.No.	Ans.	Q.No.	Ans.	Q.No.	Ans.	Q.No.	Ans.
1	False	2	True	3	True	4	True	5	False	6	False
7	True	8	False	9	True	10	True	11	True		

2.18. MACRO AND MICROPROPAGATION

2.18.1. Choose the correct answer

1. Which one of the following group lacks vascular cambium and therefore harden to propagate?
   (a) Seeds (b) Monocotyledons
   (c) Dicotyledons (d) Rhizome
2. Which one of the following crop can be propagated merely by inserting a stem in a water or moist soil?
   (a) Potato (b) Phyllanthus
   (c) Coleus (d) Sweet potato
3. Which cells can be induced to reproduce an entire plant?
   (a) Meristematic cells (b) Parenchymatus cells
   (c) Colenchymatus cells (d) Parietal cells
4. Cutting exploits the ability of plants to grow
   (a) Primary roots (b) Secondary roots
   (c) Adventitious roots (d) Root hairs
5. Bulbils are sometimes formed instead of the flowers in
   (a) Onion (b) Ginger
   (c) Garlic (d) Turmeric
6. Which one is the modified underground stem serving of an organ of vegetative reproduction?
   (a) Rhizome (b) Stem tuber
   (c) Corn (d) Root tuber
7. Which species that produce suckers for vegetative reproduction
   (a) Citrus (b) Mango
   (c) Sapota (d) Banana
8. The vegetative propagation method that is commonly practiced for apple is
   (a) Leaf cuttings (b) Stem cuttings
   (c) Grafting (d) Suckers
9. Grapes are propagated through
   (a) Stem cuttings (b) Root cuttings
   (c) Grafting (d) Layering
10. The best example for propagation through stem cutting is
    (a) Coffee (b) Peach
    (c) Tea (d) Bear
11. An example for grafting is
    (a) Apple (b) Peach
    (c) Plum (d) Poplar
12. The plant material chosen for tissue culture purpose is called as
    (a) Tissue (b) Explant
    (c) Plantlet (d) Stock plant

13. Who is the pioneer in micro propagation?
    (a) Skoog Miltan (b) Tosiba Murashige
    (c) Frederick Campion Steward (d) Cornell Redaunbaugh
14. In which year onwards the plant tissue culture is practied for micro propagation?
    (a) Late 1940s and early 1950s (b) Late 1950s and early 1960s
    (c) Late 1960s and early 1970s (d) Late 1970s and early 1980s
15. In micro propagation, the explants are surface sterilized by
    (a) Potassium permanganate (b) Sodium thiosulphate
    (c) Calcium chloride (d) Sodium hypochlorite
16. Which one of the growth regulator are used as a medium for callus induction
    (a) Auxin and Cytokinin (b) Kinetin and Zeatin
    (c) IBA and IAA (d) GA and Auxin
17. For root proliferation in tissue culture, which growth regulator is used
    (a) Cytokinin (b) Auxin
    (c) GA (d) IAA
18. Which one of the following growth regulator is applied for shoot growth in tissue culture
    (a) Cytokinin (b) Auxin
    (c) GA (d) IAA

♦ Answer Keys ♦

Q.No.	Ans.	Q.No.	Ans.	Q.No.	Ans.	Q.No.	Ans.	Q.No.	Ans.	Q.No.	Ans.
1	b	2	c	3	a	4	c	5	c	6	a
7	d	8	c	9	a	10	c	11	b	12	b
13	c	14	b	15	d	16	a	17	b	18	a

2.18.2. Fill in the blanks

1. Vegetative reproduction is a form of ....
2. A stem piece is joined onto the stem of a plant that has root is called as ....
3. .... is the practice of rapidly multiplying stock plant material to produce a large number of progeny plants using plant tissue culture
4. In horticulture, a .... is any plant material used for the purpose of plant propagation
5. Modern plant tissue culture is performed under .... conditions

♦ Answer Keys ♦

Q.No.	Ans.	Q.No.	Ans.	Q.No.	Ans.
1	Asexual reproduction	2	Root stock	3	Micro propagation
4	Propagule	5	Aseptic		

2.18.3. State True or False

1. Vegetative propagules that receives chemical treatment is capable of cellular differentiation because of paremchymatous cells
2. Vegetative propagules that receives chemical treatment is capable of cellular differentiation because of meristematic cells

♦ Answer Keys ♦

Q.No.	Ans.	Q.No.	Ans.
1	False	2	True

2.19. MATCH THE FOLLOWING

1. Question

1.	Tag colour for certified seed	A.	Deterioration of varieties
2.	Tag colour for foundation seed	B.	Seed village
3.	Breeder seed	C.	High temperature at maturation
4.	Mechanical admixture	D.	Bajra
5.	Five generation model	E.	Azur blue
6.	Compact area approach	F.	Poor seed set
7.	Hard seed	G.	White
8.	Tip sterility	H.	Viviparous
9.	*In situ* germination	I.	Progeny of nucleus seed
10.	High rainfall during flowering	J.	Groundnut

♦ Answer Keys ♦

Q.No.	Ans.	Q.No.	Ans.	Q.No.	Ans.	Q.No.	Ans.	Q.No.	Ans.	Q.No.	Ans.
1	E	2	G	3	I	4	A	5	J	6	B
7	C	8	D	9	H	10	F				

2. Question

1.	Bajra	A.	Tall plants
2.	Sorghum	B.	Objectionable weed
3.	Maize	C.	Maintainer line
4.	Cotton	D.	Partials
5.	Volunteer plants	E.	Removal of 50% fertile lines
6.	Wild okra	F.	Acid seed extraction
7.	R line	G.	Selfed bolls
8.	B line	H.	Pollen shedder
9.	Redgram	I.	Restorer line
10.	Ashgourd	J.	Shedding tassel

♦ Answer Keys ♦

Q.No.	Ans.	Q.No.	Ans.	Q.No:	Ans.	Q.No.	Ans.	Q.No.	Ans.	Q.No.	Ans.
1	D	2	H	3	J	4	G	5	A	6	B
7	I	8	C	9	E	10	F				

3. Question

1.	Onion	A.	Bolting
2.	Radish	B.	Tripping
3.	Cauliflower	C.	Non synchronize flowering
4.	Potato	D.	Hand pollination
5.	Cabbage	E.	Bulb to seed method
6.	Lucerne	F.	Emasculation and dusting
7.	Rope pulling	G.	Head to seed method
8.	Sorghum	H.	Root to seed method
9.	Sunflower	I.	Hybrid rice seed production
10.	Tomato	J.	TPS

♦ Answer Keys ♦

Q.No.	Ans.	Q.No.	Ans.	Q.No.	Ans.	Q.No.	Ans.	Q.No.	Ans.	Q.No.	Ans.
1	E	2	H	3	G	4	J	5	A	6	B
7	I	8	C	9	D	10	F				

4. Question

1.	Paddy	A.	Isogenic line
2.	Redgram	B.	CMS
3.	Maize	C.	CGMS
4.	B line	D.	GMS
5.	Cotton	E.	Panicle exertion
6.	Sorghum	F.	TGMS
7.	Castor	G.	Better than better parent
8.	Onion	H.	Detasseling
9.	GA3	I.	Thermo sensitive line
10.	Hetrobeltiosis	J.	Emasculation and dusting

♦ Answer Keys ♦

Q.No.	Ans.	Q.No.	Ans.	Q.No.	Ans.	Q.No.	Ans.	Q.No.	Ans.	Q.No.	Ans.
1	F	2	D	3	H	4	A	5	J	6	C
7	I	8	B	9	E	10	G				

5. Question

1.	Chilli	A.	500 m
2.	Tomato nursery	B.	Off-coloured seeds
3.	Groundnut	C.	Dimpled seeds
4.	Castor isolation	D.	1000 m
5.	Cucurbits isolation	E.	NAA spray
6.	Peas	F.	Negative propagation
7.	Greengram	G.	150 m
8.	Redgram	H.	Damping off
9.	Sweet potato	I.	Bleached seeds
10.	Cole crops isolation	J.	3 m

♦ Answer Keys ♦

Q.No.	Ans.	Q.No.	Ans.	Q.No.	Ans.	Q.No.	Ans.	Q.No.	Ans.	Q.No.	Ans.
1	E	2	H	3	J	4	G	5	A	6	I
7	B	8	C	9	F	10	D				

6. Question

1.	Maize isolation distance	A.	Staggered sowing
2.	Paddy	B.	Ethrel
3.	Sorghum	C.	Clipping
4.	Bajra	D.	Endosulphan pre-harvest spray
5.	Cowpea	E.	Increasing male flower in cucurbits
6.	Cotton	F.	Increasing border rows
7.	Pumpkin	G.	Barrier isolation
8.	Greengram	H.	Seed weight increase
9.	GA3 spray	I.	Jerking
10.	2% DAP spray	J.	Trimming

♦ Answer Keys ♦

Q.No.	Ans.	Q.No.	Ans.	Q.No.	Ans.	Q.No.	Ans.	Q.No.	Ans.	Q.No.	Ans.
1	F	2	G	3	A	4	I	5	C	6	J
7	B	8	D	9	E	10	H				

2.20. FIND OUT THE INCORRECT STATEMENT

1. (a) Hybrid seed production technology for redgram is CGMS system

(b) Hybrid seed production technology for cotton is emasculation and dusting

(c) Hybrid seed production technology for maize is detassaling

2. (a) High rainfall may cause poor seed set during seed production
(b) Tip sterility in bajra is due to high temperature during maturity
(c) Cool and moist climate condition is suitable for quality seed production

3. (a) Roguing is done to remove the off-type plants from the seed field
(b) Pollen shedder will cause genetic contamination
(c) Partials and pollen shedders are one and the same

4. (a) Boron deficiency causes lack of seed filling in groundnut
(b) Calcium deficiency causes production of more or single compartment pods in groundnut
(c) Calcium deficiency causes dark plumule in groundnut

5. (a) Cytoplasmic male sterility is completely stable
(b) CGMS is completely stable
(c) GMS is completely stable

♦ Answer Keys ♦

Q.No.	Ans.	Q.No.	Ans.	Q.No.	Ans.	Q.No.	Ans.	Q.No.	Ans.
1	a	2	c	3	c	4	b	5	c

2.21. ARRANGE IN ORDER

1. In seed production programme, chain of seed multiplication
Nucleus seed - Foundation seed - Breeder seed - Certified seed
Answer: Nucleus seed - Breeder seed - Foundation seed - Certified seed

2. In the seed production programme, selection should be in this order
Isolation - Land – Season – Sowing
Answer: Land – Season- Isolation – Sowing

3. The sequence to be followed during varietal development programme
ART - MLT – Release – Notification
Answer: MLT - ART - Release – Notification

4. Selection of genotype – Breeding – Advanced variety trial – Initial varietal trial – Variety – Multilocation yield trial – Variety release – Notification
Answer: Breeding – Selection of genotype – Initial varietal trial – Advanced variety trial – Multi location yield trial Variety – Varity release – Notification

5. A x R (F_1), B, R, A x B
Answer: B, AxB, R, A x R (F_1)

6. Callus formation – Plantlet – Sterilization – Explant – Shoot formation – Root formation
Answer: Explant – Sterilization – Callus formation – Root formation – Shoot formation – Plantlet

7. Explant – Somatic embryo – Sterilization – Synthetic seed – Encapsulation

Answer: Explant – Sterilization – Somatic embryo – Encapsulation – Synthetic seed

2.22. DIFFERENTIATE THE FOLLOWING

1. Differentiate between seed and grain

Seed	Grain
It should be a viable one	Need not be a viable one
It should have maximum physical and genetic quality	Not needed
Should satisfy seed standards such as maximum germination and pure seed and minimum seed moisture	No such requirements
Should be certified/truthfully labelled	No such condition
Should never be converted into grain unless warranted	May be converted as seed

2. Differentiate between certified seed and truthfully labelled seed

Certified seed	Truthfully labelled seed
Certification is voluntary	Truthfully labeling is compulsory
Applicable to notified variety only	Applicable to both notified and released varieties
It should satisfy both minimum field and seed standards	Tested for physical purity and germination
Seed Certification Officer and Seed Inspectors can take up field inspection and seed samples, respectively	Seed Inspectors alone can take samples for checking the seed quality

3. Differentiate between breeder seed and registered seed

Breeder seed	Registered seed
Obtained from the progeny of nucleus seed	Raised from nucleus, breeder or foundation seeds
Multiplied by concerned plant breeder or organization	Multiplied by seed growers
Monitered by breeder seed monitoring team	Inspected by seed certification agencies
Colour of the tag is golden yellow	Colour of the tag is purple colour
Genetic purity should be 100%	Genetic purity should be 98%

4. Differentiate between notified variety and released variety

Notified variety	Released variety
Follows release	Preceedes notification
Based on the seed demand and popularity of the variety	Based on the distinctness and superiority of the variety
Both kind and variety can be notified	Only variety is released
Normally only released varieties are notified	All released varieties are not notified
Done by the Central Sub Committee on crop standards and notification and release of varieties	Done by the Central Sub Committee or State Seed Sub Committee

5. Differentiate between hybrid and variety

Hybrid	Variety
A cross between two different varieties to get the valued attributes of each variety.	A cultivated individual which are distinguished by any characters.
Highly vigourous plant growth and development.	Standard plant growth and development
Higher yield than variety.	Standard yielding ability.
Seeds should be replaced every year	Replaced once in 3 to 4 years

6. Differentiate between genetic purity and physical purity

Genetic purity	Physical purity
True to type of the particular variety	Physically free from admixtures
Genetic contamination takes place at field level	Physical contamination takes place at post harvest management period
Genetic purity can be maintained by isolation distance and rogueing	Physical purity of the seed lot can be improved by avoiding mechanical mixtures
Genetic purity is assessed by grow out test	It can be assessed by purity analysis

7. Differentiate between physical contaminants and genetic contaminants

Physical contaminants	Genetic contaminants
It is the physical admixture of the seeds namely other crop seeds, weed seeds and inert mater	It refers the contaminants present in the seed production plot like off-types, rogues, pollen shedders, shedding tassel, partials selfbed bolls/fruits etc.
It affects the physical quality of seeds	It affects the genetic purity of the seeds

8. Differentiate between pollen shedder and shedding tassel

Pollen shedder	Shedding tassel
Presence of B line or R line plants in A line is called pollen shedder	Presence of remanants of tassel after detasseling in maize is called shedding tassel
Causes genetic contamination	Causes genetic contamination
It occurs in rice, sorghum, bajra, sunflower etc.	It occurs in maize

9. Differentiate between A line and B line with an example

A line	B line
Male sterile line and does not produce pollen	Isogeneic line of `A' line, but produces pollen grains
It is the female parent in hybrid seed production programme	It is a maintainer line of A line
It bears hybrid seed when cross pollinated with `R' line and expresses hybrid vigour	Mixing with `A' line, it is called as pollen shedders and causing genetic contamination
Eg. IR – 58025 A, IR – 62829 A	Eg. IR – 58025B, IR-62829B

10. Differentiate between genetic male sterility (GMS) and cytoplasmic male sterility (CMS) with examples

Genetic male sterility (GMS)	Cytoplasmic male sterility (CMS)
Governed by a action of genes carried in the nucleus with a particular cytoplasm	Governed by cytoplasm rather than by nuclear genes and transmitted only through the female parent
Exploited in cotton and redgram	Exploited in sorghum, paddy and bajra
50% fertile and 50% sterile	Completely sterile
50% fertile lines need to be removed	No need to remove

11. Differentiate between containerized seedling and bare rooted seedling

Containerized seedling	Bare rooted seedling
Growing seedling in the protrays	Growing seedling in the raised bed nursery
Less prone to damping off	More susceptible to damping off
Higher survival	Typically have lower survival (average 65% survival)
Higher cost per seedling	Low cost per seedling
Easier to handle seedling	More difficult to handle the seedling
Can be transported to long distance	Not possible to transport to long distance

Containerized seedling	Bare rooted seedling
Store better for longer periods	Shorter storage time
Have a wider planting window	More restrictive planting window
Lesser tolerant to deep planting	Slightly more tolerant to deeper planting
Growing media or potting mixture can be peat mass, vermiculite etc.	FYM can be applied

12. Differentiate between seed to seed method and stump to seed method with examples

Seed to seed method	Stump to seed method
Plants are allowed to produce seeds in its place of transplanting	When the crop in the first season is fully mature, the heads are examined for trueness type
May be one season seed production	May be two season seed production
Mature plants are uprooted in the transplanting method	Heads are cut just below the base by means of a sharp knife, keeping the stem with outer whorls of leaves intact
Eg : Rice	Eg : Radish

13. Differentiate between single cross hybrid and double cross hybrid with example

Single cross hybrid	Double cross hybrid
When two inbred lines or pure lines are crossed to produce the F_1 hybrid is known as single cross hybrid	When two single crosses are crossed, the resulting hybrid population is known as double cross hybrid
Eg : COH (M) 5 maize	Eg : Ganga maize, Ganga-safed-2, Deccan

14. Differentiate between triple cross hybrid and three way cross hybrid

Triple cross hybrid	Three way cross hybrid
Results from the crossing of two different three way cross hybrids	It is a cross between a single cross and an inbred to give hybrid population. Result from the cross between one parent that is an F1 hybrid and the other is from an inbred line

15. Differentiate between asexual reproduction and sexual reproduction with examples

Asexual reproduction	Sexual reproduction
Reproduction without fertilization	Reproduction by fusion of male and female gametes

Asexual reproduction	Sexual reproduction
Mitosis cell division	Meiosis cell division
Genetically identical (true to type)	Genetically not identical
Propagation by means of vegetative propagules	Propagation by means of seeds
Very short storage life	Longer storage life depending upon species
Eg. Turmeric (Rhizome), potato (Tuber)	Eg. Rice, sorghum, bajra, redgram etc. through seeds

16. Differentiate between planting ratio and border row with examples

Planting ratio	Border row
Recommended ratio in which male and female parental lines are planted in hybrid seed production	Recommended number of rows of male parental line grown on all the sides of the field
It varies from crop to crop	Act as a pollen source and prevents contamination
Eg. 8:2 in rice	Eg. 4 rows in rice

17. Differentiate between grow out test and germination test

Grow out test	Germination test
Done to determine the genetic purity status of the given seed lot	Done to predict planting value of seeds, as well as to determine whether a particular seed lot has the potential to produce a good crop or should be used for seed purpose
Observations includes the percentage of off-types	Observations include germination percentage (normal seedling including dead seeds, abnormal, diseased and damaged seedlings etc.
Conducted in the field conditions	Conducted in the controlled laboratory condition
The sample required is 100 g to 1000 g depending upon crops with a minimum of 400 plants	The number of seeds required is 400 in 4 replications

18. Differentiate between nucleus seed and breeder seed

Nucleus seed	Breeder seed
Produced by the plant breeder who breeds the variety	Produced from nucleus seed by State Agricultural Universities
Genetic purity should be 100%	It is also 100% genetically pure
Not monitored by the committee	Monitored by breeder seed monitoring committee

Nucleus seed	Breeder seed
No prescribed tag colour	The colour of the tag is golden yellow
Does not come under the perview of seed certification	Does not come under the perview of seed certification

19. Differentiate between seed and vegetative propagules

Seed	Vegetative propagules
Sexual reproduction through seed	Asexual reproduction propagated by vegetative propagules, layering, cutting, grafting and budding, tissue culture etc.
Not true to type	True to type

20. Differentiate between time isolation and barrier isolation

Time isolation	Barrier isolation
Planting the parental lines of the hybrids should be adjusted in such a way that the difference between flowering in seed production plot and the neighbouring crop does not coincide	Achieved through the physical barrier by growing taller crops like sorghum, maize, pearl millet, sugarcane
Eg. Paddy – 25 days	Eg. Paddy, the barrier crops are planted covering a distance of 30 m between hybrid seed producing plot/parental seed producing plot and other rice varieties

21. Differentiate between space isolation and time isolation with examples

Space isolation	Time isolation
Separating the seed plot from the contaminants by providing space	Separating the seed plots from contaminants by adjusting the time either prior or after the rice varieties grown nearby to start flowering
Eg. Hybrid rice seed production – 100 m	Eg. In rice hybrid seed production, heading stage should be 25 days earlier or later than other varieties grown with in 100 m around the seed crop

22. Differentiate between two line hybrid and three line hybrid

Two line hybrid	Three line hybrid
Hybrid seed production involves the use of photo period sensitive genetic male sterile (PGMS) lines	Hybrid seed production involves multiplication of cytoplasmic genetic male sterile line (A line) using

Two line hybrid	Three line hybrid
(or) Temperature sensitive genetic male sterile lines (TGMS)	maintainer line (B line)
Any normal line can serve as restorer	A restorer line (R line) is crossed with A line for production of F_1 hybrid seed (AxR)
Eg : Rice	Eg : Rice, sorghum, bajra, sunflower etc.

23. Differentiate between tripping and nicking

Tripping	Nicking
Pollen dispersal mechanism in which the staminal column spring of the keel and exposed	The receptivity of the male sterile plant to the maximum pollen load of the pollinator for cross pollination in hybrid seed production
Aids in cross pollination and good seed set	Synchronization of male and female parents in hybrid seed production for good seed set
Eg : Alfalfa	Eg : Rice, sorghum, bajra

24. Differentiate between Seed Replacement Rate (SRR) and Seed Multiplication Ratio (SMR) with examples

Seed Replacement Rate (SRR)	Seed Multiplication Ratio (SMR)
It is the percentage of area sown out of total area of crop planted in the season by using certified/quality seeds other than the farm saved seed	It is nothing but the number of seeds to be produced from a single seed when it is sown and harvested
For variety, SRR is once in 3-4years and for hybrid every season or year	For groundnut, SMR is 1:8

25. Differentiate between rogue and off-types

Rogue	Off-types
An undesirable off-type plant	A plant in a seed crop, which deviates from the norms of the cultivar, but does not belong to another cultivar or species
Rogue should be removed as soon as possible in order to protect the genetic contamination of the variety	Regularly remove the off-types from the field either by uprooting or by cutting at the ground level

26. Differentiate between pre-sowing seed treatment and pre-storage seed treatment

Pre-sowing seed treatment	Pre-storage seed treatment
Treatments given to the seeds before sowing	Treatments given to the seeds before storage
To enhance the quality of the seeds in terms of germination and vigour	To protect the seeds from storage pest and diseases which in turn increase the storability of seeds
Eg. Seed hardening, seed fortification, seed priming, seed pelleting, seed film coating, biofer-tilizer treatment, biocontrol agents etc	Eg. Seed treatment with fungicide (bavistin), insecticide (Imidacloprid), botanicals (neem leaf powder) etc.,
Different chemicals, nutrients, bio-fertilizers, biocontrol agents, insects, aids and fungicides can be used	Chemicals, insecticides and botanicals can be used

27. Differentiate between PGMS and TGMS

PGMS	TGMS
The line is sterile when the photo period (day light) exceeds 14 h and the same line becomes fertile when subjected to photo period of < 13 h	It is sterile when temperature exceeds 22°C/24°C (day/night) and becomes fertile when the tempera-ture is below 24°C (18°C day/night)
This system can be followed for rice hybrid seed production in temperate countries	This system is followed for rice hybrid seed production in tropical and sub tropical countries

28. Differentiate between self compatibility and self incompatibility

Self compatibility	Self incompatibility
Ability to set seed by the pollen produced on the same plant	It is the failure of the pollen to fertilize the same flower or other flower of the same plant
It aids self pollination	It aids cross pollination
Eg. Paddy, groundnut	Eg. Sunflower, brinjal, brassica

29. Differentiate between field inspection and seed inspection

Field inspection	Seed inspection
It is done by the seed certification officer during the seed production before harvest	It is done by seed inspectors while the seeds are in marketing
Inspection is done for minimum field standards like off-types, rogues, pollen shedders, shedding tassel, objectionable weed species, designated diseases etc.	Inspected for seed quality characters like minimum seed standards, expiry date of seed lots etc.

30. Differentiate between clipping and trimming

Clipping	Trimming
Removal of tendrils	Pruning of sympodial branches and removal of basal monopodia
It promotes early flowering to increase the seed production	Responds well to irrigation and nutrient inputs.
Eg. Cowpea	Eg. Cotton

Unit 3

SEED PROCESSING

3.1. PRINCIPLES OF SEED PROCESSING

3.1.1. Choose the correct answer

1. Dry dehiscent fruits can be extracted by
   (a) Drying (b) Maceration
   (c) Polishing (d) Threshing
2. Decortication is used for
   (a) Groundnut (b) Maize
   (c) Bhendi (d) Sunflower
3. Wet extraction is commonly used on
   (a) Bittergourd (b) Rice
   (c) Onion (d) Chillies
4. Pre cleaning mainly used to improve
   (a) Vigour (b) Viability
   (c) Physical purity (d) Genetic purity
5. Pre-cleaning of seeds is called
   (a) Stratification (b) Scarification
   (c) Scalping (d) Upgrading
6. Pre-cleaning of seeds is essential to
   (a) Increase efficiency of seed processing (b) To get more profit
   (c) To reduce labour (d) Store the seeds
7. Air is commonly used in seed processing machinery to remove
   (a) Dust (b) Dirt
   (c) Other variety seeds (d) Stones
8. Extraction of seeds or stones by removal of the fleshy part (pulp) of fruits like berries and drupes is called as
   (a) Depulping (b) Decantitation
   (c) Pulping (d) Pulp decantation
9. Method for cleaning seeds from particles with higher or lower specific density by submerging in water or other liquid is called as
   (a) Separation (b) Flotation
   (c) Desiccator (d) Humidification

10. In gas delinting of cotton, the gas used is
    (a) Dry sulphuric acid gas (b) Dry hydrochloride acid gas
    (c) Nitrogen peroxide gas (d) Nitrate nitrite gas
11. Acid used for seed extraction of tomato is
    (a) HNO_3 (b) H_2SO_4
    (c) HCl (d) KOH
12. Acid delinting of cotton is done with
    (a) HNO_3 (b) H_2SO_4
    (c) HCl (d) KOH
13. The process of removing large size particles in seed processing is
    (a) Pre-cleaning (b) Basic cleaning
    (c) Scapling (d) Upgrading
14. Removal of seed from pod in groundnut is called
    (a) Stripping (b) Decortication
    (c) Extraction (d) Threshing
15. Scapling is
    (a) Debearding (b) Grading
    (c) Pre-cleaning (d) Treating
16. Seed processing is done for improving
    (a) Genetic quality (b) Physical quality
    (c) Health quality (d) Physiological quality

♦ Answer Keys ♦

Q.No.	Ans.	Q.No.	Ans.	Q.No.	Ans.	Q.No.	Ans.	Q.No.	Ans.	Q.No.	Ans.
1	a	2	a	3	a	4	c	5	c	6	a
7	a	8	a	9	b	10	b	11	c	12	b
13	c	14	b	15	c	16	b				

3.1.2. Fill in the blanks

1. Cleaning is the .... step in processing.
2. In processing, the role of .... is rough cleaning
F. Excessive glumes in forage seeds can be removed by ....
L. Barley awns are removed by using ....
V. .... is the process of detaching grains from ear heads.
/. .... is the process of separating good seeds from a mixture of grains.
2. In barley, ... is used for the removal of appendages.
% In paddy, ... is the weight grade technique.
- .... is the organic solvent used for specific gravity grading in marigold and casuarina.
10. In peas, grading can be done based on .... property.
11. Water floatation technique is highly suitable for .... crop.
12. Expand IDS ....

♦ Answer Keys ♦

Q.No.	Ans.	Q.No.	Ans.	Q.No.	Ans.
1	Second step	2	Scalper	3	Debearder
4	Debearder	5	Threshing	6	Winnowing
7	Debearder	8	Salt floatation using egg	9	Acetone
10	Surface tension	11	Paddy	12	Incubation drying – separation

3.1.3. State True or False

1. Scapling is also called rough cleaning
2. Pre-cleaning of seeds helps in increasing the processing capacity
F. Debearding is a not a pre-treatment for seed processing
L. Depulping of fleshy fruits can be performed after some period of collection
V. Commercial H_2SO_4 is used for gas delinting of cotton
/. Delinting is a process by which the seeds are sown in the field
2. Heterogeneity of a seed lot can be increased by seed processing

♦ Answer Keys ♦

Q.No.	Ans.	Q.No.	Ans.	Q.No.	Ans.	Q.No.	Ans.	Q.No.	Ans.	Q.No.	Ans.
1	True	2	True	3	False	4	False	5	False	6	False
7	False										

3.2. SEED DRYING

3.2.1. Choose the correct answer

1. The temperature limit for artificial air drying
   (a) 100°C (b) 110°F
   (c) 110°C (d) /V°F
2. Hard seed coat can be removed by using
   (a) Stratification (b) Debearders
   (c) Aspirator (d) Scarification
3. Heated air drying limit is
   (a) 110°F (b) 110°C
   (c) -%C (d) -%F
4. If silica gel absorbs moisture which colour the gel will develop?
   (a) Blue (b) Pink
   (c) Red (d) White
5. High moistened seed should be dried
   (a) At high temperature (b) At low temperature
   (c) Under direct sun (d) Low temperature shade dried

6. Process of bringing grains or other products to a desired moisture or temperature for processing is called
   (a) Tempering (b) Grading
   (c) Processing (d) Stacking
7. Moisture content of seeds in equilibrium with atmospheric humidity at a given temperature is termed as
   (a) Inequilibrium moisture content (b) Equilibrium moisture content
   (c) Standard moisture content (d) Graded moisture content
8. Chemical compound that has a high moisture absorption affinity and can be used for desiccation or maintaining a low humidity when stored together with seeds is called as
   (a) Extractant (b) Humidiant
   (c) Absorbent (d) Desiccant

♦ Answer Keys ♦

Q.No.	Ans.	Q.No.	Ans.	Q.No.	Ans.	Q.No.	Ans.
1	b	2	d	3	a	4	b
5	b	6	a	7	b	8	d

3.2.2. Fill in the blanks

1. Caking of seed in storage can be avoided by ....
2. Optimum temperature for drying vegetable seeds is ....
F. Silica gel can absorb moisture .... more than its weight
L. The moisture content of the seed should be optimum to reduce .... during seed processing
V. Depending upon the moisture content of the seed, the .... should be adjusted while drying
/. If the initial moisture is high, use low temperature and as the seed loses moisture, drying temperature should be gradually increased and it should not exceed ....
2. Continuous drying should be avoided and a tempering period of atleast .... should be given in between two successive drying.
% Time interval provided in between two successive drying is called as ....
-. .... and .... are different methods of drying
10. .... driers are portable
11. Noon drying of seeds should be avoided to protect from ....
12. .... is the ideal temperature for mechanical drying of seeds
1F. Sun drying should not be done between .... and ....
1L. .... provides high air temperature and low relative humidity than simple sun drying
1V. .... is used for reducing moisture content mechanically

♦ Answer Keys ♦

Q.No.	Ans.	Q.No.	Ans.	Q.No.	Ans.
1	Proper drying	2	20-%0°F	3	L0%
4	Mechanical damage	5	Temperature	6	110°F
7	%hours	8	Tempering	9	Sun drying, artificial drying
10	Batch	11	UV rays	12	L0°C
13	12 noon, 2 pm	14	Solar drying	15	Drier

3.2.3. State True or False

1. Seeds dried to less than 12% moisture content resist mechanical damage
2. The maximum drying temperature for horticultural crops is - 0°F

F. Silica gel can absorb 20%/L0% moisture to its weight

L. The principle in scarification of seed and thermotherapy remains the same

V. Before using, the colour of silica gel is white

/. Wagon drying is special type batch drying with heat air find extensive use

2. Hot air is used in drying of seeds

% Drying is essential for better storage of grains and seeds

♦ Answer Keys ♦

Q.No.	Ans.	Q.No.	Ans.	Q.No.	Ans.	Q.No.	Ans.
1	False	2	False	3	False	4	True
5	True	6	True	7	True	8	True

3.3. SEED CLEANING AND GRADING

3.3.1. Choose the correct answer

1. Oblong sieve is used for grading
   (a) Paddy (b) Sorghum
   (c) Maize (d) Sunflower
2. Flat and round seeds can be separated by
   (a) Specific gravity separator (b) Spiral separator
   (c) Magnetic separator (d) Indent cylinder separator
3. What is the top screen size for wheat seed processing
   (a) 2//L" (b) %//L"
   (c) 12//L" (d) 1L//L"
4. What is the top screen size for paddy seed processing
   (a) 2//L" (b) %//L"
   (c) 12//L" (d) 1L//L"
5. What is the top screen size for moongbean seed processing?
   (a) 2//L" (b) %//L"
   (c) 12//L" (d) 1L//L"

6. What is the top screen size for castor seed processing?
   (a) 20//L″ (b) 2L//L″
   (c) F0//L″ (d) FL//L″
7. What is the top screen size for snakegourd seed processing?
   (a) 20//L″ (b) 2L//L″
   (c) F0//L″ (d) FL//L″
8. What is the bottom screen size for maize seed processing?
   (a) %/L″ (b) 10//L″
   (c) 12//L″ (d) 1///L″
9. What is the bottom screen size for redgram seed processing?
   (a) %/L″ (b) 10//L″
   (c) 12//L″ (d) 1///L″
10. What is the bottom screen size for soybean seed processing?
   (a) %/L″ (b) 10//L″
   (c) 12//L″ (d) 1///L″
11. What is the bottom screen size for cotton seed processing?
   (a) %/L″ (b) 10//L″
   (c) 12//L″ (d) 1///L″
12. What is the bottom screen size for rajmash seed processing?
   (a) %/L″ (b) 10//L″
   (c) 12//L″ (d) 1///L″

♦ Answer Keys ♦

Q.No.	Ans.	Q.No.	Ans.	Q.No.	Ans.	Q.No.	Ans.	Q.No.	Ans.	Q.No.	Ans.
1	a	2	b	3	a	4	d	5	d	6	b
7	d	8	a	9	b	10	b	11	c	12	c

3.3.2. Fill in the blanks

1. The two different systems used for size grading the seeds are .... and .... system.
2. Any other grading followed by size grading is known as ....
F. In greengram, size grading is followed by ....
L. In wheat, size grading is followed by ....
V. Size grading followed by .... is done in paddy and bhendi.

♦ Answer Keys ♦

Q.No.	Ans.	Q.No.	Ans.	Q.No.	Ans.
1	British American	2	Upgrading	3	Colour
4	Shape grading	5	Weight grading		

3.4. SEED TREATMENT AND PACKAGING

3.4.1 Choose the correct answer

1. Commonly used fungicide for seed treatment is
   (a) Furadan (b) Metaxystox
   (c) Carbendazim (d) Atrazine
2. Which one of the following chemical is used against bacteria?
   (a) Carbendazim (b) Mancozeb
   (c) Carboxyl (d) Streptomycin
3. Which one of the following organics is commonly used for seed treatment?
   (a) Pungam leaf powder (b) Neem leaf powder
   (c) Maize leaf powder (d) Cumbu leaf powder
4. What is the dosage of bavistin for seed treatment?
   (a) 1 g (b) 2 g
   (c) F g (d) L g
5. What is the dosage of neem leaf powder for seed treatment?
   (a) V g (b) 10 g
   (c) 1V g (d) 20 g
6. The systemic fungicides are used to control
   (a) Soil borne organism (b) Soil borne nematodes
   (c) Seed borne pests (d) Seed borne pathogen
7. The most successful application method of seed treatment to protect environmental hazarad should be
   (a) Free from dust-off (b) Easy sticking capacity
   (c) Poor loading efficiency (d) Poor seed damage
8. What is the dosage of imidacloprid used for seed treatment?
   (a) 1 ml/kg of seed (b) 2 ml/kg of seed
   (c) F ml/kg of seed (d) L ml/kg of seed

♦ Answer Keys ♦

Q.No.	Ans.	Q.No.	Ans.	Q.No.	Ans.	Q.No.	Ans.
1	c	2	d	3	b	4	d
5	b	6	d	7	a	8	a

3.4.2. Fill in the blanks

1. The fungicide chemical used for seed treatment is ....
2. .... refers to the application of insecticides and fungicides to disinfect the pathogen and insects

F. The trade name of carboxin is ....

L. The control of disease organism from the seed surface is known as ....

V. The elimination of a pathogen that has penetrated into the living cells of seed is termed as ....

/. Expand PCNB....................

2. NSKE expands ....

% One of the desirable quality of seed treating chemical is ....

-. .... treatment is more efficient than dry treatment

♦ Answer Keys ♦

Q.No.	Ans.	Q.No.	Ans.	Q.No.	Ans.
1	Bavistin	2	Seed treatment	3	Vitavax
4	Disinfestations	5	Disinfection	6	Pentachloronitrobenzene
7	Neem Seed Kernel Extract	8	Non phytotoxic	9	Slurry

3.4.3. State True or False

1. Fumigants are used for seed treatment

♦ Answer Keys ♦

Q.No.	Ans.
1	False

3.5. SEED QUALITY ENHANCEMENT

3.5.1. Choose the correct answer

1. What is the temperature range at which the seeds are given with seed hardening treatment?
   (a) 10-1V°C (b) 1V-20°C
   (c) 1V-2V°C (d) 1V-F0°C
2. Which one of the following crop can be given with seed hardening treatment using 2% dihydrogen phosphate?
   (a) Cotton (b) Safflower
   (c) Blackgram (d) Sorghum
3. Which one of the following crop can be given with seed hardening treatment using 2% potassium chloride?
   (a) Cotton (b) Safflower
   (c) Blackgram (d) Sorghum
4. Which one of the following crop can be given with seed hardening treatment using 100 ppm zinc sulphate?
   (a) Cotton (b) Safflower
   (c) Blackgram (d) Sorghum
5. Halopriming involves
   (a) Water soaking (b) Osmotic solution soaking
   (c) Salt solution soaking (d) Mixing with inorganic carrier

6. What do you mean by solid matrix priming?
   (a) Water soaking (b) Osmotic solution soaking
   (c) Salt solution soaking (d) Mixing with inorganic carrier
7. Which one of the following organism is used for biopriming of seed?
   (a) Liquid biofertilizer (b) *Trichoderma viride*
   (c) *Xanthomonas campestris* (d) *Bacillus thuriengiensis*
8. Who has coined the different terms like osmopriming, halopriming etc. ?
   (a) Heydecker (b) Henkel
   (c) Hiltner (d) Hegarty
9. What is the temperature at which the seed priming is done?
   (a) V-10°C (b) 10-1V°C
   (c) 1V-20°C (d) 20-2V°C
10. Which one of the following biofertilizer is used for biopriming?
    (a) *Rhizobuim* (b) *Pseudomonas*
    (c) *Azospirillum* (d) *Azolla*
11. Up to which moisture level, the seeds are soaked while fortification of seeds?
    (a) 1V-20% (b) 20-2V%
    (c) 2V-F0% (d) F0 FV%
12. What is the volume of solution used for soaking for seed fortification treatment?
    (a) Less than equal volume (b) More than equal volume
    (c) Equal volume (d) 1/F volume
13. The use of biocontrol agent in the seed pelleting is called as
    (a) Inoculant pelleting (b) Herbicide pelleting
    (c) Protective pelleting (d) Nutrient pelleting
14. The use of *Rhizobium* in seed pelleting is known as
    (a) Inoculant pelleting (b) Herbicide pelleting
    (c) Protective pelleting (d) Organic pelleting
15. The use of DAP in seed pelleting is termed as
    (a) Inoculant pelleting (b) Herbicide pelleting
    (c) Protective pelleting (d) Nutrient pelleting
16. What is the optimum dose of polykote for seed coating?
    (a) 2 g/kg of seed (b) F g/kg of seed
    (c) L g/kg of seed (d) Vg/kg of seed
17. Controlled hydration is
    (a) Fortification (b) Hardening
    (c) Priming (d) Infusion
18. Magnesium salts are used for fortification to increase
    (a) Amino acid content (b) Respiration
    (c) Mitochondrial activity (d) Chlorophyll

19. Seed fortification denotes
(a) Pelleting (b) Nutrient soaking
(c) Incubation (d) Drying

20. Chemical which is more suitable for seed priming is
(a) KCl (b) PEG /000
(c) HCl (d) H_2SO_4

21. The chemical used for seed hardening is
(a) KCl (b) Water
(c) IBA (d) CCC

22. Chemical which is more suitable for acid delinting is
(a) KCl (b) PEG /000
(c) HCl (d) H_2SO_4

23. The treatment of seeds with an osmotic solution i.e., polyethylene glycol to initiate germination and then dried to get more uniform and rapid germination of certain vegetable seeds is called as
(a) Priming (b) Stratification
(c) Hardening (d) Fortification

24. A seed treatment to enhance the market value
(a) Seed priming (b) Polykote treatment
(c) Seed dipping (d) Seed fortification

25. PEG is commonly used in
(a) Halo priming (b) Osmopriming
(c) Hydro priming (d) Solid matrix priming

26. The chemical used for creating osmotic stress is
(a) Sodium hypochloride (b) Chlorohydrins
(c) Mannitol (d) Silica gel

27. .... is the growth regulator used in ragi for seed hardening
(a) Triacontanol (b) Brassinolide
(c) GA_F (d) IAA

28. Pelleting technique is used for enhancing
(a) Seed germination (b) Seed storability
(c) Seed viability (d) Seed longevity

29. Process of enclosing the seed with a small quantity of inert material to produce a globular unit is
(a) Seed coating (b) Seed pelleting
(c) Seed fortification (d) Seed hardening

30. Procedure by which individual seeds are provided with an envelope of adhesive material containing e.g. nutrients, microsymbiont inoculants and/or pesticides is known as
(a) Grading (b) Stacking
(c) Crushing (d) Pelleting

31. Major seed enhancement technique used to increase the market value of seed is
    (a) Fluid drilling (b) Seed colouring
    (c) Seed priming (d) Seed hardening

♦ Answer Keys ♦

Q.No.	Ans.	Q.No.	Ans.	Q.No.	Ans.	Q.No.	Ans.	Q.No.	Ans.	Q.No.	Ans.
1	c	2	d	3	a	4	c	5	c	6	d
7	b	8	a	9	b	10	c	11	b	12	c
13	c	14	a	15	d	16	b	17	c	18	b
19	b	20	b	21	a	22	d	23	a	24	b
25	b	26	c	27	b	28	a	29	b	30	d
31	b										

3.5.2. Fill in the blanks

1. The hardened seeds are dried back to original ....
2. In the seed hardening process, there is an .... in germination during hydration.

F. During dehydration of seed hardening, the germination advancement activity ....

L. The hardened seed when sown in the field, it .... the germination activity where its stopped with the available moisture.

V. For greengram seed hardening, ... ppm zinc sulphate is used.

/. The use of biocontrol agents for priming of seeds is called as ....

2. Seed pelleting is done to seed to facilitate .... planting.

% Pelleting is defined as the deposition of inert materials that may result in a substantial .... increase.

-. During seed pelleting process, an .... is used as a sticking material.

10. The use of magnesium carbonate in seed pelleting is named as .... coating.

11. In aerial seedling of pelleted seed, the .... property of seed helps in precision sowing.

12. The seed coating does not increase the size, weight and shape of seed is called as ....

1F. The polykote film coating prevents the .... of the pesticides that are mixed with polykote for control.

1L. .... improved seed quality without altering seed moisture content

1V. .... is an impregnation of required substances for invigourating the seed for improved production.

1/. .... technique helps to avoid the damage caused to the seed due to soaking in water.

12. .... treatment imparts resistance against stress conditions in dry land agriculture

1% Soaking of seeds in salt solution is ....

1-. The chemical recommended for seed hardening for blackgram is ....

20. The chemical used for seed hardenting of greengram is ....

21. .... grade H_2SO_4 is used for acid delinting of cotton

22. For dryland sowing of sorghum, ... is recommended as a pre-sowing management to drought.

2F. Hardening treatment using seed hydration, dehydration techniques are physiological methods for improving for plant behaviour under .... conditions.

2L. The growth regulators like NAA, IAA, IBA, ..., ... can be used for seed hardening.

2V. Controlled hydration is ....

2/. Soaking of seeds in water is ....

22. .... are the salts used to prepare osmotic solution.

2% Soaking the seeds with biological agents like bacteria is called ....

2-. The pre-sowing treatment suitable for vegetable during cold seasons is ....

F0. .... is a pre-sowing seed treatment which aids in precision sowing

F1. In smaller seeds, ... is adopted as a pre-sowing management to increase the size of seed for easy sowing

F2. .... is singling of seeds

FF. The seed enhancement technique most suitable for precision sowing is ....

FL. .... is the process of enclosing a seed with small quantity of inert material just large enough to produce a unit standard size.

FV. During pelleting, ... is added to the pelleting mixture as a last pelleting material.

F/. The most commonly used adhesive for seed pelleting is ....

F2. Enclosing pre-germinated seed in viscous gel is ....

F% A thin layer of coating which does not alter shape and size of seed is called as ....

F-. .... enables brand identification and gives the seed a distinct and attractive colour.

L0. In oxygen coating, seeds are coated with peroxides of ....

L1. In a .... thin film of biodegradable polymer are used in seed treatment

L2. By film coating treatment, ... can be avoided while adding dry chemicals

LF. .... is a commercial product utilized for film coating

LL. .... is prepared by combining various seed treatments in a sequential manner.

♦ Answer Keys ♦

Q.No.	Ans.	Q.No.	Ans.	Q.No.	Ans.
1	Moisture/weight	2	Advancement	3	Stopped
4	Remembers	5	100	6	Biopriming
7	Precision	8	Weight	9	Adhesive
10	Hydrophilic	11	Ballistic	12	Polykote film coating
13	Dusting off	14	Seed infusion	15	Seed fortification
16	Seed infusion	17	Seed hardening	18	Halopriming
19	$MnSO_4$	20	$ZnSO_4$	21	Commercial
22	Hardening (2% KH_2PO_4 for 1/ h)	23	Dryland	24	CCC; GA_3
25	Seed priming	26	Hydropriming	27	Osmoticants
28	Biopriming	29	Priming	30	Seed pelleting
31	Pelleting	32	Seed pelleting	33	Pelleting
34	Seed pelleting	35	Fungicide	36	Carboxylmethyl Cellulose
37	Fluid drilling	38	Seed coating	39	Seed colouring
40	Zinc and calcium	41	Seed film coating	42	Dusting off
43	Polymer	44	Designer seed		

3.5.3. State True or False

1. Cow's urine can be used for seed hardening.

2. On sowing of hardened seeds, the seed does not remember its germination activity where it stopped

F. Soaking of seed in *Xanthomonas* is called as biopriming.

L. Biopriming using biocontrol agents like *Pseudomonas* is called as biopriming.

V. The fortified seeds are dehydrated to original moisture content/weight.

/. The pelleting of seed does not increase the original shape and size of seed.

2. The seed pelleting may increase the shape and size of seed.

% The polykote seed coating increases the size, weight and shape of seed.

-. The prevention of dusting off of the pesticides helps in the environmental protection

10. The objective of pre-sowing seed treatment is to improve germination only

11. Seed enhancement techniques are improving the germination and vigour of seed
12. Alternate wetting and drying of soybean seeds causes natural crushing
1F. The utility of organic solvents in infusion treatment is much felt because of deeper penetration of the solvent rather than uniform distribution
1L. The best pre sowing seed treatment for rice is 20 hrs soaking in 1% KCl followed by an equal period of incubation of seeds
1V. Seed soaking treatment does more harm than good in most of the pulse seeds
1/. Seed hardening is a mid-storage treatment
12. The best dry farming technique for rainfed sorghum is seed hardening with 2% KNO_3 for 2Lh
1% Seed priming techniques involve controlled hydration of seeds
1-. Osmotic priming is done with polyethylene glycol (PEG)
20. Temperature suggested during priming is 10°-1V°C
21. Seed pelleting improves seed size
22. Seed pelleting helps in precision sowing of seeds

♦ Answer Keys ♦

Q.No.	Ans.	Q.No.	Ans.	Q.No.	Ans.	Q.No.	Ans.	Q.No.	Ans.	Q.No.	Ans.
1	True	2	False	3	False	4	True	5	False	6	False
7	True	8	False	9	True	10	False	11	True	12	True
13	True	14	True	15	True	16	False	17	False	18	True
19	True	20	True	21	True	22	True				

3.6. SEED PROCESSING MACHINES

3.6.1. Choose the correct answer

1. For which crop, the specific gravity separator separated seeds are in positive association with vigour
   (a) Cotton (b) Sunflower
   (c) Casuarina (d) Cashewnut
2. Dodder mill separates seed based on
   (a) Surface texture (b) Colour
   (c) Size (d) Shape
3. Air screen cleaner used for
   (a) Scalping and grading (b) Grading
   (c) Separating (d) Cleaning
4. Round and flat seeds in a seed lot can be separated by
   (a) Draper belt (b) Indent cylinder separator
   (c) Specific gravity separator (d) Debearder

5. Clusters of seeds can be separated by
   (a) Magnetic separator (b) Debearder
   (c) Huller (d) Decorticator
6. In processing, the role of scalper is
   (a) Scarification (b) Debearding
   (c) Dewinging (d) Rough cleaning
7. Seeds of same size can be separated by
   (a) Disc separator (b) Specific gravity separator
   (c) Spiral separator (d) Air screen cleaner
8. Scarifier is used to give a .... on the seed surface
   (a) Cut (b) Damage
   (c) Scratch (d) Polish
9. When the mass difference exists between the seeds and impurities, the following separation is used
   (a) Inclined belt separator (b) Spiral separator
   (c) Specific gravity separator (d) Magnetic separator
10. Magnetic separator uses
   (a) Calcium chloride powder (b) Iron powder
   (c) Magnesium chloride powder (d) Chalk powder
11. When the difference in surface texture exists between the seeds and impurities, the following separation is used
   (a) Inclined belt separator (b) Spiral separator
   (c) Roller separator (d) Magnetic separator
12. When the difference in length exists between the seeds and impurities, the following separation is used
   (a) Inclined belt separator (b) Length separator
   (c) Roller separator (d) Magnetic separator
13. When the difference in shape (round and flat) exists between the seeds and impurities, the following separation is used
   (a) Spiral separator (b) Length separator
   (c) Roller separator (d) Magnetic separator
14. When the difference in colour exists between the seeds and impurities, the following separation is used
   (a) Length separator (b) Colour separator
   (c) Roller separator (d) Magnetic separator
15. For conveying seeds vertically, the following equipment is used
   (a) Screw conveyor (b) Bucket elevator
   (c) Belt conveyor (d) Vertical conveyor
16. The equipment used to apply chemicals that involves suspension and wettable powder treatment material in water is
   (a) Slurry treaters (b) Panogen treaters
   (c) Direct treaters (d) Mist-o-matic treaters

17. To separate smooth seed from the rough seeds, ... equipment is used.
    (a) Indent cylinder (b) Roll mill
    (c) Inclined draper (d) Magnetic separator
18. Roller mill is used to separate
    (a) Smooth and rough seeds (b) Small and big seeds
    (c) Mature and immature seeds (d) Crop and weed seeds
19. To separate smooth seed from rough seeds, the machines used is
    (a) Indent cylinder (b) Roll mill
    (c) Inclined draper (d) Magnetic separator
20. The removal of husk from the grasses is done by
    (a) Huller (b) Sheller
    (c) Scarifier (d) Debearder
21. Which one of the following machine separates the higher inert mater and adulterant from the crop seed?
    (a) Screen cleaner (b) Air screen cleaner
    (c) Aspirator (d) Seed cleaner
22. The weed seeds are easily separated by
    (a) Magnetic separator (b) Indent cylinder separator
    (c) Spiral separator (d) Disc separator
23. The air screen cleaner cum grader separates the seeds based on the
    (a) Shape (b) Size
    (c) Shape & Size (d) Weight
24. The gravity separator works on the principle of
    (a) Gravity principle (b) Floatation principle
    (c) Scalping principle (d) Grading principle

♦ Answer Keys ♦

Q.No.	Ans.	Q.No.	Ans.	Q.No.	Ans.	Q.No.	Ans.	Q.No.	Ans.	Q.No.	Ans.
1	b	2	a	3	a	4	a	5	c	6	d
7	c	8	c	9	c	10	b	11	c	12	b
13	a	14	b	15	b	16	a	17	b	18	a
19	b	20	a	21	c	22	d	23	b	24	a

3.6.2. Fill in the blanks

1. Soybean mixed with kidney beans can be separated using ....
2. For density grading, ... separator is used
F. Seed grading with .... is known as basic grading
L. The machine used for grading sunflower is ....
V. Air screen separator works on the principle of ....
/. Air screen cleaner works based on the principle of ....

2. Dodder mill separates seeds based on ....
% Magnetic separator separates seeds based on their ....
-. Electrostatic separator separate seeds based on their ....
10. In processing, the role of scalper is ....
11. Scalper favors the cleaning the seeds which contain ....
12. Huller is used in processing to separate ....
1F. Troublesome seed appendages can be removed by ....
1L. Debearder is used to break the seed pods of ....
1V. Seeds differing in density can be separated by ....
1/. Seed differing in size can be separated by ....
12. Round and flat seeds in a seedlot can be separated by ....
1% Seed which differ in colour can be separated by ....
1-. .... is used for separation of seeds based on surface tension.
20. .... is used for separation of seeds based on colour.
21. .... is used for grading seed based on weight.
22. Indented cylinder is used for grading seeds based on ....
2F. The weight grading technique followed in sunflower is done by ....
2L. The machine used for colour grading is ....
2V. .... separator is used for separating lucerne seed from Dodder seed.

♦ Answer Keys ♦

Q.No.	Ans.	Q.No.	Ans.	Q.No.	Ans.
1	Indented cylinder separator	2	Specific gravity	3	Air screen cleaner
4	Specific gravity separator	5	Size and weight	6	Size and density
7	Surface texture	8	Affinity towards liquid	9	Difference in electrical conductivity
10	Rough cleaning	11	Light trash	12	Cluster of seeds
13	Hammer mills	14	Lucerne	15	Specific gravity separator
16	Air screen cleaner	17	Draper belt	18	Electronic colour sorter
19	Magnetic separator	20	Electronic colour sorter	21	Specific gravity separator
22	Shape	23	Specific gravity separator	24	Electronic colour separator
25	Magnetic				

3.6.3. State True or False

1. Mechanical depulper is mostly used for fruits with relatively hard seed coats or stones
2. Specific gravity separator separate the seeds based on the density of seeds
F. During mechanical grading, the scalper stand always be the bottom most screen
L. Maize sheller cylinder shafting speed range is LV0-V00 rpm
V. Bag sewing machine is used for closing of bags
/. Vibrating conveyor (or) shakers move material through a metal at horizontal movement
2. Processing screens are used for scalping and grading operations
% In spiral separator, the seeds are separated based on size
-. Electrostatic separator separates the seeds based on electrical properties
10. Centrifugal force is the working principle of Boerner divider
11. The machine that separates seeds based on length is known as precision grader
12. The recommended seed upgrading machine for sunflower is air screen cleaner
1F. Conveying equipments are not used in seed processing industries
1L. Screw conveyor is used for conveying the seeds both in horizontal direction and mild slope
1V. Indented cylinder is used for the separation of seeds based on the density
1/. Fluid drilling denotes breaking the seeds in water
12. The disc separator separates the round seeds
1% Specific gravity separator works on the principle of size of the seeds.

♦ Answer Keys ♦

Q.No.	Ans.	Q.No.	Ans.	Q.No.	Ans.	Q.No.	Ans.	Q.No.	Ans.	Q.No.	Ans.
1	True	2	True	3	False	4	True	5	True	6	True
7	False	8	False	9	True	10	False	11	False	12	False
13	False	14	True	15	False	16	False	17	False	18	False

3.7. SEED QUALITY MAINTENANCE

3.7.1. Choose the correct answer

1. What is the concentration of fast green solution used for testing seed coat cracking?
   (a) 0.2V% (b) 0.V0%
   (c) 0. 2V% (d) 1.0%

2. Which one of the following test is used for assessing the seed coat damage of large seeded legumes?
   (a) Acetocarmine test (b) Indoxyl acetate test
   (c) Fast green test (d) Ferric chloride test
3. Which one of the following test is used for assessing the seed coat damage of large seeded grasses?
   (a) Acetocarmine test (b) Indoxyl acetate test
   (c) Fast green test (d) Ferric chloride test
4. What is the packaging material used for paddy seed packing?
   (a) Super bag (b) Hitech bag
   (c) Gunny bag (d) Aluminium foil pouches
5. What is the packaging material used for vegetable seed packing?
   (a) Super bag (b) Hitech bag
   (c) Gunny bag (d) Aluminium foil pouches
6. What is the unit of packaging for paddy seed?
   (a) 1VKg (b) 2VKg
   (c) FVKg (d) LVKg
7. What is the packaging material for groundnut seed packaging?
   (a) Tyvec bag (b) Hitech bag
   (c) Gunny bag (d) Polythene bag
8. Which one of the following container is moisture resistant container?
   (a) Gunny bag (b) Cloth bag
   (c) Aluminium foil bag (d) 200 gauge polythene bag
9. Which one of the following container is used for long ter storage?
   (a) Gunny bag (b) Cloth bag
   (c) F00 gauge polythene bag (d) 200 gauge polythene bag
10. Which one of the following container is a moisture pervious container?
   (a) Gunny bag (b) Aluminium foil pouches
   (c) F00 gauge polythene bag (d) 200 gauge polythene bag
11. Which one the following container is a moisture impervious container?
   (a) Gunny bag (b) Aluminium foil pouches
   (c) F00 gauge polythene bag (d) 200 gauge polythene bag

♦ Answer Keys ♦

Q.No.	Ans.	Q.No.	Ans.	Q.No.	Ans.	Q.No.	Ans.	Q.No.	Ans.	Q.No.	Ans.
1	a	2	b	3	c	4	a	5	d	6	b
7	a	8	c	9	c	10	d	11	d		

3.7.2. Fill in the blanks

1. The two layers of polyethylene bag with .... in between them form the super bag which is the packaging material for paddy seed

♦ Answer Keys ♦

Q.No.	Ans.
1	Vaccum

3.8. MATCH THE FOLLOWING

1. Question

1.	Scalping	A.	Groundnut
2.	Deawning	B.	Length grading
F.	Dehusking	C.	Weight grading
L.	Haulm cutting	D.	Removal of material higher than seed size
V.	Decortication	E.	Density grading
/.	Air screen cleaner	F.	Cotton
2.	Indented cylinder separator	G.	Potato
%	Specific gravity separator	H.	Grasses
-.	Water flotation technique	I.	Size grading
10.	Ginning	J.	Maize

♦ Answer Keys ♦

Q.No.	Ans.	Q.No.	Ans.	Q.No.	Ans.	Q.No.	Ans.	Q.No.	Ans.	Q.No.	Ans.
1	D	2	H	3	J	4	G	5	A	6	I
7	B	8	C	9	E	10	F				

2. Question

1.	Ferric chloride test	A.	PEG
2.	Pelleting	B.	Artificial drying
F.	Hardening	C.	Surface texture
L.	Seed priming	D.	HCl gas
V.	Seed coating	E.	Debearders
/.	Batch drier	F.	Mechanical damage
2.	Magnetic separator	G.	Increasing seed size
%	IDS method	H.	Does not increase seed size
-.	Gas delinting	I.	Drought resistance
10.	Pre-cleaning	J.	Bhendi

♦ Answer Keys ♦

Q.No.	Ans.	Q.No.	Ans.	Q.No.	Ans.	Q.No.	Ans.	Q.No.	Ans.	Q.No.	Ans.
1	F	2	G	3	I	4	A	5	H	6	B
7	C	8	J	9	D	10	E				

3.9. FINDOUT THE INCORRECT STATEMENT

1. (a) Seed pelleting is a seed treatment by which the size of the seed is increased

(b) In smaller seeds, pelleting is adopted as a pre-sowing management technology

(c) For rainfed sowing, pelleting is recommended as a pre-sowing management technology

2. (a) Electrostatic separator separates the seeds based on electrical properties

(b) Indented cylinder separator is used for the separation of seeds based on the seed length

(c) Specific gravity separator separate the round seeds

♦ Answer Keys ♦

Q.No.	Ans.	Q.No.	Ans.
1	c	2	c

3.10. ARRANGE IN ORDER

1. Seed processing sequence of barely

   Debearding – Cleaning – Scalping – Grading

   Answer: Debearding –Scalping – Cleaning – Grading

2. Seed processing of cotton

   Grading – Cleaning – Delinting – Ginning

   Answer: Ginning – Delinting – Cleaning – Grading

3. Sequence of seed processing in maize

   Shelling – Sorting cobs – Cleaning - Grading

   Answer: Sorting cobs – Shelling – Cleaning – Grading

4. Drying – Precleaning – Threshing – Upgrading – Size grading – Seed packing – Seed treatment

   Answer: Threshing – Drying – Precleaning – Size grading – Upgrading – Seed treatment – Seed packing

5. Soaking – Dry seed – Sun dry – Hardened seed – Shade dry

   Answer: Dry seed – Soaking – Shade dry – Sun dry – Hardened seed

6. Polykote coating – Hardening or priming – Insecticides – Nutrients – Biofertilizers – Designer seeds – Fungicide

   Answer: Hardening or priming – Polykote coating – Insecticides – Nutrients – Biofertilizers –Fungicide - Designer seed

3.11. DIFFERENTIATE THE FOLLOWING

1. Differentiate between hydration and dehydration

Hydration	Dehydration
Absorption of water by the seeds	Loss of water from the seeds
Eg. Hardening and priming	Eg. Drying back to original moisture or weight after hydration in the case of hardening and priming
Hydration process is very quick process	Dehydration process is slow process
An advancement in metabolic activity and germination occurs	While drying, it stops the metabolic activity and germination process

2. Differentiate between size grading and weight grading with examples

Size grading	Weight grading
Separation of seeds based on the size and shape of the seeds	Separation of seeds based on the weight and density of seeds
Principle used is gravitational, centrifugal and product pressure	Principle employed here is floatation pressure
Eg. Air screen cleaner	Eg. Grading separator stoners, Specific gravity separator
Weed seeds can be removed	Badly damaged, deteriorated, diseased, chaffy, ill-filled, insect-damaged seeds, contaminating crop or weed seeds can be removed

3. Differentiate between aspirator and roll mill

Aspirator	Roll mill
Used for general cleaning, coarse grading and specific separations	Separation of smooth seeds from rough seed coat seed with shrivelled seeds, broken, chopped or damaged seeds, rough or irregularly shaped inert materials
Use the movement of air to divide seeds according to their terminal velocities	Separates seeds according to difference in texture of seed coat
The fan is at discharging end and induces a vacuum which allows the atmospheric pressure to force the air through separator	
Eg. Lifting the seeds of meadow foxtail from seeds of *Alfa ficus*	Eg. Clover

4. Differentiate between wet seed treatment and dry seed treatment

Wet seed treatment	Dry seed treatment
A dilute solution of a volatile pesticide (both fungicide and insecticide) applied to the seed and is thoroughly mixed with them	The treating material is applied to the seeds or seedlings in the form of a dust
Used to treat small batches of seeds of valuable strains of wheat, barley and millet that are infected by smut	This method is used to disinfect seeds of tubers and grain, vegetable, bulb and root crops
Labour and time consuming process	Relatively less time and labour saving process
More adherence to the seed surface and materials are not dispersed	
Less health hazard	Health hazard is high due to dusting-off

5. Differentiate between inoculant pelleting and nutrient pelleting

Inoculant pelleting	Nutrient pelleting
Coating the seeds with biofertilizers and biocontrol agents	Coating the seeds with macro and micro nutrients
Eg. *Rhizobium, Azospirillum, Trichoderma, Pseudomonas* etc.	Eg. DAP, gypsum, micronutrient mixture
Should consider the viability periods of bioinoculants (shorter period)	Coated seeds can able to provide nutrient for longer period
Sowing should be taken within 2Lh after treatment.	Seeds can be stored 10-1V days based on the crop
Can be stored for few weeks against pathogens	Can be stored for few months
It provides protection to seeds against pathogen. It fixes atmospheric N as well as involved in mobilization of nutrients	It supplies nutrients to the seeds

6. Differentiate between pre-cleaning and cleaning

Pre-cleaning	Cleaning
It is the removal of larger matter such as remenants of leaves, twigs, trashes, stones, clods etc.	Separation of empty seed, undesirable material, inert matter, weed seeds, light and chaffy seeds, off-sized or deteriorated seeds etc., besides uniform grading based on seed size

Pre-cleaning	Cleaning
Scalper, debearders can be used for pre-cleaning the seeds	Air screen cleaner cum grader can be used for cleaning the seeds
No pre-cleaning is required for hand harvested and winnowed seeds	Cleaning should be done for all kind of seeds

7. Differentiate between scalping and grading

Scalping	Grading
Separation of materials which are larger than the required seed size like trash, clods, plant parts, etc.	The seeds of the variety that are under sized undergo grading operation
They do not pass through the top screen	They pass through the top screen, but do not pass through the bottom screen and collected in a separate outlet
Collected in a separate outlet	Undersized seed, damaged seed and the inert matters that are smaller in size than seed size pass through the bottom screen and collected in a separate outlet

8. Differentiate between seed disinfection and seed disinfestation

Seed disinfection	Seed disinfestation
Eradication of fungal spores that have become established within the seed coat or more deep seated tissues	Destruction of surface borne organisms that have contaminated the seed surface, but not infected the seed source
The pathogen is internally seed borne	The pathogen is externally seed borne
For effective control, the fungicidal treatment must actually penetrate the seed in order to kill the fungus that is present internally	Chemical dips, soaks, fungicides, applied as dust, slurry or liquid have been found successful to prevent seed disinfestation
Systemic fungicide must be used for seed treatment	Non-systemic fungicide can be used

9. Differentiate between seed coating and seed pelleting

Seed coating	Seed pelleting
Application of nutrient/fungicide over the seeds using binding materials without modifying the seed size	Seed pelleting is the process of enclosing a seed with a small quantity of inert material just large enough to produce a globular unit of standard size

Seed coating	Seed pelleting
Polycote or polymer is used for thin film coating	Fevicol or gum Arabica or maida gruel can be used as an adhesive (Binding agent)
No filler or inert material is used	Filler materials like gypsum or botanical powder are used
No increase in size, shape and weight of seeds	A considerable increase in seed size, shape and weight
To keep the identity of the seed firm	To keep away the seed destroying or damaging agents especially birds after sowing
Not suitable for aerial seeding	Aerial seeding is possible because of increased ballistic property of seeds (Free from blowing off seeds while aerial seeding)
Ecofriendly because it prevents dusting off	Not ecofriendly because free dusting off is injurious to human health and degrading of environment as well
It does not enable precision planting	It enables the precision planting

10. Differentiate between seed priming and seed hardening

Seed priming	Seed hardening
It is the process of controlled hydration of seeds to a level that permits pre-germinative metabolic activity to proceed, but that prevents actual emergence of the radicle	It is the process of hydrating the seeds in shorter period than priming to initiate the pre-germinative metabolism
Soaking should be done at 10-1V°C	Hydration should be done at 1V-2V°C
Drying back to original moisture content is required	Drying back to original moisture content is necessary
Inorganic salts, sugars and polyethylene glycol, liquid biofertilizers and biocontrol agents can be used	Potassium dihydrogen phosphate, potassium chloride and other suitable chemical salts can be used

11. Differentiate between seed fortification and seed infusion

Seed fortification	Seed infusion
An approach to improve the germinability and vigour of the seeds by infusion of bioactive chemicals into the seed by soaking in growth regulators and nutrient solutions	Infusion of nutrients, growth promoting substances can be done with organic solvent such as acetone and dichloromethane

Seed fortification	Seed infusion
Drying back is necessary	Drying back is not necessary

12. Differentiate between pre harvest management and post harvest management of seeds

Pre harvest management	Post harvest management
Management techniques adopted before harvesting of the seed crop	Management techniques adopted after harvesting of the seed crop
To increase the seed yield and quality and to maintain the true to type of the variety	To improve the quality of the seeds
To control the field carry over fungi and insects	To size grade and upgrade the seed for quality and for fetching higher prices in the market
Preharvest spraying of 0.1% endosulphan to control bruchids in pulses	Size grading of seeds to improve the physical purity of seeds

13. Differentiate between indent cylinder separator and magnetic separator

Indent cylinder separator	Magnetic separator
Helps to separate the seeds according to the seed length	Separates the seed according to its surface texture or related characteristics
Small seeds are passed into the indents by centrifugal force and can be removed	Treated seeds are passed over a revolving magnetic drum and separated from smooth, uncoated seed
The larger seeds flows in the centre of the cylinder and is discharged by gravity	Effectiveness of magnetic separation depends on the components of the seed lot and the powder and water used in treating operation

14. Differentiate between air screen cleaner and specific gravity separator

Air screen cleaner	Specific gravity separator
Separates seeds based on seed size	Separates seeds based on weight and density of seed
Uses a combination of air, gravity and screens to separate seed	This machine separates the seeds based on air force
Air streams remove the trash and impurities	This machines effective at removing stones from the seed

Air screen cleaner	Specific gravity separator
It's a basic cleaning	It is a upgrading of seeds
Principle is scalping and grading	Principle is air blowing

15. Differentiate between osmopriming and halopriming

Osmopriming	Halopriming
Soaking of seeds in osmotic solutions	Soaking of seeds in inorganic salt solutions
Eg. PEG	Eg. KNO_3, KCl etc.
Relatively high cost	Low cost
PEG is chemically inert and does not toxic to seed	At high concentration, it is toxic to seed
Ecofriendly	Not ecofriendly and causes hazards to environment

16. Differentiate between hydropriming and halopriming

Hydropriming	Halopriming
Soaking the seeds in water	Soaking the seeds in chemicals salt solutions
The process must be arrested at a specific time to prevent the onset of phase III of germination	High ion concentration in the embryo may affect germination adversely when inorganic salts are used
No cost compared to halopriming	High cost compared to hydropriming
Ecofriendly	Environmental hazards
No phytotoxic effect	Phytotoxic, if concentration increases

17. Differentiate between osmopriming and solid matrix priming

Osmopriming	Solid matrix priming
Soaking the seeds in osmotic solutions	Mixing of seeds with an organic or inorganic carrier and water for a period of time
PEG is commonly used because, it is chemically inert and does not have adverse effects on the embryo	Seed water potential is regulated by matrix potential of the seed and during priming the water is largely held by the carrier
Mannitol can also be used	Peat or vermiculite can also be used
Not ecofriendly	Ecofriendly

18. Differentiate between acid delinting and gas delinting

Acid delinting	Gas delinting
Removal of fuzz from the cotton seed using commercial sulphuric acid	Removal of fuzz from cotton seeds using HCl gas
Drying back is necessary	Cotton seed is in dry state throughout the process. No need of drying
Requires no specified equipments and can be done by manually	Requires specific equipments
Hazard to environment	Pollution free
Handling of seed is maximum and there is possibility for seed damage	Minimum handling of seeds, hence less damage to the seeds
Culminates the pathogen present in the seed coat	Requires fungicide and insecticide treatment to eliminate the pathogens

19. Differentiate between manual threshing and mechanical threshing

Manual threshing	Mechanical threshing
Threshing can be manually by beating the bundles of cut crop	Threshing the harvested produce using a tractor or power operated machines
Never over dry the crop, as over drying will cause excessive shattering	Timing is most important factor for mechanical threshing
Never thresh on barren floor as this will lead to the introduction of foreign matter such as stones, soil particles etc	Immediate threshing reduces the exposure of crop to insects, birds and rodents and prevents the onset of disease and mould
A worker can obtain 1V to L0 kg of rice products per hour by manual threshing	In machine threshing, about /L0 kg of rice can be obtained per day
Winnowing can be done manually for separation of seeds	Both threshing and winnowing are done in the machine itself
Involves more labour work, time consuming and cumbersome process	Easy and less time consuming

20. Differentiate between acid method and alkali method of seed extraction with examples

Acid method	Alkali method
Hydrochloric acid is used for seed extraction	Sodium bicarbonate or sodium hydroxide is used
Eg. Tomato – 2V ml/kg of fruit pulp	Eg. Tomato –0.V% sodium bicarbonate or 2.V% to V% NaOH

Acid method	Alkali method
The acid and pulp mixture should be kept for shorter period for about V-/0 minutes Eg. Tomato – F0 min.	The acid and pulp mixture should be kept comparatively longer periods ie., 1V min. – one day depends on nature of alkali used Eg. Tomato
Seed recovery percentage is higher Seed recovery of tomato PKM-1 variety by this method is about 1.2%	Seed recovery percentage is lower than acid method Seed recovery of tomato PKM-1 variety by this method is 0.- %
Seeds are also bright in colour with good germinability and free from fungal attack Tomato – PKM – 1 variety -V% germination	Seeds recorded lesser germination compared to acid method

21. Differentiate between acid method and fermentation method of seed extraction with examples

Acid method	Fermentation method
Pulps are mixed with commercial hydrochloric acid	Fruit pulps are kept as such without any chemicals for fermentation
The acid and pulp mixture should be kept for shorter period Eg. Tomato – F0 minutes	Allowed for fermentation for two to three days Eg. Tomato
Corrosiveness of the acid removes the mucilage adhering to the seed and makes the seed free of pulp	The seeds detached from the adhering pulp and settles to the bottom of the container due to fermentation
Seed recovery is higher 1.2% for tomato	Seed recovery is less 1.0% for tomato
Seeds are bright in colour with good germinability and free from fungal attack	Seeds become dull colour with lower germination due to fermentation of the pulp and the fungal growth on the seeds
It situ germination is not occurred Germination percentage --V% in tomato	*In situ* germination may occur due to long period of fermentation Germination percentage – 21% in tomato

Unit 4

SEED QUALITY CONTROL

4.1. SEED LEGISLATION

4.1.1. Choose the best answer

1. The pioneer organization involved in seed production and quality control is
   (a) NSC (b) ISTA
   (c) CSTL (d) SSFC
2. The National Seed Research and Training Centre is located at
   (a) Varanasi (b) Delhi
   (c) Mumbai (d) Kolkata
3. The Seed Act has
   (a) 2V sections (b) 20 sections
   (c) V0 sections (d) /0 sections
4. New Seed Policy was implemented during the year
   (a) 1-/% (b) 1-%%
   (c) 1-2V (d) 1-%V
5. Government of India enacted Seed Act during the year
   (a) 1-/% (b) 1-/-
   (c) 1-// (d) 1-%0
6. Restriction of export and import of seeds was envisaged through
   (a) Seed Act, 1-/% (b) Seed Control Order, 1-%F
   (c) New Seed Policy (d) Seed Bill
7. The tremendous progress has been made in the development of HYV by
   (a) AICRP (b) NSP
   (c) NSC (d) NADP
8. The Indian Seed Act was passed by Govt. of India to regulate quality of seeds in the year
   (a) 1-/% (b) 1-/F
   (c) 1-/2 (d) 1-//
9. National Seeds Corporation was established to develop Indian Seed Industry in the year
   (a) 1-/% (b) 1-/F
   (c) 1-/2 (d) 1-/L

10. The act that provides for protection of farmers and plant varieties is
    (a) Plant Varieties Protection and Farmers Rights Act, 2001
    (b) Essential Commodities Act
    (c) Seed Control Order
    (d) Environmental Protection Act

11. Seed control order insists on
    (a) Licensing (b) Export
    (c) Processing (d) Import

12. Orange certificate is issued by
    (a) ISTA (b) OECD
    (c) AOSA (d) UPOV

13. The total set of rules and protocols related to variety development and release, seed production, quality control and delivery is called
    (a) Variety regulation (b) Plant regulation
    (c) Field regulation (d) Seed regulation

14. Seed Rules was implemented in the year
    (a) 1-V/ (b) 1-/-
    (c) 1-2/ (d) 1-%/

15. Seed quality in India is regulated through
    (a) Seed schemes (b) Seed Act and Rules
    (c) Seed Control Order (d) Seed Mission

16. Seed control order was enacted during the year
    (a) 1-%F (b) 1-VF
    (c) 1-/F (d) 1-2F

17. The system developed in India by integrating the rights of breeder, farmers and village communities is
    (a) TRIPS (b) CBD
    (c) *Sui genesis* (d) PPV&FRA

18. The year in which Indian Seed Act was implemented is
    (a) 1-// (b) 1-/2
    (c) 1-/% (d) 1-/-

19. The seed rules of the Seed Act were notified in
    (a) August 1-/% (b) September 1-/%
    (c) October 1-/% (d) November 1-/%

20. Seed act was implemented in India from
    (a) Jan 1 200V (b) Dec 2, 1-//
    (c) October 2 1-/- (d) 2-th Dec 1-//

21. Seed rule was implemented in India from
    (a) Jan 1 200V (b) Vth October, 1-/-
    (c) 2nd October 1-/- (d) 2-th Dec 1-//

22. Seed Control Order, 1983
    (a) Promotes export and import of seeds
    (b) Restricts export and import of seeds
    (c) Controls export and import of seeds
    (d) Regulates export and import of seeds

♦ Answer Keys ♦

Q.No.	Ans.	Q.No.	Ans.	Q.No.	Ans.	Q.No.	Ans.	Q.No.	Ans.	Q.No.	Ans.
1	a	2	a	3	a	4	b	5	c	6	b
7	a	8	d	9	b	10	a	11	a	12	a
13	d	14	b	15	b	16	a	17	c	18	d
19	b	20	d	21	c	22	a				

4.1.2. Fill in the blanks

1. Under Seeds Act, 1-// .... is voluntary
2. In the year 1-%F, Government passed .... and seeds have been declared as an essential commodity

F. Seed quality in India is legally controlled by ....

L. .... is enforced by Government of India during 1-// for seed quality control

V. The Indian Seeds Act was passed by Govt. of India to regulate quality of seeds in the year ....

/. Seed Act was passed during ....

2. The act that provides for protection of farmers and plant varieties is ....

% Announcement of .... in 1-%% liberalized the seed trade in India

-. New Seed Policy was implemented during the year ....

10. The Indian Seed Act was enacted in the year ....

11. When was new seed policy passed? ....

12. Seed control order was implemented in ....

1F. High yielding variety programme was launched in the year ....

1L. The two main features of Seed Act, 1-// are .... and ....

1V. Supreme body under Seeds Act is ....

1/. Seed Act has .... sections

12. The seed rules are enacted in the year ....

1% National Seeds Corporation was established in the year ....

1-. National seed policy was proposed in the year ....

20. The total number of clauses in the Seed Bill 200L is ....

21. The Indian Minimum Seed Certification standards manual is published ....

22. Seed Inspectors are appointed under section number .... of Seeds Act 1-//

♦ Answer Keys ♦

Q.No.	Ans.	Q.No.	Ans.	Q.No.	Ans.
1	Seed certification	2	Seed Control Order	3	Seed Act, 1-//
4	Seed Act	5	1-/F	6	1-//
7	Protection of Plant Varieties and Farmers Right	8	New Seed Policy	9	1-%%
10	1-//	11	1-%%	12	1-%F
13	1-/V	14	Voluntary Certification, compulsory labelling	15	Central Seed Committee
16	2V	17	1-/%	18	1-/F
19	2002	20	L-	21	1-%%
22	1F				

4.1.3. State True or False

1. Seed Rules was enacted during 1966

2. Seed Act is enforced only on the notified varieties of certified seeds

F. The containers without proper label are allowed for sale in the market

L. Preventing of seed inspector from taking seed sample at sale point is not an offence

V. After the implementation of Seed Control Order (1-%F), seed is considered as an essential commodity

/. Seed Act 1-// is formulated to bring the seed production under legal umbrella

2. The aim of New Seed Policy is to improve the export and import of seeds

% New seed policy envisages import of improved varieties/parental lines of hybrids in small quantities from any country

-. Seed control order was formed insisting on compulsory licensing of the dealers

10. The seed rules were passed in the year 1-//

11. The power to notify the kind or variety falls Vth section of the seed act

12. New seed policy was enacted in the year 1--V

1F. New seed policy was enacted in 1-%/

1L. Seed rules were framed in the year 1-/%

♦ Answer Keys ♦

Q.No.	Ans.	Q.No.	Ans.	Q.No.	Ans.	Q.No.	Ans.	Q.No.	Ans.	Q.No.	Ans.
1	True	2	True	3	False	4	False	5	True	6	True
7	True	8	False	9	True	10	False	11	True	12	False
13	False	14	True								

4.2. SEED CERTIFICATION

4.2.1. Choose the best answer

1. Minimum germination requirement for castor seed
   (a) /V% (b) %0%
   (c) 20% (d) 2V%
2. The inseparable other crop plant of oats is
   (a) Barley (b) Paddy
   (c) Sorghum (d) Bajra
3. Breeder seed tag colour is
   (a) Golden yellow (b) Azhur blue
   (c) White (d) Opal green
4. Only these kinds of crops are eligible for seed certification
   (a) Newly released varieties (b) Newly released hybrids
   (c) All notified varieties (d) All released varieties
5. Labeling of seed packages under Seed Act is
   (a) Voluntary (b) Compulsory
   (c) Need based (d) Demand based
6. Minimum germination standard for foundation seed of paddy is
   (a) %0 (b) 100
   (c) /V (d) -%
7. Objectionable weed plant in bhendi is
   (a) *Convolvulus arvensis* (b) Cuscuta
   (c) Wild okra (d) *Argemone maxicana*
8. Minimum requirement of germination as per IMSCS for cauliflower is
   (a) %0% (b) 20%
   (c) /0% (d) V0%
9. Isolation distance required for foundation seed production of cabbage is
   (a) 1/00 m (b) %00 m
   (c) 1000 m (d) V00 m
10. Genetic contamination is caused by
    (a) Other crop plants (b) Weed plants
    (c) Diseased plants (d) Off type plants
11. Seed production which comes under seed certification
    (a) Nucleus seed (b) Breeder seed
    (c) Foundation seed (d) TFL seed
12. Maximum area per unit in seed certification is
    (a) Vac (b) 10 ac
    (c) 20 ac (d) 2Vac

13. Pearson's square law is used in
    (a) Seed sampling (b) Seed blending
    (c) Seed dividing (d) Seed mixing
14. The pure seed fraction recommended for groundnut is
    (a) --% (b) -%%
    (c) -/% (d) -2%
15. Blending is permitted in case of
    (a) Different varieties of same species (b) Same varieties
    (c) Composites (d) Synthetic
16. Without phytosanitary certificate for smut disease, the onion seeds cannot be exported to
    (a) Thailand (b) Malaysia
    (c) Singapore (d) Pakistan
17. Orange certificate is
    (a) Phytosanitary certificate (b) Seed sample certificate
    (c) Seed lot certificate (d) Seed passport
18. The maximum area could be registered for seed certification is
    (a) L ac (b) V ac
    (c) L ha (d) 10 ha
19. Seed lot certificate colour is
    (a) Blue or White (b) White or Yellow
    (c) Purple or Red (d) Orange or Green
20. Pure seed fraction according to seed standards in *Abelmoschus esculentus* should be
    (a) -%% (b) -V%
    (c) --% (d) -F%
21. Certified seed of cotton should have minimum germination of
    (a) V0% (b) /0%
    (c) 20% (d) %0%
22. A procedure developed as a means of assuring that the seeds have a high standard of purity and quality is called
    (a) Quarantine (b) Registration certificate
    (c) Seed certification (d) Patent certificate
23. Cob sorting in maize is done to identify
    (a) Diseased cobs (b) Off type cobs
    (c) Mature cobs (d) Big sized cobs
24. Crop varieties which are eligible for certification are
    (a) Non notified varieties (b) Notified varieties
    (c) Denotified varieties (d) Newly released varieties
25. Validity period for certification tag is
    (a) / months (b) - months
    (c) 12 months (d) 1V months

26. Genetic purity in the field is maintained by
    (a) Seed testing (b) Roguing
    (c) Seed sampling (d) Field inspection
27. Breeder seed production plot is inspected by
    (a) Seed Certification Officers (b) Monitoring team
    (c) University Scientists (d) Farmers
28. Selfed bolls/fruits are
    (a) Physical contaminants (b) Genetic contaminants
    (c) Health contaminants (d) Growth contaminants
29. Objectionable weed in rice is
    (a) *Cuscuta* sp (b) *Striga* sp
    (c) *Circium* sp (d) *Oryza sativa* var. *fatua*
30. Genetic purity of the foundation seed should be
    (a) --.0% (b) --.V%
    (c) -%0% (d) -%V%
31. The minimum germination of groundnut as per the prescribed seed standard is
    (a) /V (b) 20
    (c) 2V (d) %0
32. Seed certification in India is
    (a) Compulsory (b) Voluntary
    (c) Depends on state Government (d) Depends on crop varieties
33. The certification tag attached with foundation seed is
    (a) White colour (b) Blue colour
    (c) Golden yellow (d) Opal green
34. This class of seed is not certified by the Seed Certification Agency
    (a) Breeder seed (b) Certified seed
    (c) Foundation seed (d) Registered seed
35. Pod verification is done in
    (a) Groundnut (b) Blackgram
    (c) Cowpea d.Redgram
36. The unit of the DNA ladder is based on
    (a) Microgram (b) Base pair
    (c) Micro litre (d) Voltage current
37. Probe based markers are
    (a) Microsatellite (b) RAPD
    (c) SSR (d) ISSR
38. What is the maximum permitted limit for loose smut disease in wheat foundation seed production?
    (a) 0.0V% (b) 0.10%
    (c) 0.2V% (d) 0.V0%

39. What is the maximum permitted limit for head smut disease in sorghum foundation seed production?
(a) 0.0V% (b) 0.10%
(b) 0.2V% (d) 0.V0%

40. What is the maximum permitted limit for early blight and leaf spot of tomato in certified seed production?
(a) 0.0V% (b) 0.10%
(c) 0.2V% (d) 0.V0%

41. What is the maximum permitted limit for watermelon mosaic disease in summer squash certified seed production?
(a) 0.0V% (b) 0.10%
(c) 0.2V% (d) 0.V0%

42. What is the maximum permitted limit for anthracnose disease in the case of chilli foundation seed production?
(a) 0.0V% (b) 0.10%
(c) 0.2V% (d) 0.V0%

43. Maximum number of offtypes permitted for chickpea foundation seed production
(a) 0.0V% (b) 0.10%
(c) 0.1V% (d) 0.20%

44. Maximum number of offtypes permitted for barley foundation seed production
(a) 0.0V% (b) 0.10%
(c) 0.1V% (d) 0.20%

45. What is the maximum standards for inseparable other crop plants for certified seed production of wheat?
(a) 0.01% (b) 0.1F%
(c) 0.0V% (d) 0.10%

46. What is the maximum permitted limit for objectionable weeds at field level in the case of rice certified seed production?
(a) 0.01% (b) 0.1F%
(c) 0.0V% (d) 0.10%

47. What is the maximum permitted limit for objectionable weeds field level in the case of lucerne certified seed production?
(a) 0.01% (b) 0.1F%
(c) 0.0V% (d) 0.10%

48. .... of cuscuta species is allowed in lucerne foundation seed production
(a) 0% (b) V%
(c) 10% (d) 1V%

♦ Answer Keys ♦

Q.No.	Ans.	Q.No.	Ans.	Q.No.	Ans.	Q.No.	Ans.	Q.No.	Ans.	Q.No.	Ans.
1	c	2	a	3	a	4	c	5	b	6	a
7	c	8	b	9	a	10	d	11	c	12	d
13	b	14	c	15	b	16	d	17	c	18	d
19	d	20	c	21	b	22	c	23	b	24	b
25	b	26	b	27	b	28	b	29	d	30	b
31	b	32	b	33	a	34	a	35	a	36	b
37	b	38	b	39	a	40	d	41	d	42	b
43	b	44	a	45	c	46	a	47	c	48	b

4.2.2. Fill in the blanks

1. In Indian seed quality control system, certification is voluntary and labeling is ....

2. Trueness to variety is ....

F. .... varieties only eligible for certification

L. The validity period as per certification is ....

V. In seed quality control system, golden yellow tag represents ....

/. The maximum area registered for single certification is ....

2. The tag colour of TFL seed is ....

% .... is exempted from certification

-. National organization involved in seed production is ....

10. .... is the legally sanctioned system for seed quality control

11. State level organization empowered to conduct seed certification works is ....

12. .... and .... are coming under the preview of seed certification

1F. Under the Seeds Act and Rules .... is voluntary and .... is compulsory

1L. Breeder seeds are affixed with .... colour tag

1V. What is the sequence of generation systems adopted in paddy seed production? ....

1/. Purpose of Pearson's square law is ....

12. The FS and CS seed tags contain informations on .... and .... details

1% Two main features of Seed Act 1-// are .... and ....

1-. Blending high and low germinable seed lots in sorghum is done based on ....

20. .... tag is used for truthfully labelled seed

21. The prescribed distance which separates a cross pollinated seed crop from the contaminants is called ....

22. Seed certification agency can certify only .... variety

2F. Minimum genetic purity for certified seed is ....

2L. A class or seed that does not come under the purview of seed certification ....

2V. .... is directly controlled by the originating or the sponsoring breeder or institution

2/. Post harvest verification for genetic purity is done in ....

22. TMV is designated disease of ....

2% Mention the minimum seed certification standards for germination of cucumber seed ....

2-. The objectionable weed plant in rice is ....

F0. Breeder seeds are affixed with .... colour tag

F1. Volunteer plants in the seed plot occur from ....

F2. Certification standards are classified into 2 broad categories namely .... and specific seed certification standard.

FF. In India, seed certification standards were initially developed by ....

FL. The seed certification standards are developed by ....

FV. National commission on Agriculture was established in ....

F/. Seed quality in India is legally controlled by ....

♦ Answer Keys ♦

Q.No.	Ans.	Q.No.	Ans.	Q.No.	Ans.
1	Compulsory	2	Genetic purity	3	Notified
4	- months	5	Breeder seed	6	2V acres
7	Opal green	8	Breeder seed	9	National Seed Corporation
10	Seed certification	11	State Seed Certification Agency	12	Foundation, Certified seed
13	Certification, Truthfully labelling	14	Golden yellow	15	BS-FS-CS1-CS2-Commercial
16	Seed lot blending	17	Kind, Variety	18	Compulsory labelling, Voluntary Certification
19	Pearson Square Law	20	Opal green	21	Isolation distance
22	Notified	23	- - %	24	TFL
25	Nucleus/Breeder seed	26	Groundnut	27	Tobacco
28	/0%	29	Wild rice	30	Golden yellow
31	Previous crop plants	32	General Seed Certification Standard	33	National Seeds Corporation Limited
34	Central Seed Committee	35	1-22	36	Seed act 1-//

4.2.3. State True or False

1. The truthful labelling concept has been allowed to control and prevent the deterioration of minimum standards by the seed industry
2. Breeder seed is exempted from seed certification

F. The minimum number of seed samples taken by the seed inspector is five

L. Foundation seed is not under the preview of seed certification

V. Registered seed is the source of breeder seed

/. Separate sowing report is necessary, if the distance between two seed field is F0 m

2. Validation period for certification tag is - months

% Seed Certification Officer is the person responsible for protection of quality of seed sold in market

-. Certification is applicable for notified varieties only

10. The minimum germination standard for paddy is %0 per cent
11. Truthful labeling is applicable for notified varieties only
12. Validation period for certification is / months

1F. Seed should be compulsorily certified

1L. A crop which passes the field standards, but does not satisfy for minimum seed standard can be sold as seed

1V. Seed certification is compulsory and labelling of seed is voluntary in India

1/. As per the Seed Act 1-//, notification of a variety is compulsory

12. Seed certification is voluntary, but labelling is compulsory

1% The validation period is %/- months for all the crops from the date of harvest/testing

1-. Revalidation is possible for lots which posses minimum germination percentage even if the genetic purity is below the seed standard

20. Seed quality control is maintained by field inspection, sample testing and enforcement of minimum seed standards
21. Certification is compulsory and Truthful labeling is compulsory
22. Standards for seed certification and truthfully labeling are same

2F. Seed inspector checks the quality of seed at certification

2L. Border rows are to be provided for hybrids, but not for varieties

2V. For quality seed production, seed certification is compulsory

2/. Blue color tag is used for foundation class of seeds

22. All varieties can be certified

2% Breeder seed is produced from foundation seed

2-. Through field inspection, the genetic purity of the crop can be ensured

F0. Harvest time inspection is the last inspection conducted for a seed crop

F1. The breeder seed production field is monitored by seed certification department

F2. Truthfully labeled seeds are self certified seeds

FF. Generation system of seed multiplication is followed for maintaining the genetic purity of crop varieties

FL. The genetic purity of the breeder seed plot is monitored by certification agency

FV. Seed certification is compulsory as per Seed Act, 1-//

♦ Answer Keys ♦

Q.No.	Ans.	Q.No.	Ans.	Q.No.	Ans.	Q.No.	Ans.	Q.No.	Ans.	Q.No.	Ans.
1	True	2	True	3	False	4	False	5	False	6	False
7	True	8	False	9	True	10	True	11	True	12	False
13	False	14	False	15	False	16	False	17	True	18	False
19	False	20	True	21	False	22	True	23	False	24	True
25	False	26	False	27	False	28	False	29	True	30	False
31	False	32	True	33	True	34	False	35	False		

4.3. FIELD INSPECTION

4.3.1. Choose the correct answer

1. Number of field inspections is more in
   (a) Self pollinated crops (b) Cross pollinated crops
   (c) Often cross pollinated crops (d) Hybrids

2. In cotton, the number of plants/count for field inspection is
   (a) 100 (b) 1000
   (c) V00 (d) 2V0

3. It is necessary to take a minimum of five counts upto
   (a) 2 acres (b) 10 acres
   (c) Vacres (d) %acres

4. Normally, the number of field inspection made for a self pollinated seed crop is
   (a) 2 (b) F
   (c) L (d) V

5. Genetic contaminants present in hybrid rice field are
   (a) Shedding tassel (b) Selfed bolls
   (c) Pollen shedders (d) Off-types

6. Number of plants counted per single count during field inspection of bhendi, brinjal and chilli is
   (a) 100 plants (b) 200 plants
   (c) V00 plants (d) 1000 plants

7. In seed certification, field inspection is taken up by
   (a) Seed Certification Officer (b) Seed Inspector
   (c) Seed Analyst (d) Agricultural Officer

8. What is the maximum permitted limit for loose smut disease in wheat foundation seed production?
   (a) 0.0V% (b) 0.10%
   (c) 0.2V% (d) 0.V0%
9. What is the maximum permitted limit for head smut disease in sorghum foundation seed production?
   (a) 0.0V% (b) 0.10%
   (c) 0.2V% (d) 0.V0%
10. What is the maximum permitted limit for early blight and leaf spot of tomato in certified seed production?
   (a) 0.0V% (b) 0.10%
   (c) 0.2V% (d) 0.V0%
11. What is the maximum permitted limit for watermelon mosaic disease in summer squash certified seed production?
   (a) 0.0V% (b) 0.10%
   (c) 0.2V% (d) 0.V0%
12. What is the maximum permitted limit for anthracnose disease in the case of chilli foundation seed production?
   (a) 0.0V% (b) 0.10%
   (c) 0.2V% (d) 0.V0%
13. If the size of the field is upto 2 hectares, how many counts you will make in field inspection of seed crop?
   (a) V (b) /
   (c) 2 (d) %
14. If the area of the field is 6.1 to 8 acres, how many counts you will make in field inspection of a seed crop?
   (a) V (b) /
   (c) 2 (d) %
15. How many plants or earheads to be observed in one count of wheat seed crop field inspection?
   (a) 100 plants/earheads (b) V00 plants/earheads
   (c) 1000 plants/earheads (d) 1V00 plants/earheads
16. In mustard, how many number of plants/earheads to be observed in one count of field inspection
   (a) 100 plants/earheads (b) V00 plants/earheads
   (c) 1000 plants/earheads (d) 1V00 plants/earheads
17. In maize, how many number of plants/earheads to be observed in one count of field inspection?
   (a) 100 plants/earheads (b) V00 plants/earheads
   (c) 1000 plants/earheads (d) 1V00 plants/earheads
18. How many field inspection to be made for potato?
   (a) 1 (b) 2
   (c) F (d) L

19. How many field inspection to be made for radish?
    (a) 1 (b) 2
    (c) F (d) L
20. How many field inspections to be made for rape seed?
    (a) 1 (b) 2
    (c) F (d) L
21. What is the isolation distance for the wheat crop infected with loose smut for certified seed production?
    (a) 100 m (b) 1V0 m
    (c) 200 m (d) 2V0 m
22. Maximum number of offtypes permitted for chickpea foundation seed production
    (a) 0.0V% (b) 0.10%
    (c) 0.1V% (d) 0.20%
23. Maximum number of offtypes permitted for barley foundation seed production
    (a) 0.0V% (b) 0.10%
    (c) 0.1V% (d) 0.20%
24. What is the maximum standards for inseparable other crop plants for certified seed production of wheat?
    (a) 0.01% (b) 0.1F%
    (c) 0.0V% (d) 0.10%
25. What is the maximum permitted limit for objectionable weeds at field level in the case of rice certified seed production?
    (a) 0.01% (b) 0.1F%
    (c) 0.0V% (d) 0.10%
26. What is the maximum permitted limit for objectionable weeds at field level in the case of lucerne certified seed production?
    (a) 0.01% (b) 0.1F%
    (c) 0.0V% (d) 0.10%

♦ Answer Keys ♦

Q.No.	Ans.	Q.No.	Ans.	Q.No.	Ans.	Q.No.	Ans.	Q.No.	Ans.	Q.No.	Ans.
1	d	2	a	3	c	4	b	5	c	6	a
7	a	8	b	9	a	10	d	11	d	12	b
13	a	14	b	15	c	16	b	17	a	18	d
19	a	20	c	21	b	22	b	23	a	24	c
25	a	26	c								

4.3.2. Fill in the blanks

1. Field inspection should be done without prior notice in ....
2. The objectionable weed in paddy seed production is ....

F. Stages of roguing in chilli is ....

L. In a self pollinated crop, inspection of crop to verify the field standards is not performed during ....

♦ Answer Keys ♦

Q.No.	Ans.	Q.No.	Ans.
1	Self pollinated crop	2	*Oryza sativa var fatua*
3	Flowering, maturity	4	Vegetative stage

4.3.3. State True or False

1. The number of plants to be observed per count in maize is 1000

♦ Answer Keys ♦

Q.No.	Ans.
1	False

4.4. GENETIC PURITY

4.4.1. Choose the correct answer

1. Genuineness of cultivars refers to
   (a) Genetic purity (b) Physical purity
   (c) Physiological quality (d) Health quality
2. The pre and post control test of genetic purity is
   (a) Grow out test (b) Chemical test
   (c) Morphological test (d) Vigour test
3. The minimum population observed for varietal purity testing of certified class seed
   (a) L00 plants (b) L000 plants
   (c) L0 plants (d) 200 plants
4. In grow out test, varieties/hybrids should be tested to assess the
   (a) Genetic purity (b) Germination
   (c) Vigour (d) Moisture
5. Which one of the following crop requires 500 g of submitted sample for grow out test
   (a) Maize (b) Sorghum
   (c) Bajra (d) Wheat
6. For bajra, the size of submitted sample for grow out test is
   (a) 100 g (b) 2V0 g
   (c) V00 g (d) 1000 g
7. For cotton, the size of submitted sample for grow out test is
   (a) 100 g (b) 2V0 g
   (c) V00 g (d) 1000 g

8. Which one of the following crop needs 1000 g submitted sample for grow out test
   (a) Wheat (b) Paddy
   (c) Maize (d) Sorghum
9. For paddy, the size of submitted sample for grow out test is
   (a) 100 g (b) 2V0 g
   (c) V00 g (d) 1000 g
10. Grow out test (GOT) is practiced to ensure
   (a) Field emergence (b) Genetic purity
   (c) Seed vigour (d) Physical purity
11. The maximum number of ODV/kg of seed permitted for maize composite is
   (a) V (b) 10
   (c) 1V (d) 20
12. The maximum number of ODV/kg of seed permitted for carrot is
   (a) V (b) 10
   (c) 1V (d) 20
13. The maximum number of ODV/kg of seed permitted for rice hybrid is
   (a) V (b) 10
   (c) 1V (d) 20
14. Recommended size of the sample submitted for grow out test of *Beta vulgaris* is
   (a) 100 g (b) 2V0 g
   (c) V00 g (d) 1000 g
15. Recommended size of the sample submitted for grow out test of groundnut is
   (a) 100 g (b) 2V0 g
   (c) V00 g (d) 1000 g
16. Recommended size of the sample submitted for grow out test of wheat is
   (a) 100 g (b) 2V0 g
   (c) V00 g (d) 1000 g
17. Recommended size of the sample submitted for grow out test of pearl millet is
   (a) 100 g (b) 2V0 g
   (c) V00 g (d) 1000 g
18. The number of plants required to conduct grow out test, if maximum permissible off type percentage of 0.1% is
   (a) L000 (b) 2000
   (c) %00 (d) L00

19. The number of plants required to conduct grow out test, if maximum permissible off type percentage of 0.5% is

(a) L000 (b) 2000

(c) %00 (d) L00

20. The number of plants required to conduct grow out test, if maximum permissible off type percentage of 0.2% is

(a) L000 (b) 2000

(c) %00 (d) L00

21. The number of plants required to conduct grow out test, if maximum permissible off type percentage of 1.00% and above is

(a) L000 (b) 2000

(c) %00 (d) L00

♦ Answer Keys ♦

Q.No.	Ans.	Q.No.	Ans.	Q.No.	Ans.	Q.No.	Ans.	Q.No.	Ans.	Q.No.	Ans.
1	a	2	a	3	a	4	a	5	b	6	a
7	d	8	c	9	c	10	b	11	d	12	b
13	d	14	b	15	d	16	c	17	a	18	a
19	c	20	b	21	d						

4.4.2. Fill in the blanks

1. The number of plants tested for genetic purity of certified seed class is ....
2. Grow out test is practiced to ensure ....
F. The term ODV stands for ....
L. Maximum number of ODV allowed for rice variety is ....
V. For hybrid cotton grow out test, space between rows is .... cm.
/. Plant to plant spacing of .... cm is needed for grow out test of maize.
2. The sample with genetic purity less than MSCS is ....

♦ Answer Keys ♦

Q.No.	Ans.	Q.No.	Ans.
1	L00	2	Genetic purity
3	Other Distinguishing Varieties	4	10 ODV/kg
5	LV	6	2V
7	Rejected		

4.4.3. State True or False

1. Grow out test should be made as an integrated part of seed testing
2. Genetic purity refers to trueness to type
F. Grow out test is followed for genetic purity assessment of seed corps
L. Grow out test is applicable for notified varieties

V. Trueness to type referred as physical purity

/. Grow out test is must for cotton hybrids

2. Grow out test is practiced to ensure genetic purity

% As on date, genetic purity is conducted as a post control quality test for cotton

-. Standard sample is must for comparison of test samples in grow out test

10. Genetic purity test is the one of the routine tests conducted by STL

♦ Answer Keys ♦

Q.No.	Ans.	Q.No.	Ans.	Q.No.	Ans.	Q.No.	Ans.	Q.No.	Ans.
1	True	2	True	3	True	4	True	5	False
6	True	7	True	8	False	9	True	10	False

4.5. SEED TESTING

4.5.1. Choose the correct answer

1. The sample of seed submitted to a seed testing laboratory is known as
   (a) Original sample (b) Submitted sample
   (c) Test sample (d) Primary sample

2. The optimum germination temperature for a wide range of seeds is between
   (a) 2V and F0°C (b) F0 and FV°C
   (c) 10 and 1V°C (d) FV and L0°C

3. The person responsible for quality control in the seed market channel is
   (a) Seed Inspector (b) Certification Officer
   (c) Seed Analyst (d) Agricultural Officer

4. The Boerner divider works on the principle of
   (a) Gravitational force (b) Centrifugal force
   (c) Magnetic force (d) Electrical force

5. ODV test is mainly conducted for
   (a) Rice (b) Cotton
   (c) Onion (d) Castor

6. Seed testing is the science of evaluating seed for its
   (a) Planting value (b) Vigour
   (c) Contamination of seed lot (d) Health status

7. Nobbe trier is useful in
   (a) Sampling (b) Dividing
   (c) Weighing (d) Purity analysis

8. The submitted sample size for paddy is
   (a) 100 g (b) L0 g
   (c) L00 g (d) 10 g

9. The seed samples sent for analysis by the farmer
   (a) Certified sample (b) Official sample
   (c) Service sample (d) Bulk sample

10. Seed sampling in cotton is done using
   (a) Stick trier (b) Nobbe trier
   (c) Spoon method (d) Hand method

11. In paddy, first and final count days for seed germination test are
   (a) L, 10 (b) L, 21
   (c) L, 2 (d) V, 1L

12. In purity analysis, seeds less than half size is considered as
   (a) Other crop seed (b) Inert matter
   (c) Pure seed (d) Weed seed

13. The RH required in a germination room is
   (a) 20% (b) %V%
   (c) -V± 2% (d) 2V%

14. The working sample size for tomato is
   (a) 10 g (b) 100 g
   (c) L0 g (d) 2 g

15. For estimation of moisture content, the seeds are packed in containers which are
   (a) Moisture pervious (b) Moisture resistant
   (c) Moisture impervious (d) Moisture vapour proof

16. As per ISTA, working sample should contain a minimum of
   (a) 2V00 seeds (b) L000 seeds
   (c) 2V0 seeds (d) L00 seeds

17. The light intensity required for germination of non-dormant seed is
   (a) 1V0 lux (b) LV0 lux
   (c) 2V0 lux (d) FV0 lux

18. Seed testing is to be done under
   (a) Controlled condition (b) Ambient condition
   (c) Cold atmosphere (d) At field condition

19. Germination test results are to be reported in
   (a) One decimal places (b) Whole number
   (c) Two decimals (d) Three decimals

20. Samaras without true seed in purity analysis is known as
   (a) Pure seed (b) Inert matter
   (c) Other crop seed (d) Fruit

21. The submitted sample size of maize for grow out test
   (a) 1000 g (b) V00 g
   (c) 100 g (d) 2V0 g

22. Duties of seed analyst is detailed in
    (a) Section Vof Seed Act (b) Section 12 of Seed Act
    (c) Section 11 of Seed Act (d) Section 2Vof Seed Act

23. At market level, seed quality is monitored by
    (a) Seed Certification Officer (b) Seed Inspector
    (c) Agricultural Officer (d) Seed Analyst

24. The duty of the seed analyst is
    (a) Seed testing (b) Certification
    (c) Seed marketing (d) Seed quality

25. The seed sample taken by seed inspector and submitted to seed testing laboratory
    (a) Certified sample (b) Official sample
    (c) Service sample (d) Primary sample

26. NSRTC is located in
    (a) New Delhi (b) Katrain (HP)
    (c) Kashmir Valley (d) Varanasi

27. Seed testing is carried out on
    (a) Primary sample (b) Working sample
    (c) Submitted sample (d) Composite sample

28. Submitted sample size for maize
    (a) %00 g (b) 1200 g
    (c) 1000 g (d) V00 g

29. Hand sampling is followed in
    (a) Tomato (b) Sorghum
    (c) Maize (d) Paddy

30. Germinated seed is
    (a) Pure seed (b) Inert matter
    (c) Weed seed (d) Other crop seed

31. The submitted sample size for groundnut is
    (a) V00 gm (b) 2V0 gm
    (c) -00 gm (d) 1000 gm

32. ODV test is being done for
    (a) Groundnut (b) Sorghum
    (c) Maize (d) Paddy

33. In seed testing, permitted deviation from the standard is called
    (a) Objection (b) Tolerance
    (c) Deviation (d) Critical difference

34. The temperature maintained at low constant moisture content is
    (a) 10F ± 2°C (b) 11/°C
    (c) 1F0°C (d) 120°C

35. The submitted sample size for amaranthus is
    (a) 20 g (b) 10 g
    (c) 20 g (d) 100 g

36. In a seed sample, the proportion of all nonliving matter, disease and insect infested seeds and broken seeds is called as
    (a) Garbage (b) Genetic purity
    (c) Pure matter (d) Inert matter

37. Half gram is working sample for purity analysis for
    (a) Tomato (b) Onion
    (c) Berseem (d) Tobacco

38. While sampling from a seed lot of 50 bags, the minimum number of primary samples should be
    (a) V (b) 10
    (c) 1V (d) 20

39. The test which determines the percentage of seeds belonging to the variety under certification is called
    (a) TZ test (b) Germination test
    (c) Purity test (d) Vigour test

40. A device that divides a seed lot and puts sub samples directly into a various number of planting envelopes is called as
    (a) Seed divider (b) Stacking machine
    (c) Seiver (d) Differentiation machine

41. Seeds of a particular crop gathered at one time and likely to have similar germination rates and other characteristics are called as
    (a) Variety lot (b) Plant lot
    (c) Field lot (d) Seed lot

42. The process of evaluation of seeds in terms of purity and germination is called
    (a) Seed testing (b) Plant testing
    (c) Propagule testing (d) Plots testing

43. Mix of several primary samples taken from different parts of a seed lot is called as
    (a) Uniform sample (b) Composite sample
    (c) Grade sample (d) Specified sample

44. At the end of the germination test period, neither hard nor fresh seeds that have not produced seedlings, are classified as
    (a) Dead seeds (b) Obsolete seed
    (c) Recalcitrant seed (d) Calcitrant seed

45. The officer appointed by State Government to protect the quality of seed sold in the market is
    (a) Seed Inspectors (b) Seed Analysts
    (c) Seed Certification Officers (d) Seed Testing Officers

46. The seed sample sent for analysis by the seed inspector is
(a) Certified sample (b) Service sample
(c) Official sample (d) Working sample

47. Planting value of the seed is evaluated through
(a) Field inspection (b) Seed testing
(c) Seed inspection (d) Grow out test

48. The RH requirement for germination room is
(a) LV% (b) %0%
(c) -V% (d) 2V%

49. Boerner divider is otherwise called as
(a) Conical divider (b) Gamet divider
(c) Centrifugal divider (d) Soil divider

50. The final count day for cotton is
(a) 10 (b) 12
(c) 1L (d) %

51. The final count for paddy germination test is
(a) 1L days (b) 1/ days
(c) %days (d) 10 days

52. The germination requirement for paddy is
(a) %0 (b) 20
(c) /V (d) -0

53. Generally seed samples meant for testing has to be packed and sent in
(a) Polythene bag (b) Cloth bag
(c) Gunny bag (d) Paper cover

54. Air blower is meant for the purpose of separation of
(a) Broken seeds (b) Inert matter
(c) Soil particles (d) Other crop seeds

55. First seed testing station in Thrandt of Saxony, Germany was started during
(a) 1%/- (b) 1%2/
(c) 1-2L (d) 1-10

56. The trier which cannot be used for seed sampling is
(a) Nobbe trier (b) Bin sampler
(c) Stick trier (d) Theif trier

57. Maximum prescribed seed lot size for paddy is
(a) 10000 kg (b) 1/000 kg
(c) 20000 kg (d) L0000 kg

58. The number of seeds in a working sample size will approximately
(a) 2V00 (b) 2000
(c) F000 (d) 1/00

59. The quantity of seed sample that to be sent for moisture estimation which needs to be ground
    (a) V0 gm (b) 2Vgm
    (c) 100 gm (d) Vgm
60. Submitted sample size for tomato
    (a) F0 gm (b) 20 gm
    (c) 100 gm (d) 1V0 gm
61. ODV stands for
    (a) Other distinguishable variety (b) Optimum density variation
    (c) Other dominant variety (d) Other distant variety
62. The weight of submitted sample is usually
    (a) Vtimes more than the weight of working sample
    (b) 10 times more than the weight of working sample
    (c) 1Vtimes more than the weight of working sample
    (d) 20 times more than the weight of working sample
63. The working sample subject to a minimum of 0.5 g and maximum of 1000 g for quality tests should contain
    (a) 1V00 seeds (b) 2000 seeds
    (c) 2V00 seeds (d) F000 seeds
64. When the first STL was started in the world?
    (a) 1%/ - (b) 1%2-
    (c) 1%% (d) 1%-
65. In which country, the first STL was started?
    (a) USA (b) UK
    (c) Germany (d) Paris
66. Who has started the first STL is the world?
    (a) Franklin Bonner (b) E.H. Roberts
    (c) Woodstick (d) Friedrich Nobbe
67. For germination test, pH of the water should be
    (a) L-V.2 (b) V-/.2
    (c) /-2.2 (d) 2-%2
68. As per ISTA standard how many number of seeds can be used for germination test
    (a) 200 (b) L00
    (c) /00 (d) %00
69. What is the medium used for groundnut germination test?
    (a) Vermiculite (b) Top of paper
    (c) Soil (d) Sand
70. What is the medium used for cluster bean germination test?
    (a) Between paper (b) Top of paper
    (c) Soil (d) Sand

71. What is the medium used for sesame germination test?
(a) B/W paper (b) Top of paper
(c) Soil (d) Sand

72. What is the day to first count for testing germination of cucumber seed?
(a) F (b) L
(c) V (d) /

73. What is the day to first count for testing germination of mustard seed?
(a) F (b) L
(c) V (d) /

74. What is the day to first count for testing germination of guinea grass seed?
(a) 2 (b) %
(c) - (d) 10

75. What is the day to the final count for testing germination of jute seed?
(a) V (b) /
(c) 2 (d) %

76. What is the day to the final count for testing germination of cowpea seed?
(a) V (b) /
(c) 2 (d) %

77. What is the day to the final count for testing germination of parsley seed?
(a) 20 (b) 21
(c) 2/ (d) 2%

78. What is the day to the final count for testing germination of kodo millet?
(a) 20 (b) 21
(c) 2/ (d) 2%

79. In sugar beet, more than one seedling is produced by a seed that is known as multigerm. while counting normal seedlings, it is counted as one seedling or more
(a) One seedling (b) Two seedlings
(c) Three seedlings (d) Four seedlings

80. If the seeds are kept in 1-5 containers, what is the sampling intensity?
(a) Each container, from atleast Vpositions
(b) Each container, from atleast / positions
(c) Not less than Vsamples
(d) Atleast one container in every Vpositions

81. What is the sampling intensity when the seeds are stored in 561 or more containers?
(a) Atleast one container in every five
(b) Not less than 10 containers
(c) Not less than %0 containers
(d) Atleast one sample in every 2 containers

82. In which seed lot size, not less than 10 samples to be taken for seed testing?
(a) V1-1V00 kgs (b) 1V01 – F000 kgs
(c) F001- V000 kgs (d) V001-20000 kgs

83. In which containers, the seeds should be packed and sent to STL for moisture estimation?
(a) Moisture pervious container (b) Moisture proof container
(c) Cloth bag (d) Gunny bag

84. What is the quantity of submitted sample for the species that has to be ground and sent to STL for moisture determination?
(a) 2Vg (b) V0 g
(c) 2Vg (d) 100 g

85. What is quantity of submitted sample for the species that does not need grounding, to be submitted to STL for moisture?
(a) 2Vg (b) V0 g
(c) 2Vg (d) 100 g

86. What is the submitted sample size for guinea grass?
(a) 2Vg (b) V0 g
(c) 2Vg (d) 100 g

87. What is the submitted sample size for lucerne?
(a) 2Vg (b) V0 g
(c) 2Vg (d) 100 g

88. What is the working sample size for groundnut?
(a) 2V0 g (b) V00 g
(c) 2V0 g (d) 1000g

89. What is the working sample size for bottle gourd?
(a) 2V0 g (b) V00 g
(c) 2V0 g (d) 1000 g

90. The temperature used in high constant temperature method for estimation of moisture in rice seed is
(a) 1L0°C (b) 1F0°C
(c) 120°C (d) 110°C

91. A high constant temperature method is used for estimation of seed moisture in
(a) Chickpea (b) Groundnut
(c) Soybean (d) Cotton

92. A low constant temperature method is used for estimation of seed moisture in
(a) Chickpea (b) Peas
(c) Castor (d) Cotton

93. The weight of sample of tomato used for moisture determination in universal moisture meter is
(a) 2Vg (b) F0 g
(c) V0 g (d) /0 g

94. The weight of sample of sunflower used for moisture determination in universal moisture meter is
(a) 2Vg (b) F0 g
(c) V0 g (d) /0 g

95. The weight of sample of soybean used for moisture determination in universal moisture meter is
(a) 2Vg (b) F0 g
(c) V0 g (d) /0 g

96. The volume cup for wheat used in moisture determination in universal moisture meter is
(a) A (b) B
(c) C (d) D

97. The volume cup for oat used in moisture determination in universal moisture meter is
(a) A (b) B
(c) C (d) D

98. The volume cup for okra used in moisture determination in universal moisture meter is
(a) A (b) B
(c) C (d) D

♦ Answer Keys ♦

Q.No.	Ans.	Q.No.	Ans.	Q.No.	Ans.	Q.No.	Ans.	Q.No.	Ans.	Q.No.	Ans.
1	b	2	a	3	a	4	a	5	a	6	a
7	a	8	c	9	c	10	d	11	d	12	b
13	c	14	d	15	d	16	a	17	c	18	a
19	b	20	a	21	a	22	b	23	b	24	a
25	b	26	d	27	b	28	c	29	a	30	a
31	d	32	d	33	b	34	a	35	a	36	d
37	d	38	b	39	c	40	a	41	d	42	a
43	b	44	a	45	a	46	c	47	b	48	c
49	a	50	b	51	a	52	a	53	b	54	b
55	a	56	d	57	c	58	a	59	c	60	b
61	a	62	b	63	c	64	a	65	c	66	d
67	c	68	b	69	d	70	a	71	b	72	b

Q.No.	Ans.	Q.No.	Ans.	Q.No.	Ans.	Q.No.	Ans.	Q.No.	Ans.	Q.No.	Ans.
73	c	74	d	75	a	76	d	77	d	78	a
79	a	80	a	81	d	82	c	83	b	84	d
85	b	86	a	87	b	88	d	89	b	90	b
91	a	92	c	93	a	94	b	95	d	96	a
97	b	98	c								

4.5.2. Fill in the blanks

1. .... is the father of seed testing

2. The sample submitted to the seed testing laboratory is ....

F. The world's first Seed Testing Station was established by .... in Tharandt, Saxony

L. The first Vegetable Seed Testing Station was established in ....

V. Working sample size of paddy is ....

/. Method of sampling for fuzzy seed is ....

2. Hand sampling is followed for .... seeds

% The germination room temperature should be ....

-. The germination % is calculated to the nearest ....

10. The drying temperature for moisture estimation of oilseeds should be ....

11. A minimum of .... seeds are required for germination test

12. Head quarter of ISTA ....

1F. Journal published by ISTA ....

1L. Journal published by AOSA ....

1V. .... publish seed abstract.

1/. Triphenyl formazon is extracted using ....

12. The germination of deep water paddy can be improved by coating with ....

1% Cool germination test is employed in .... crop

1-. Baffle type divider is generally known as ....

20. Cardinal temperature requirement of crop is determined using ....

21. Maximum quantity in a seed lot of wheat is ....

22. Tensile strength for germination paper should be ....

2F. The minimum number of seeds for physical purity analysis is ....

2L. Concentration of mercuric chloride solution used for surface sterilization of seeds is ....

2V. Most susceptible seed to soaking injury is ....

2/. Vigour and germination is predicted by ....

22. Seed emptiness is determined by ....

2% Different seed fractions in purity analysis ...., other crop seeds and inert matter

2-. Different methods used for estimating seed moisture ...., electronic moisture meter and toluene distillation method

F0. In a Seed Testing Laboratory the moisture percentage of a seed is estimated and reported on ....

F1. The working principle involved in Boerner divider is .... force

F2. Differences in seed quality observed within the seed lot is known as ....

FF. Skotoblastic seeds require .... for germination

FL. The formula for speed of germination was formulated by ....

FV. Dr. Maguire has formulated the formula for speed of germination during ....

F/. Temperature and RH requirement in the germination room is ....

F2. The components of purity analysis are ....

F% The working principle involved in Gamet divider is ....

F-. Germination % and purity % are added up together to arrive at ....

L0. The submitted sample size should not be less than .... for maize

L1. The sample taken at one point is called .... sample

L2. In rice, germination counts are made .... after sowing

LF. The personnal responsible for seed quality testing in seed testing laboratory is ....

LL. .... is the method of assessing the planting value of seed

LV. The final count day for paddy is ....

L/. .... test is used to detect the glucose exudates from non viable seed

L2. A common moisture absorbent used in seed testing is ....

L% Damaged, decayed and deformed seedlings are collectively called as .... seedlings

L-. .... is the contrast agent in X-radiography

V0. .... is a mixing and dividing equipment for seeds

V1. The equipments used for seed sampling is known as ....

V2. PLS stands for ....

VF. The size of submitted sample of hybrid tomato is ....

VL. First and second counts for germination test for chilli are .... respectively

VV. EST stands for ....

V/. Medium used for finger millet germination test is ....

V2. Medium used for maize germination test is ....

V% Medium used for red gram germination test is ....

V-. Medium used for safflower germination test is ....

/0. Medium used for guinea grass germination test is ....

/1. Medium used for cole crop germination test is ....

/2. The optimum temperature for germination of crop seed is ....

/F. Pearl millet seed moisture is determined by ....

/L. Mustard seed moisture is determined by ....

/V. Seed lot with moisture content more than the MSCS is recommended for ....

//. Another name for germination is ....

/2. In STL, the physical purity is tested using ....

/% The objectionable weed for paddy is ....

/-. The objectionable weed for wheat is ....

20. The objectionable weed for sunflower is ....

21. The objectionable weed for lucerne is ....

22. The objectionable weed for bitter gourd is ....

2F. The objectionable weed for okra is ....

2L. The objectionable weed for fenugreek is ....

♦ Answer Keys ♦

Q.No.	Ans.	Q.No.	Ans.	Q.No.	Ans.
1	Friedrich Nobbe	2	Submitted sample	3	Friedrich Nobbe
4	IARI, New Delhi	5	L0 g	6	Hand method
7	Fuzzy	8	2V± 1°C	9	Whole number
10	10/°C	11	L00	12	Switzerland
13	Seed Science and Technology	14	Journal of Seed Technology	15	CAB–(Commonwealth Agricultural Bureau)
16	Methyl cellosolve	17	Calcium Peroxide	18	Cotton
19	Boerner divider	20	Two way thermo gradient plate	21	20,000 kg
22	High	23	2V00	24	0.01%
25	Soybean	26	Germination test	27	X-ray test
28	Pure seed	29	Oven method	30	Wet basis
31	Gravitational	32	Heterogeneity	33	Darkness
34	Maguire	35	1-/2	36	2V°C±1°C and - V± 2%
37	Pure seed, other crop seeds and inert matter	38	Centrifugal	39	Pure Live Seed
40	1000 g	41	Primary	42	1L days
43	Seed Analyst	44	Germination test	45	1Lth day
46	Urine sugar analysis paper	47	Silica gel	48	Abnormal

Q.No.	Ans.	Q.No.	Ans.	Q.No.	Ans.
49	Barium chloride	50	Borner divider	51	Triers
52	Pure Live Seed	53	2 g	54	2th and 1Lth day
55	Essentially Sequenced Tags	56	Top of paper	57	Between paper/sand
58	Between paper/sand	59	Top of paper/ between paper	60	Top of paper
61	Top of paper	62	20 to 2V°C	63	High constant temperature
64	Low constant temperature	65	Drying	66	Sprouting
67	Purity work board	68	Wild rice	69	Wild morning glory
70	Wild helianthus	71	Dodder	72	Balsam apple
73	Wild okra	74	Senji		

4.5.3. State True or False

1. Seed testing is the method of assessing the planting value of the seed
2. Physical purity refers to trueness to type
F. Working sample is derived from primary sample
L. Standard sample is must for comparison of test samples in grow out test
V. The person who is analyzing the seed sample is called seed analyst
/. Working sample will be always 1/V of the submitted sample
2. Maximum size (quantity) of seed lot for maize will be 20000 kgs
% The mixed version of the primary sample is called as submitted sample
-. The sample drawn and sent by the seed certification officer is called as official sample
10. Delinted cotton seeds will come under the category of free flowing sample
11. Estimation of moisture content by oven method is inaccurate method
12. As the moisture content in the seed increases resistant decreases
1F. Seeds of most tropical crops are more specific to germination conditions than the temperate ones
1L. For good germination, the moisture holding capacity of the medium should be more than /0%
1V. Referee testing results can also be questioned
1/. The minimum limit of germination prescribed for cotton is /V%
12. A seed lot is a physically identifiable quantity of seed which is homogenous
1% The chemical used for testing the viability of seed is Selenium chloride
1-. In a moisture meter, the seed moisture content and resistance are inversely proportional
20. Too much of intensive mixing may separate more than its blends

21. Groundnut seed without seed coat is an inert matter
22. The minimum number of seeds used for germination test in a seed testing lab is L00 for all crops
2F. Moisture estimation by electronic moisture meter is more accurate than the conventional hot air oven method
2L. Seeds containing volatile oils are tested using low constant temperature method for moisture estimation
2V. More than two lots can be blended using Pearson square law method
2/. The maximum size of seed lot for maize is L0,000 kg
22. The working sample contains minimum of FV00 seeds
2% The moisture content of the seed is directly proportional to EC
2-. The final count days for paddy is 12 days
F0. Broken seeds exactly half is considered as a pure seed
F1. The germination standard fixed for hybrid cotton is /0%
F2. Trueness to type referred as physical purity
FF. For paddy, the submitted sample size is - 00 gm
FL. Centrifugal force is the working principle of Boerner divider
FV. A separate seed sample in polythene bag is necessary for moisture testing
F/. Triers are used to take primary samples of cotton seeds
F2. Physical purity is not essential for a quality seed
F% A sample formed by combining and mixing all the primary samples taken from the lot is working sample
F-. Gamet divider makes use of centrifugal force to scatter the seeds over the dividing surface
L0. Fluff is the seed inert matter
L1. In seed testing, twisted seedlings are considered as normal seedlings
L2. Kari fisher reagent method is used to determine the moisture content of seeds
LF. For moisture estimation of seeds, the commonly used method is freezing method

♦ Answer Keys ♦

Q.No.	Ans.	Q.No.	Ans.	Q.No.	Ans.	Q.No.	Ans.	Q.No.	Ans.	Q.No.	Ans.
1	True	2	False	3	False	4	True	5	True	6	False
7	False	8	False	9	False	10	True	11	False	12	True
13	False	14	False	15	False	16	True	17	True	18	False
19	True	20	True	21	True	22	True	23	False	24	True
25	True	26	False	27	False	28	True	29	False	30	True
31	False	32	False	33	False	34	False	35	True	36	False
37	False	38	False	39	True	40	False	41	False	42	False
43	False										

4.6. QUICK VIABILITY TEST

4.6.1. Choose the best answer

1. The enzyme associated with tetrazolium test is
   (a) Catalase (b) Super oxide dismutase
   (c) Dehydrogenase (d) Amylases
2. TZ test was not a suitable method for seed vigour estimation
   (a) Woodstock (b) Moore
   (c) Vanderlip (d) Perry
3. Chemical used in quick viability test is
   (a) Potassium chloride
   (b) Gibberellic acid
   (c) Sodium chloride
   (d) 2,F-VTriphenyl tetrazolium chloride
4. In TZ test, colour of living tissues of seed changed to
   (a) Yellow (b) Green
   (c) Red (d) Blue
5. Topographical tetrazolium test is
   (a) Mechanical injury test (b) Quick viability test
   (c) Vigour test (d) Stress test
6. Tetrazolium test was first formulated by
   (a) Lakon, 1-L2 (b) Roberts, 1-/-
   (c) Hiltner 1-/- (d) Woodstock, 1-2/
7. Quantification of formazan is done for measuring the activity of
   (a) GADA (b) Amylase
   (c) Peroxidase (d) Dehydrogenase
8. The enzyme involved in the red colour staining in living tissue of seed is
   (a) Amylase (b) Peroxidase
   (c) Catalase (d) Dehydrogenase
9. How many replications are used for tetrazolium testing
   (a) 2 (b) L
   (c) / (d) %
10. How many seeds are used for one replication while testing viability/ tetrazolium test?
    (a) V (b) 100
    (c) 1V0 (d) 200
11. The reduction reaction due to dehydrogenase enzymes changes the colourless solution into
    (a) Remains colourless (b) No colour change
    (c) Red colour (d) Mottled colour

12. The pH of the tetrazolium chloride should be

(a) /.V-2.V (b) V.V.../.V

(c) L.V-V.V. (d) F.V-L.V

♦ Answer Keys ♦

Q.No.	Ans.	Q.No.	Ans.	Q.No.	Ans.	Q.No.	Ans.	Q.No.	Ans.	Q.No.	Ans.
1	c	2	c	3	d	4	c	5	b	6	a
7	d	8	d	9	b	10	b	11	c	12	a

4.6.2. Fill in the blanks

1. Quick viability test is .... test
2. .... is used for determining the viability of seeds very fast and quick

F. .... chemical is used to conduct quick viability test

L. Tetrazolium test is a .... test

V. Areas of viable and non viable seeds in individual seed is determined by ....

/. The chemical name of tetrazolium is ....

2. In tetrazolium test, a reduction reaction takes place and .... released.

% The colourless solution of 2,F,V-triphenyl tetrazolium reacts with dehydrogenases and releases hydrogen by .... process.

-. During tetrazolium test, the colourless solution changed to a red, stable and non defusable substance named as ....

10. In tetrazolium staining, the reduction process takes place in .... tissues

♦ Answer Keys ♦

Q.No.	Ans.	Q.No.	Ans.
1	TZ	2	Tetrazolium test
3	2,F,VTriphenyl tetrazolium chloride	4	Quick viability
5	TZ test	6	2,F,V-triphenyl tetrazolium chloride
7	Hydrogen	8	Reduction
9	Triphenyl formazan	10	Living

4.6.3. State True or False

1. The scientist Lakon proved the possibility of TZ test for vigour estimation
2. Paddy kernel lacking aleurone are more suitable for viability testing

F. TZ test will indicate the viability potential of both dormant and non-dormant seeds

L. TZ bromide/chloride is the salt used in quick viability test

V. Final product of TZ test is Triphenyl Formazan

/. The enzyme associated with TZ test is catalase

♦ Answer Keys ♦

Q.No.	Ans.	Q.No.	Ans.	Q.No.	Ans.
1	True	2	False	3	True
4	False	5	True	6	False

4.7. SEED VIGOUR

4.7.1. Choose the correct answer

1. The ability of the seed to perform under unfavourable condition is
   (a) Seed vigour (b) Seed viability
   (c) Seed dormancy (d) Seed senescence
2. Brick gravel test is a
   (a) Vigour test (b) Viability test
   (c) Varietal identification test (d) Quick viability test
3. Cold test is considered as best method of estimating vigour in
   (a) Corn (b) Paddy
   (c) Wheat (d) Soybean
4. The scientists closely associated with vigour index calculation is
   (a) Abdul-Baki and Anderson (b) Shrivastva and Sareen
   (c) Roberts (d) Ellies
5. Cold test was recommended for
   (a) Pulses (b) Corn
   (c) Pea (d) Wheat
6. ISTA seed vigour definition was suggested by
   (a) Perry (b) Ihseen
   (c) Pollock (d) Neels
7. Degree of aliveness of seed is called as
   (a) Seed vigour (b) Hybrid vigour
   (c) Viability (d) Heterosis
8. The chemicals used for osmotic stress test
   (a) KH_2PO_4 (b) Jalasakthi
   (c) Glycerol & Sucrose (d) PEG & Mannitol
9. The size of the brick in the brickgravel test is
   (a) Fmm (b) F.Vmm
   (c) 2.Vmm (d) Lmm
10. The concept of seed vigour originally given by .... in 1911
    (a) Hiltner Inssen (b) Perry
    (c) Isley (d) Moore
11. Who suggested X-ray radiography test for vigour analysis
    (a) Simek and Gustafsson (b) Hasegowa
    (c) Reddy (d) Stathio

12. Cold test was recommended for corn in the year
(a) 1-/-
(b) 1-L0
(c) 1-2V
(d) 1-2-

13. Stress and funk test for vigour estimation was suggested by
(a) Mc Donald
(b) Verhay
(c) Pollock and Roos
(d) Delouche and Caldmeu

14. The scientist closely associated with speed of germination is
(a) Maguire
(b) Woodstock
(c) Roberts
(d) Ellies

15. Accelerated ageing technique was identified as vigour test by
(a) Fritz, 1-/V
(b) Roberts, 1-/-
(c) Hiltner, 1-/-
(d) Woodstock, 1-2/

16. The quality of the seed responsible for rapid and uniform germination, extended storability and good field emergence is called
(a) Plant vigour
(b) Field vigour
(c) Seed vigour
(d) Plantlets vigour

17. Progression of cytological and biochemical events which ultimately leads to the death of the seed is defined as
(a) Ageing
(b) Activation
(c) Germination
(d) Death progression

18. Vigour test method in which the ability of germination to overcome physical stress is evaluated by germinating seeds under a layer of crushed brick stone gravel is called as
(a) Gardner test
(b) Hiltner test
(c) Vilmorin test
(d) Health test

19. Which of the following does not come under direct vigor test?
(a) Brick gravel test
(b) Paper piercing test
(c) Cold test
(d) Exhaustion test

20. The term rate or speed of germination for vigour test is proposed by?
(a) Maguire
(b) Stahi
(c) Roberts
(d) Ellies

21. Paper piercing technique was identified as vigour test by
(a) Fritz, 1-/V
(b) Roberts, 1-/-
(c) Hiltner, 1-/-
(d) Woodstock, 1-//

22. The instrument used to identify the respiration rate of seeds is
(a) Nanometer
(b) Psychrometer
(c) Lactometer
(d) Hygrometer

23. In which crop, the cold test is performed
(a) Cotton
(b) Wheat
(c) Corn
(d) Rice

24. Which one of the following is a physical test of seed vigour?
    (a) Brick gravel test (b) Seed size
    (c) First count (d) Tetrazolium test
25. Which one of the following is a stress test?
    (a) Brick gravel test (b) Seed size
    (c) First count (d) Tetrazolium test
26. Which one of the following is the biochemical test?
    (a) Brick gravel test (b) Seed size
    (c) First count (d) Tetrazolium test
27. Which one of the following is the physiological test?
    (a) Brick gravel test (b) Seed size
    (c) First count (d) Tetrazolium test
28. The scientist who invented paper piercing test is
    (a) Delouche (b) Fritz
    (c) Hiltner (d) Sareen
29. In pathogen infected soil test, which one of the pathogen is used
    (a) *Phythium* (b) *Penecillium*
    (c) *Aspergillus* (d) *Rhizophus*
30. For bick gravel test, the size of granules used is
    (a) 1 -F mm (b) 2 -F mm
    (c) F -L mm (d) L -V mm
31. For testing vigour using osmotic stress, the chemical used is
    (a) NaCl (b) Jalasakthi
    (c) Mannitol (d) Magnesium carbonate
32. Which one the following scientist is responsible for TZ test?
    (a) Moore (b) Woodstock
    (c) Roberts (d) Ellis

♦ Answer Keys ♦

Q.No.	Ans.	Q.No.	Ans.	Q.No.	Ans.	Q.No.	Ans.	Q.No.	Ans.	Q.No.	Ans.
1	a	2	a	3	a	4	a	5	b	6	a
7	c	8	d	9	a	10	a	11	a	12	b
13	c	14	a	15	d	16	c	17	a	18	b
19	d	20	a	21	a	22	a	23	c	24	b
25	a	26	d	27	c	28	b	29	a	30	b
31	c	32	a								

4.7.2. Fill in the blanks

1. Controlled deterioration test for seed vigour is evolved by ....
2. Cold test is an accepted vigour test for .... crop
F. Seed vigour committee of .... formulates the rules and regulations for seed vigour testing

L. An example for stress vigour test is ....

V. Hiltner, 1-11 is associated with .... test

/. Cold test was recommended for ....

2. The RQ is the ratio of .... evolved per unit time to the amount of O_2 consumed.

% GADA expands ....

-. For accelerated ageing the seeds are subjected to extreme temperature of ....°C and RH of - V±V%

10. SME stands for ....

11. ME stands for ....

12. Brick gravel test is otherwise called as .... test.

♦ Answer Keys ♦

Q.No.	Ans.	Q.No.	Ans.	Q.No.	Ans.
1	Mathews	2	Maize	3	ISTA
4	Brick gravel test	5	Brick gravel	6	Corn
7	CO_2	8	Glutamic acid decarboxylase activity	9	L0±1
10	Seed metabolic efficiency	11	Mobilization efficiency	12	Hiltner

4.7.3. State True or False

1. Vigour test is the routine test conducted in the seed testing laboratory

2. Seed vigour is correlated with field emergence

F. Seed vigour is physiological stamina of seed

L. Vigour test is more sensitive index of seed quality than germination test

V. Seed vigour is a single measurable attribute

/. Seed vigour test is complementary and not a substitute to standard germination test

2. Sum total of all seed attributes that favour stand establishment under varying field condition is seed vigour

% Speed of germination is physiological vigour test

-. In accelerated ageing test, low temperature and low relative humidity is used for test the vigour

10. Paper piercing technique was suggested as vigour test

11. A test to determine field emergence potential is brick gravel

12. Manometer is used to identify the respiration of seed

1F. Roberts proposed brick gravel method as vigour test

1L. KOH and sulphuric acid are used to create the required RH in accelerated ageing test

♦ Answer Keys ♦

Q.No.	Ans.	Q.No.	Ans.	Q.No.	Ans.	Q.No.	Ans.	Q.No.	Ans.	Q.No.	Ans.
1	False	2	True	3	True	4	True	5	False	6	True
7	True	8	True	9	False	10	True	11	True	12	False
13	False	14	True								

4.8. VARIETAL IDENTIFICATION

4.8.1. Choose the correct answer

1. A hybridization based marker is
   (a) RFLP (b) ISSR
   (c) SSR (d) SDS
2. More number of targets can be analyzed simultaneously in
   (a) RT-PCR (b) Multiplex PCR
   (c) Southern blot technique (d) ELISA
3. Proteins can be visualized by
   (a) Ethidium Bromide (b) Silver staining
   (c) Acridine orange (d) Acetocarmine
4. Temperature probe is an essential component of
   (a) Centrifuge (b) Autoclave
   (c) Vortex shaker (d) pH meter
5. Which one of the following method cannot be used for quantitative assay of transgenes?
   (a) RT-PCR (b) Southern blot technique
   (c) PCR (d) Q-PCR
6. Which of the following does not involve radio activity?
   (a) AFLP (b) ELISA
   (c) PCR (d) SDS
7. Individual loci controlling a complex trait is known as
   (a) Qualitative Trait Locus (b) Quantitative Trait Locus
   (c) Alleles (d) Single locus
8. Which one of the following is used for amplifying DNA samples?
   (a) Gel documentation (b) PCR
   (c) Speed vac (d) Thermo mixture comfort
9. Role of mercaptoethanol in DNA finger printing technology is
   (a) To break cell well
   (b) To distrupt cell membrane
   (c) To prevent phenolic oxidation
   (d) To prevent lipid perioxidation
10. TE stands for
    (a) Tri ethylene (b) Tris ethylene
    (c) Tris EDTA (d) Triester

11. pH stands for
    (a) Power of helium (b) Power of hydrosylion
    (c) Power of hydrogen (d) Power of hexane
12. For DNA quantification, the equipment used is
    (a) Nano drop (b) Thermo mixture
    (c) Speed vac (d) PCR
13. The solution in the electrode of pH meter is
    (a) KNO_F (b) KCl
    (c) Mercury (d) Methane
14. The term TDS stands for
    (a) Total digested solids (b) Total digested salts
    (c) Total dissolved solids (d) Total dissolved salts
15. The extraction buffer used in analysis of seed protein by PAGE method is
    (a) Tris HCl (b) SDS
    (c) Mercapto ethanol (d) Tri ethylene
16. Temperature probe is an essential component of
    (a) Centrifuge (b) Autoclave
    (c) Vortex shaker (d) pH meter
17. For evaporating liquid from DNA samples, the following equipment is used
    (a) Speed vac (b) Water bath
    (c) Microwave oven (d) Nano drop
18. For DNA extraction, the suitable method is
    (a) Liquid nitrogen (b) cTAB method
    (c) Minipep method (d) Nano drop
19. Easy method for detection of GM crop is
    (a) ELISA method (b) Strip test method
    (c) Finger printing method (d) Visual method
20. ELISA plates are coated with
    (a) Enzymes (b) Lipids
    (c) Antibody (d) Antigen
21. Dongle key is an essential part of
    (a) Gel documentation unit (b) Fume hood
    (c) Magnetic stirrer (d) Incubator
22. The specific test recommended for varietal identification in sorghum is
    (a) KOH bleach test (b) Phenol test
    (c) Fluorescent test (d) Vigour test
23. Phenol test is recommended for varietal identification of
    (a) Paddy (b) Groundnut
    (c) Blackgram (d) Brinjal

24. The primary products of genes specific to each variety
(a) Proteins (b) Enzymes
(c) Lipids (d) Carbohydrate

25. The seed appendages are useful for
(a) Varietal identification (b) Seed dispersal
(c) Seed development (d) Seed maturation

26. Phenol color test is used for identification of
(a) Purity analysis (b) Mechanical damage
(c) Varietal difference (d) Seed health

27. The chemical used in modified phenol test
(a) $CuSO_4$ (b) NaOH
(c) KOH (d) KCl

28. Seed coat peroxidase test is mainly used for
(a) Soybean (b) Maize
(c) Cumbu (d) Paddy

29. KOH bleach test is recommended for
(a) Sorghum (b) Maize
(c) Cumbu (d) Paddy

30. Garden/Fodder pea can be easily distinguished by
(a) Potassium dichromate test (b) $FeSO_4$
(c) K_2SO_4 (d) $MnSO_4$

31. The specific test used for varietal identification of *Lupinus* spp
(a) Alkaloid test (b) Tanin test
(c) KOH test (d) $FeSO_4$

32. The scientist first evolved the phenol test is
(a) Piper (1-20) (b) Walls (1-/V)
(c) Roberts (1-2%) (d) Lakon (1-LF)

33. Seedling illumination test is used for varietal identification of
(a) Oats (b) Wheat
(c) Sorghum (d) Bajra

34. Hypocotyl colour can be used for varietal identification of
(a) Beet root (b) Sugar beet
(c) Carrot (d) Turnip

35. In hybrid cotton, the stem pigmentation is governed by
(a) R gene (b) Poly gene
(c) Single gene (d) Multiple genes

36. Root length inhibition takes place in
(a) 2, L-D test (b) NaOH test
(c) KOH test (d) TZ test

37. Guaiacol solution is used for
(a) Peroxidase test (b) KOH test
(c) $FeSO_4$ test (d) Phenol test

38. The enzyme involved in phenol test is
(a) Tgrosinase (b) Peroxidase
(c) Catalase (d) Reductase

39. Image analysis of seed for varietal identification is based on
(a) Morphological features (b) Physiological features
(c) Biochemical characteristics of seeds
(d) Histological features of seeds

40. KOH chemical test is useful for
(a) Varietal identification (b) Viability testing
(c) Assessing vigour potential (d) Mechanical damage

41. The name of the protein, which is much useful in varietal identification is called as
(a) Tocophenol (b) Prolamine
(c) ABA (d) Auxin

42. KOH bleach test is applied for
(a) Varietal identification (b) Improving germination
(c) Improving vigour (d) Improving viability

43. Electrophoresis is a test for the determination of
(a) Storage potential (b) Age of the seeds
(c) Varietal identification (d) Viability of seed

44. A hybridization based marker is
(a) RFLP (b) ISSR
(c) SSR (d) SNP

45. In GM crops, many number of the transgenes can be analyzed by
(a) Isoenzyme analysis (b) Urea PAGE
(c) Agarose gel (d) Q-PCR

46. The function of cTAB is to
(a) Distrupt the cell membrane (b) Separate proteins
(c) Separate RNA (d) Prevent phenolic oxidation

47. The methods which reveal genetic variability in protein composition are
(a) Electrophoresis (b) HPLC and Electrophoresis
(c) Peroxidase test (d) Peroxidase test and HPLC

48. The biochemical test which is used to distinguish the varieties of soybean is
(a) Peroxidase test (b) Phenol test
(c) Sodium hydroxide test (d) Phenol & Sodium hydroxide test

49. The genuinity of the seed marketed by the private seed dealers is checked by
(a) Seed law enforcement officers (b) Seed inspectors
(c) Seed certification officers (d) Assistant Director of seed certification

50. What is the final count day for germination test in legumes?
(a) V (b) 2
(c) 10 (d) 12

51. The characters which are influenced by environmental factors
(a) Biochemical (b) Morphological
(c) Protein (d) DNA

52. Chemical test used for varietal identification of rice is
(a) Phenol test (b) Modified phenol test
(c) Peroxidase test (d) KOH bleach test

53. Seed coat peroxidase test is used for varietal identification of
(a) Wheat (b) Sorghum
(c) Soybean (d) Peas

54. Presence or absence of tannic acid in sorghum can be identified by
(a) Phenol test (b) KOH bleach test
(c) Modified phenol test (d) Alkaloid test

55. Identification of red rice varieties by
(a) Phenol test (b) KOH test
(c) Peroxidase test (d) Alkaloid test

56. Chemical used for intensification of coleoptile colour in cereals is
(a) NaCl (b) KNO_3
(c) KCl (d) Phenol

57. What is the concentration of phenol used for variety testing of wheat?
(a) 0.V% (b) 1.0%
(c) 1.V% (d) 2.0%

58. Which one of the following crops is exposed to the UV light for variety testing?
(a) Wheat (b) Barely
(c) Triticale (d) Oat

59. For identification of red rice seeds by KOH test, the concentration used is
(a) 0.V% (b) 1.0%
(c) 1.V% (d) 2.0%

60. Which pigmentation is verified for identification of seedling?
(a) Chlorophyll (b) Xanthophylls
(c) Anthocyanin (d) Fluorescence

61. Which one of the following crop is tested for varietal identification using ammonia fluorescent test
(a) *Festuca rubra* (b) *Lollium perenne*
(c) *Melilotus alba* (d) *Melilotus officinalis*

62. Which one of the following crop is tested for varietal identification using $CuSO_4$ ammonia fluorescent test
    (a) *Festuca rubra* (b) *Lolium perenne*
    (c) *Melilotus alba* (d) *Oriyza sativa var fatua*
63. Which one of the following crop is tested for varietal identification using KOH test
    (a) *Festuca rubra* (b) *Lollium perenne*
    (c) *Melilotus alba* (d) *Oriyza sativa* var *fatua*

♦ Answer Keys ♦

Q.No.	Ans.	Q.No.	Ans.	Q.No.	Ans.	Q.No.	Ans.	Q.No.	Ans.	Q.No.	Ans.
1	a	2	b	3	b	4	d	5	c	6	c
7	b	8	b	9	c	10	c	11	c	12	c
13	c	14	c	15	a	16	d	17	a	18	b
19	b	20	c	21	b	22	a	23	a	24	a
25	a	26	c	27	a	28	a	29	a	30	a
31	a	32	a	33	a	34	a	35	a	36	a
37	a	38	a	39	a	40	a	41	b	42	a
43	c	44	a	45	d	46	a	47	b	48	a
49	b	50	b	51	b	52	a	53	c	54	b
55	b	56	a	57	b	58	d	59	d	60	c
61	a	62	c	63	d						

4.8.2. Fill in the blanks

1. Cytological test is observed in .... crop for varietal identification
2. Peroxidase test is used in the identification of .... (varietal identification)

F. Pigment formed during phenol test ....

L. PCR stands for ....

V. Varietal identification is used to test the .... of the variety.

/. .... and .... characters are used for varietal identification.

2. .... and .... have been extensively used in the characterization of crop varieties.

% .... and .... are used to determine the varietal characters of groundnut.

-. .... is used for verification of oat varieties at seed level.

♦ Answer Keys ♦

Q.No.	Ans.	Q.No.	Ans.	Q.No.	Ans.
1	Cereal	2	Soybean	3	Melanin
4	Polymerase Chain Reaction	5	Genuinity	6	Quantitative qualitative
7	Protein isozymes	8	KOH bleach test modified phenol test	9	Fluorescence test

4.8.3. State True or False

1. Isozymes are biochemical markers which are not influenced by the environment

2. 2D-PAGE analysis cannot be used to study Post Translational modification of proteins

F. Interval analysis for QTL mapping needs a genetic map of mapping population

L. SDS is otherwise known as sodium dodecyl sulphate

V. In SDS-PAGE analysis, proteins will be separated by both charge and molecular weight

/. DNA fingerprinting can be done by ELISA

2. For defating procedure in the electrophoresis method, chloroform is used

% The electrodes of the pH meter should be immersed in distilled water when not in use

-. RNAse is added during the process of DNA extraction for estimation of RNA content in the plant sample

10. Gel documentation unit is used for calculating the time taken for running the gel

11. The extracted DNA from the plant sample is negatively charged

12. The electrode buffer solution should be prepared freshly each time

1F. The Rm of the protein band denotes the molar weight of the protein bands

1L. For DNA quantification, the best method used is agarose gel method

1V. Tris E is otherwise known as Tris EDTA

1/. Ethylene can be used as a defeating reagent in the electrophoresis method

12. The property of molecular movement based on size and net electrical charge utilized in electrophoresis

1% Phenol colour reaction is widely used for varietal identification of wheat varieties

1-. Protein with higher charge density migrate slower than the protein with lower charge density

20. Sodium hydroxide test is recommended by ISTA for varietal identification

21. The phenol colour reaction is an index of poly phenol oxidase activity

22. Varietal identification based on seed physiomorphic characters are more dependable than seedling or floral characters

2F. Plant height is a definite character for varietal identification

2L. Chromosome counts method is used for the varietal identification

2V. Electrophoresis is used for the varietal identification

2/. DNA finger printing is not used for the varietal identification

22. GLC and HPLC are used for the varietal identification

2% Nucleic acid analysis is used for the varietal identification

2-. Hypocotyl colour can be used for varietal identification of sugar beet and beet root

F0. RFLP is a hybridization based marker

F1. In GM crops, many number of the transgene (s) can be analyzed by Q-PCR

F2. Varietal identification is pre-requisite for establishing plant breeders right

FF. Image analysis technique is one of the varietal identification test

♦ Answer Keys ♦

Q.No.	Ans.	Q.No.	Ans.	Q.No.	Ans.	Q.No.	Ans.	Q.No.	Ans.	Q.No.	Ans.
1	False	2	False	3	True	4	True	5	False	6	False
7	True	8	False	9	False	10	False	11	True	12	True
13	False	14	True	15	True	16	False	17	True	18	True
19	True	20	True	21	True	22	False	23	False	24	True
25	True	26	False	27	True	28	True	29	True	30	True
31	True	32	True	33	True						

4.9. ISTA

4.9.1. Choose the correct answer

1. The head quarters of ISTA is located in
   (a) Zurich (Switzerland) (b) Connecticut (USA)
   (c) Thavandt (Germany) (d) Glasgow (UK)

2. Orange International Certificate given by ISTA refers to
   (a) Seed sample certificate
   (b) Seed lot certificate
   (c) No linkage between seed sample and seed lot
   (d) Phytosanitary certificate

3. The blue certificate issued by ISTA is applicable for
   (a) Seed sample
   (b) Seed lot
   (c) Linkage between seed sample and seed lot
   (d) Phytosanitary certificate

4. When ISTA was established?
   (a) 1-0L (b) 1-1L
   (c) 1-2L (d) 1-FL

5. AOSA is involved in standardization of seed testing procedures for
   (a) USA and UK (b) USA and Mexico
   (c) USA and Canada (d) USA and Germany

6. Where UPOV is located?
   (a) Germany (b) Switzerland
   (c) Paris (d) UK
7. Where OECD is established?
   (a) Germany (b) Switzerland
   (c) Paris (d) France

♦ Answer Keys ♦

Q.No.	Ans.	Q.No.	Ans.	Q.No.	Ans.	Q.No.	Ans.
1	a	2	b	3	a	4	c
5	c	6	a	7	d		

4.9.2. Fill in the blanks

1. Most important committee of ISTA is .... Committee (2 elected ISTA members)
2. ISTA issues .... types of certificate.
F. ISTA issues blue, orange and .... certificates.
L. The green certificate issued by ISTA is applicable for ....
V. ISTA stands for ....
/. AOSA stands for ....
2. UPOV stands for ....
% OECD expands for ....

♦ Answer Keys ♦

Q.No.	Ans.	Q.No.	Ans.	Q.No.	Ans.
1	Executive	2	Three	3	Green
4	Seed lot	5	International Seed Testing Association	6	Association of Official Seed Analysts
7	International Union for the Protection of New Varieties of Plants	8	Organization for Economic Cooperation and Development		

4.9.3. State True or False

1. ISTA has formed during the year 1 - 2L
2. Accredited stations of ISTA alone are empowered to issue blue tag for international trade
F. The main objective of ISTA is to facilitate international movement of certified seeds

♦ Answer Keys ♦

Q.No.	Ans.	Q.No.	Ans.	Q.No.	Ans.
1	True	2	False	3	False

4.10. MATCH THE FOLLOWING

1. Question

1.	Physical purity	A.	Assistant Director of Seed Certification
2.	Breeder seed	B.	Not < V primary samples
F.	Field inspection	C.	Hand sampling
L.	Certified seed registration	D.	L00 seeds
V.	Sowing report	E.	Purity work board
/.	Off-type	F.	Seed Certification Officer
2.	Upto V containers	G.	Genuineness of varieties
%	Fuzzy seeds	H.	Exempted from certification
-.	Germination test	I.	Plant differs in morphological characters
10.	Electrophoresis	J.	Reach 1V days before flowering

♦ Answer Keys ♦

Q.No.	Ans.	Q.No.	Ans.	Q.No.	Ans.	Q.No.	Ans.	Q.No.	Ans.	Q.No.	Ans.
1	E	2	H	3	F	4	A	5	J	6	I
7	B	8	C	9	D	10	G				

2. Question

1.	Seed Act	A.	1-%%
2.	Seed Inspector	B.	Hybrids
F.	Seed Analyst	C.	Section V of Seed Act
L.	Seed Control Order	D.	Self pollinated crops
V.	New seed policy	E.	Section 1F of the Seed Act
/.	Notification	F.	Cross pollinated crops
2.	Four inspections	G.	1-//
%	Two inspections	H.	1000 heads
-.	Three inspections	I.	Section 12 of the Seed Act
10.	Paddy	J.	1-%F

♦ Answer Keys ♦

Q.No.	Ans.	Q.No.	Ans.	Q.No.	Ans.	Q.No.	Ans.	Q.No.	Ans.	Q.No.	Ans.
1	G	2	E	3	I	4	J	5	A	6	C
7	B	8	D	9	F	10	H				

3. Question

1.	L.F.R	A.	International Seed Testing Association
2.	Bengalgram	B.	/ months
F.	Brinjal	C.	Genetic purity

L.	ODV	D.	Liable for Rejection
V.	ISTA	E.	Vcounts
/.	Validity period	F.	V00 plants
2.	Re-validation period	G.	- counts
%	Grow Out Test	H.	100 plants
-.	Upto V acres	I.	- months
10.	21-2V acres	J.	Other distinguishable variety

♦ Answer Keys ♦

Q.No.	Ans.	Q.No.	Ans.	Q.No.	Ans.	Q.No.	Ans.	Q.No.	Ans.	Q.No.	Ans.
1	D	2	F	3	H	4	J	5	A	6	I
7	B	8	C	9	E	10	G				

4. Question

1.	Moisture estimation	A.	Triers
2.	NSC	B.	Constant temperature for germination
F.	ISTA	C.	High constant temperature method
L.	Free flowing seeds	D.	Air oven method
V.	Friedrich Nobbe	E.	Alternate temperature for germination
/.	Tropical seeds	F.	Low constant temperature method
2.	Temperate seeds	G.	National Seed Corporation
%	Soybean moisture estimation	H.	Minimum of 2V00 seeds
-.	Wheat moisture estimation	I.	1-2L
10.	Working sample	J.	Father of Seed Testing

♦ Answer Keys ♦

Q.No.	Ans.	Q.No.	Ans.	Q.No.	Ans.	Q.No.	Ans.	Q.No.	Ans.	Q.No.	Ans.
1	D	2	G	3	I	4	A	5	J	6	B
7	E	8	F	9	C	10	H				

5. Question

1.	Phenol colour test	A.	Hiltner
2.	Maize	B.	Woodstock
F.	Brick gravel test	C.	Switzerland
L.	R/Q ratio	D.	France
V.	Accelerated ageing test	E.	First Seed Testing Station

/.	Molecular test	F.	Varietal identification
2.	ISTA	G.	Varanasi
%	UPOV	H.	Biochemical vigour test
-.	Tharandt	I.	Cold test
10.	NSRTC	J.	Electrophoresis

♦ Answer Keys ♦

Q.No.	Ans.	Q.No.	Ans.	Q.No.	Ans.	Q.No.	Ans.	Q.No.	Ans.	Q.No.	Ans.
1	F	2	I	3	A	4	H	5	B	6	J
7	C	8	D	9	E	10	G				

6. Question

1.	Breeder seed	A.	Opal green
2.	Foundation seed	B.	1-21
F.	Certified seed	C.	1-2V
L.	TFL seeds	D.	200L
V.	TDC	E.	Golden yellow
/.	NSC	F.	Mau
2.	ISST	G.	Azure blue
%	NSP	H.	1-/-
-.	New Seed Bill	I.	1-/F
10.	Directorate of Seed Research	J.	White

♦ Answer Keys ♦

Q.No.	Ans.	Q.No.	Ans.	Q.No.	Ans.	Q.No.	Ans.	Q.No.	Ans.	Q.No.	Ans.
1	E	2	J	3	G	4	A	5	H	6	I
7	B	8	C	9	D	10	F				

7. Question

1.	Sample for moisture estimation	A.	Centrifugal force
2.	Submitted sample	B.	Physiological vigour test
F.	Germination	C.	TZ test
L.	Boerner divider	D.	Stress test
V.	Gamete divider	E.	Polythene bag
/.	Soil divider	F.	Red formazan
2.	Seedling length test	G.	Cloth bag
%	Paper piercing test	H.	Gravitational force
-.	Quick viability test	I.	2V°C + - V% RH
10.	Tetrazolium test	J.	Non free flowing seeds

♦ Answer Keys ♦

Q.No.	Ans.	Q.No.	Ans.	Q.No.	Ans.	Q.No.	Ans.	Q.No.	Ans.	Q.No.	Ans.
1	E	2	G	3	I	4	H	5	A	6	J
7	B	8	D	9	C	10	F				

8. Question

1.	Grow out test	A.	Planting value
2.	NaCl test	B.	Species that require grinding
F.	KOH test	C.	*Osmaticum*
L.	Fast green test	D.	Fresh ungerminated seed
V.	Germination test	E.	Included in normal seedlings
/.	Moisture estimation sample size V0 g	F.	Assessment of genetic contamination
2.	Moisture estimation sample size 100 g	G.	Stress test
%	PEG	H.	Species that does not need grinding
-.	Dormant seed	I.	Varietal identification
10.	Hard seed	J.	Mechanical damagc

♦ Answer Keys ♦

Q.No.	Ans.	Q.No.	Ans.	Q.No.	Ans.	Q.No.	Ans.	Q.No.	Ans.	Q.No.	Ans.
1	F	2	G	3	I	4	J	5	A	6	H
7	B	8	C	9	D	10	E				

4.11. FIND OUT THE INCORRECT STATEMENT

1. (a) The tag colour of TFL seed is opal green
(b) The tag colour of breeder seed is golden yellow
(c) The tag colour of certified seed is white

2. (a) The germination room temperature should be 2V± 1°C
(b) A minimum of 2V00 seeds are required for germination test
(c) The germination percentage is calculated to the nearest whole number

3. (a) Seed vigour is a single measurable attribute
(b) Seed vigour test is complementary and not a substitute to standard germination test
(c) Seed vigour is correlated with field emergence

4. (a) Breeder seed is certified by SCA
(b) Foundation seed is certified by SCA
(c) Certified seed is certified by SCA

♦ Answer Keys ♦

Q.No.	Ans.	Q.No.	Ans.	Q.No.	Ans.	Q.No.	Ans.
1	c	2	b	3	a	4	a

4.12. ARRANGE IN ORDER

1. The order of seed sampling for seed quality analysis is
   Working sample – Composite sample – Primary sample – Submitted sample
   Answer: Primary sample – Composite sample – Submitted sample – Working sample –
2. The Tetrazolium test is to proceed
   Incubating – Evaluation – Preparation – Pre-conditioning
   Answer: Pre-conditioning – Preparation – Incubating – Evaluation
3. Process of seed certification
   Verification of seed source – Field inspection – Application for registration – Grant of certificate
   Answer: Application for registration – Verification of seed source – Field inspection – Grant Certificate
4. The sequence of assigning seed lot number
   Seed processing plant code – Location code – Month/year – Seed producer code
   Answer: Location code – Month/year – Seed processing plant code – Seed producer code
5. The order of seed production under seed quality control system
   Field inspection – Selection of variety – Offer for seed certification – Seed testing
   Answer: Selection of variety – Offer for seed certification – Field inspection – Seed testing
6. Composite sample – Submitted sample – Primary sample – Working sample
   Answer: Primary sample – Composite sample – Submitted sample – Working sample
7. Submission – Dividing – Mixing – Submitted sample
   Answer: Submitted sample – Submission – Mixing – Dividing
8. Germination test – Report – Moisture estimation – Seedling evaluation – Purity analysis
   Answer: Purity analysis – Moisture estimation – Germination test – Seedling evaluation – Report
9. Flowering stage – Harvest stage – Vegetative stage – Preflowering stage – Preharvest stage – Field inspection
   Answer: Field inspection – Vegetative stage–Preflowering stage – Flowering stage – Preharvest stage – Harvest stage
10. Preflowering stage– Flowering stage – Sowing – Post flowering stage – Preharvest stage – Harvest stage

Answer: Sowing – Preflowering stage– Flowering stage– Post flowering stage– Pre harvest stage– Harvest stage

11. Health purity – Physical purity – Genetic purity

    Answer: Physical purity – Health purity – Genetic purity

12. Receipt of sample – Identification – Physical examination – Chemical test – Reporting of results

    Answer: Receipt of sample – Physical examination – Chemical test – Identification – Reporting of result

13. Field inspection – Receipt of application – Sampling – Secrutining – Processing

    Answer: Receipt of Application – Secrutining – Field inspection – Processing – Sampling

14. Sample receipt – Biochemical test – Molecular test – Seed examination – Seedling examination – Field examination

    Answer: Sample receipt – Seed examination – Seedling examination – Biochemical test– Molecular test– Field examination

15. Physiological test – Stress test – Physical test – Biochemical test – Seed vigour test methods

    Answer: Seed vigour test method – Physical test – Physiological test – Stress test – Biochemical test

4.13. DIFFERENTIATE THE FOLLOWING

1. Differentiate between normal seedling and abnormal seedling

Normal seedling	Abnormal seedling
Seedling which has the capacity for continued development into normal plant when grown in favourable conditions of soil, oxygen, water, temperature and light	Seedling which do not show the capacity for continued development into normal plant when grown in favourable condition of soil, oxygen, water, temperature and light
In root system, the primary root may be intact or shows acceptable defects includes discolouration or necrotic spots, healed cracks and splits, superficial cracks and splits	In root system, the primary root is defective if it is stunned or stubby, tap root missing, broken, split from the tip, trapped in the seed coat, shows negative geotropism; constricted, spindly, glassy, decayed as a result of primary infection
Terminal and surrounding tissue is intact and are not of usually viable	In shoot system, the hypocotyl is defective if it is too short and thick, deeply cracked or broken, split right through, missing, bent over or forming a loop, is tightly twisted forming a spiral, constricted, spindy, glossy, diseased as result of primary infection

Normal seedling	Abnormal seedling
Cotyledons shows an acceptable defects upto V0% of tissue not functioning normally, only one cotyledons (may intact)	Terminal of surrounding tissue is defective and usually not viable
	The cotyledons are defective, if they are defective to such an extent that less than V0% of the original tissue not functioning normally, swollen or curled, deformed, broken or otherwise damaged, separate or missing, discoloured, necrotic, glossy. Decayed as a result of primary infection

2. Differentiate between pure seed and inert matter

Pure seed	Inert matter
Pure seed means the seeds of kind/species stated by the sender or found to predominate in the test	Inert matter includes seed units and all other matter and structures that are not defined as pure seed
It includes all botanical varieties and varieties of that kind/species	Inter matter means all materials other than seed that includes seed like structures, sterile florets, chaff, fungal bodies & stones as determined by methods defined by ISTA rules
Pure seed means seed exclusive of inert matter and all other seeds not being considered as determined by methods defined by rule	It includes broken seeds, if the size is one half of the original size
It also includes broken seeds, if the size is >1/2 of the original size.	Seeds without seed coat are recorded as inert matter
In leguminaceae and curciferae, seeds should have intact seed coat	

3. Differentiate between deformed seedling and decayed seedling

Deformed seedling	Decayed seedling
Weak or unbalanced development of essential structures such as spirally twisted or stunted plumule or hypocotyl or epicotyl, swollen shoot, stunted roots etc.	Seedlings with any one of the essential structures showing diseased symptoms as a result of primary infection from the seed which prevents the development of the seedlings

4. Differentiate between primary sampling and composite sampling

Primary sampling	Composite sampling
A small portion of seed sample or handful of seed sample taken from different places of a seed lot	A mixture of all the primary samples taken from the seed lot
It is a representative sample of a seed lot	From this sample, submitted sample can be derived
Number of sampling depends upon the seed lot size	Weight of composite sample varies depending on kind and crop species

5. Differentiate between hard seeds and fresh ungerminated seeds

Hard seeds	Fresh ungerminated seeds
Seeds which do not absorb moisture till the end of the test period and remain hard	Seeds which are neither hard nor have germinated, but remain firm and apparently viable at the end of the test period
If dropped down, metallic sound is noticed	While dropping no such sound.
It is due to hard seed coat	It is due to morphophysiological dormancy
Broken by mechanical and acid scarification	Broken by after ripening or stratification
Eg. Seeds of leguminaceae and Malvaceae	Eg. Apple

6. Differentiate between hard seeds and dead seeds

Hard seeds	Dead seeds
Seeds with do not absorb moisture till the end of the test period and remain hard. Eg. Seeds of leguminaceae and Malvaceae	Seeds at the end of the test period are neither hard nor fresh or have not produced any part of a seedling. Often dead seeds collapse and milky paste comes out when pressed at the end of the test
It needs treatment for breaking dormancy	Already dead seeds

7. Differentiate between submitted sample and working sample

Submitted sample	Working sample
Seed sample is reduced from composite sample which is usually much larger in quantity than test	A reduced quantity of sample taken from submitted sample in the laboratory for use in a given quality

Submitted sample	Working sample
requirements. The size of the submitted sample should be as large as specified in the ISTA rule.	test
The samples are submitted to the testing laboratory.	A minimum of 2V00 seeds are needed and obtained by mixing and dividing equipments
Eg. Paddy – L00g	Eg. Paddy – L0g

8. Differentiate between metaxenia and xenia with examples

Metaxenia	Xenia
Effect of foreign pollen on the endosperm colouration	Direct and immediate visible effects of the foreign pollen on the seed coat colour during the formation of seed.
Eg. Maize	Eg. Maize

9. Differentiate between working sample and official sample

Working sample	Official sample
A reduced quantity of sample taken from submitted sample in the laboratory	Official sample means any sample taken and designated as official by the department
Used for seed quality testing like purity analysis, moisture determ-ination and seed germination	Quality testing is done based on the request

10. Differentiate between pure seed and live seed

Pure seed	Live seed
Pure seed means the seeds of kind/species stated by the sender and found to predominate in the test	Percentage of viable seeds in a seed lot
Free from physical contaminants and ensure seed standards	Refers to germinable seed
Purity test is employed to determine the pure seeds	Germination test is done to determine the live seeds
Purely seed fraction is considered	Germination percentage is directly correlated with live seeds
All the pure seeds are not live seeds	All the live seeds are pure seeds
Purity test is conducted in the working sample	Germination test can be conducted from the pure seeds and pure live see-ds can be calculated using the formula Purity% x germination%

11. Differentiate between brick gravel test and cold test

Brick gravel test	Cold test
This test is used to differentiate vigour levels in cereals seeds	Used to evaluate the vigour of maize
Required tools are germination box, aluminium tray, sand, sand marker, brick gravel of 2-F mm size, germinator, seed sample	Tools required are aluminium tray, field soil, sand marker, germinator, seed samples
Principles employed here is the weak seedlings and are not able to generate enough force to overcome the pressure of brick gravels	The principle adopted here is the weak seeds that do not germinate and establish. Therefore to stimulate the actual field conditions witnessed at the time of corn planting, cold test have been developed
Percentage of emerged seedlings are used to compare seed vigour of different lots	The germination percentage is computed by counting the number of normal seedlings as in germination test
100 seeds are required to perform this test	V0 seeds are needed to carry out this test

12. Differentiate between paper piercing test and Hiltner test

Paper piercing test	Hiltner test
Principle employed in this test is similar to that of the brick gravel test	Principle adopted here is that the weak seedlings are not able to generate enough force to overcome the pressure of brick gravels
High vigour seed lots are expected to produce strong seedlings which can pierce a particular type of paper, while seedlings of poor vigour lots may not able to pierce the paper	Healthy seeds are able to generate enough force to overcome the pressure
Seed lot having maximum number of seedlings coming out of paper is considered to be most vigourous	Seedlings which have emerged through the brick layer are counted

13. Differentiate between indirect vigour test and direct vigour test

Indirect vigour test	Direct vigour test
This test measures certain physiological attributes of seeds	This test stimulates pertinent unfavourable field conditions on a laboratory scale
The advantage is that the variables can be precisely controlled, allowing	Advantage of the test is that it simultaneously evaluates all the

Indirect vigour test	Direct vigour test
reproducibility of result, less time consuming, less complex and required less equipment and also allows direct vigour comparisons over a wide geographical area.	factors affecting seed vigour
Do not simultaneously evaluate all vigour factors particularly injuries and morphological abnormalities	Difficult in standardization of test, inconsistency of test results and time consuming and more complex

14. Differentiate between seed vigour and germination

Seed vigour	Germination
It is the physiological stamina of a seed lot	It is the emergence and development of seed embryo of those essential structures
It is the ability of the seeds to perform in a wide range of field conditions	It indicates the ability to develop into normal plant under favourable conditions
Used to predict stand – producing potential of seed lots	Used to predict the field planting value of the seeds and used to compare the value of seed lots
Loss of vigour appears before loss of germination	Loss of germination occurs next to vigour loss

15. Differentiate between seed analyst and seed inspector

Seed analyst	Seed inspector
Seed analyst means a seed analyst appointed under section 12 of Seed act, 1-//	Seed inspector means a seed inspector appointed under section of 1F of Seed act, 1-//
Duties of the seed analyst is to analyze the sample at the state seed testing laboratory and deliver in prescribed form	Duties of the seed inspector is to take samples of any seed of any notified kind/variety which are in sale outlet
Seed analyst have to send the report of the result of the analysis to the seed certification officer and another copy to the person from whom the sample has been taken	Send the samples for analysis to the seed analysts for the area within which such sample has been taken
The report sent by the central seed laboratory under sub-section shall super side the report given by the seed analyst under sub section	Can examine any record register document or any other material Deliver one sample to the person from whom it has been taken

Seed analyst	Seed inspector
	Retain the remaining sample in the prescribed manner for production in case any legal proceedings are taken or for analysis by the central seed testing laboratory

16. Differentiate between validity period and revalidation

Validity period	Revalidation
Validity period of certified seed is - months from the date of certification for certified seeds	After reanalysis of the seed sample , if seed is found to confirm to prescribed standards, the certification agency shall extend the validity of seed for a further period of six months from the date of test
Submit the sample for revalidation to STL	The date of test and period of validity and name of certification agency who has extended the validity period must be rubber stamped on the tags affixed on the seed containers

17. Differentiate between constant temperature and alternate temperature

Constant temperature	Alternate temperature
In germination chamber, the temperature indicated should be recorded as a maximum of 2V°C and variation should not be more than ±1°C	In alternate temperatures, the lower temperature of 20°C should be maintained for 1/h and higher of F0°C for %h
Maintain the constant temperature until the test ends	If the alternating temperatures are not controlled over week ends or holidays, the test should be maintained at the lower temperatures
Example. All tropical seeds	Example. All temperate seeds

18. Differentiate between hot air oven method and electronic method of moisture estimation

Hot air oven method	Electronic method
In this method, seed moisture is removed by drying at a specified temperature for a specified duration	Moisture meters estimate seed moisture quickly, but the estimation is not as precise as by the air-oven method
Temperature maintained for moisture estimation here should be	Principles involved in the moisture meter is that the wet-seeds are good

Hot air oven method	Electronic method
1F0±1°C and 10F±2°C for high constant and low constant temperature methods, respectively	conductors, while dry seeds are less conductors of electricity
Moisture content is expressed as a percentage of the original weight (wet weight basis)	Meters should be calibrated and standardized against the airoven method
Air currents that might cause loss of moisture must be recorded to a minimum and time consuming	Use of moisture meter is a very convenient and rapid way for determining seed moisture content
Containers of glass or non-corrosive metal (stainless steel) should be used	The seed is weighted by a built in balance and heated by an infrared lamp or electric heating elements
Moisture content may be calculated by the formula as $\frac{\text{Loss in weight}}{\text{Initial weight of seed}} \times 100$	Moisture content is directly proportional to electrical resistance (I/C) and direct constant of the sample

19. Differentiate between orange certificate and green certificate issued by ISTA

Orange certificate	Green certificate
This certificate is issued when the sample is drawn officially from the seed lot and tests carried out under the authority of the same member station and the procedure followed identified the certificate with the seed lot	The certificate is issued when the sample is drawn officially from the seed lot under the authority of one authorized member station, but the tests are carried out at another authorized member station and only responsible for testing the sample as submitted
The certificate is orange coloured	The certificate is green coloured

20. Differentiate between ISTA and AOSA

ISTA	AOSA
ISTA stands for International Seed Testing Association	AOSA stands for Association of Official Seed Analyst
It is located at Zurich, Switzerland	It is located at Wichita, Kansas, USA
ISTA rules are applicable for testing of seeds in European and Asian countries	Rules are applicable for USA and Canada
Publish the scientific journal, Seed Science and Technology	Publishes AOSA Newsletter and Proceedings

21. Differentiate between field standard and seed standard

Field standard	Seed standard
In seed certification, there is minimum standards for isolation distance and land requirements and number of field inspections. Likewise maximum permissible level is fixed for off-types, inseparable other crop plants, objectionable weed plants and plants affected by designated diseases. These are called field standards	In seed certification process, the seed should meet out minimum standards for germination per cent, pure seeds, inert matter, other crop seeds and weed seeds and maximum permissible limit for objectionable weeds, diseased seeds, ODV and moisture content for ordinary and vapour proof pack. These are called seed standards
Seed certification officer inspect all those field standards during field inspection	The seeds are analyzed by seed analyst in the seed testing laboratory
The seed crop deviating from this standards is liable for rejection	Seeds failed meet out these standards are not passed

Unit 5

SEED STORAGE

5.1. REQUIREMENTS AND TYPES OF SEED STORAGE

5.1.1. Choose the correct answer

1. Example for recalcitrant seed
   (a) Mango (b) Paddy
   (c) Bajra (d) Blackgram
2. The seeds having short life
   (a) Mango (b) Bajra
   (c) Rice (d) Greengram
3. The seeds withstand long term storage are
   (a) Orthodox seeds (b) Recalcitrant seeds
   (c) Intermediate seed (d) Difficult recalcitrant
4. Which one of the following seed is a recalcitrant type?
   (a) Rice (b) Soybean
   (c) Onion (d) Jack
5. Microbiotic seed
   (a) Onion (b) Paddy
   (c) Maize (d) Tomato
6. The viability of dry recalcitrant fruits can be maintained by
   (a) Moist storage (b) Drying
   (c) Depulping (d) Dry storage
7. Neem is the best example for
   (a) Orthodox seed (b) Recalcitrant seed
   (c) Intermediate seed (d) Difficult recalcitrant seed
8. The storage period for commercial seed ranges
   (a) For three to five years
   (b) For many years
   (c) One year to one and a half year
   (d) From few days to eight to nine months
9. The storage period of carryover seeds is
   (a) $1-1\frac{1}{2}$ years (b) %- months
   (c) Many years (d) 20-2L months

10. The first segment of the storage period represents the time interval between
    (a) Physiological maturity and harvest maturity
    (b) Morphological maturity and physiological maturity
    (c) Harvest and processing
    (d) Harvest maturity and processing
11. In short term storage, seed is stored up to
    (a) L-%months (b) V-%months
    (c) /-%months (d) 2-%months
12. An example for long lived seed is
    (a) Lupine (b) Mild rice
    (c) River maple (d) Mango
13. Which one of the following crop seed is a orthodox seed?
    (a) Mango (b) Jamun
    (c) Jack (d) Pineapple

♦ Answer Keys ♦

Q.No.	Ans.	Q.No.	Ans.	Q.No.	Ans.	Q.No.	Ans.	Q.No.	Ans.	Q.No.	Ans.
1	a	2	a	3	a	4	d	5	a	6	a
7	c	8	d	9	a	10	a	11	c	12	a
13	d										

5.1.2. Fill in the blanks

1\. The advanced method used for seed storage is ....

2\. Hermatic storage is commercially adopted for ....

F. The largest germplasm storage bank in the world is ....

L. Dr. H.F. Chin is known for his work in .... seeds

V. Cocoa is a .... seed

/. A seed which cannot be dried to a safe moisture level and stored longer is called .... seed

2\. The seeds that can be dried to a safe moisture level and stored for a reasonably longer period is called ....

% A seed having a short life span is known as ....

-. Onion is a .... lived seed

10\. Storing in .... is permafrost storage

11\. An example for short lived seed is ....

12\. The storage period for carry-over seeds ranges from ....

1F. Foundation seeds can be stored upto ....

1L. Short term period of storage is otherwise called as ....

1V. Most recalcitrant seeds are killed at .... temperature

1/. .... from Calcutta University is the scientists who proposed "Mid storage correction technique, Hydration-Dehydration

♦ Answer Keys ♦

Q.No.	Ans.	Q.No.	Ans.
1	Cryopreservation	2	High value low volume seeds
3	Colorado University, USA	4	Recalcitrant
5	Recalcitrant	6	Recalcitrant
7	Orthodox	8	Microbiotic
9	Short	10	Ice caves
11	Onion	12	1 to 1 ½ year
13	Three to five years	14	Microbiotic seed
15	Sub zero	16	Dr RNBasu

5.1.3. State True or False

1. As the moisture content in the seed increases, resistant to different stress decreases
2. Sorghum seeds with the moisture content of 1F-1L% can be stored for a long time

F. Seed storage starts in plant itself

L. Nutmeg seed is a recalcitrant one

V. Intensive research on recalcitrant seeds is done in Pertanian University, Malaysia

/. Soil seed banks act as dispersal agents

2. *In-situ* germplasm conservation is more suitable than *ex-situ* germplasm

% The long term storage is also known as mesobiotic.

-. Neem is a recalcitrant seed

10. Avocado is a recalcitrant seed

♦ Answer Keys ♦

Q.No.	Ans.	Q.No.	Ans.	Q.No.	Ans.	Q.No.	Ans.	Q.No.	Ans.
1	True	2	False	3	True	4	True	5	True
6	False	7	True	8	False	9	False	10	True

5.2. SEED STORAGE FACTORS

5.2.1. Choose the correct answer

1. The prescribed seed moisture content for storing paddy seeds under open storage is
   - (a) 10 per cent
   - (b) 12 per cent
   - (c) 1L per cent
   - (d) %per cent
2. The seeds, which can be dried to low moisture content and stored at low temperatures without loosing their viability are called as
   - (a) Naked seed
   - (b) Basic seed
   - (c) Synthetic seed
   - (d) Orthodox seed

3. The two most important factors influencing the life span of seeds under storage are
   (a) RH and temperature (b) Climate and temperature
   (c) Rainfall and RH (d) Seed size and moisture content
4. For safe storage, the required moisture content of rice seed is below
   (a) %% (b) 10%
   (c) 1L% (d) 12%
5. For every reduction in one per cent moisture content, the seed life is
   (a) Increased (b) Decreased
   (c) Not affected (d) Doubled
6. Recalcitrant seeds are desiccation
   (a) Tolerant (b) Sensitive
   (c) Resistant (d) Non-sensitive
7. The most effective means of maintaining seed quality in storage is
   (a) Decreasing temperature and seed moisture
   (b) Increasing temperature and seed moisture
   (c) Decreasing temperature and increasing seed moisture
   (d) Increasing temperature and decreasing seed moisture
8. The period of viability of seed decreases with
   (a) Increase in carbondioxide pressure
   (b) Increase in oxygen pressure
   (c) Decrease in oxygen pressure
   (d) Decrease in carbondioxide pressure
9. For sealed storage, the safe moisture content is
   (a) F-%% (b) L-%%
   (c) V-%% (d) /-%%
10. The seeds can be stored for two years with a moisture content of
   (a) 11-1F% (b) 10-12%
   (c) --11% (d) %-10%
11. According to Harrington's rule, the storage potential of the seed is doubled when there is
   a. 1 per cent decrease in moisture content and 10°F decrease in temperature
   b. 1 per cent decrease in moisture content and 10°F increase in temperature
   c. 1 per cent increase in moisture content and 10°F increase in temperature
   d. 1 per cent increase in moisture content and 10°F decrease in temperature
12. Seeds having moisture content of 10-12% can be stored for a period of ....
   (a) ½ year (b) 1 year
   (c) 2 year (d) L year

13. Which among the following is a serious factor in affecting the seed viability and vigour at physiological maturity stage
    (a) Weather (b) Mineral nutrient deficiencies
    (c) Soil conditions (d) Disease and insect damage

14. Which kind of seeds store better for longer period
    (a) Proteinaceous (b) Oily seeds
    (c) Starchy seeds (d) Oil seeds

15. .... seeds are intermediate in longevity
    (a) Onion (b) Watermelon
    (c) Radish (d) Tomato

16. For storing the cereals in ordinary conditions for twelve to eighteen months, seeds should be dried upto ....
    (a) 10% (b) %8%
    (c) 1V% (d) 1/%

17. Seed internal factors that influence the seed viability loss in storage is
    (a) Rainfall (b) Termperature
    (c) Initial seed quality (d) Container

18. At what moisture range, the heating occurs in storage?
    (a) %-10% (b) 10-12%
    (c) 12-1L% (d) 1%-20%

19. At 4% moisture, the seed undergoes?
    (a) Autooxidation (b) Dehydration
    (c) Rehydration (d) Absorption

20. At what moisture range, the mold growth is higher in storage?
    (a) %-% (b) 10-12%
    (c) 12-1L% (d) /-%8%

21. What type of injury takes place when the seeds are stored at high moisture level?
    (a) Freezing injury (b) Soaking injury
    (c) High respiration (d) Germination

22. At what RH, the fungi and bacteria are very active during storage?
    (a) /2% (b) 2V%
    (c) <2V% (d) >2V%

23. The fungi and bacterial activity in storage is controlled only by
    (a) Temperature (b) Oxygen
    (c) Relative humidity (d) Deep freezing

24. For grain cum seed storage of pulses, the seeds are treated with
    (a) Activated clay (b) Red earth
    (c) Carbendazim (d) Bavistin

♦ Answer Keys ♦

Q.No.	Ans.	Q.No.	Ans.	Q.No.	Ans.	Q.No.	Ans.	Q.No.	Ans.	Q.No.	Ans.
1	b	2	d	3	a	4	c	5	d	6	b
7	a	8	b	9	b	10	c	11	a	12	b
13	a	14	c	15	c	16	a	17	c	18	d
19	a	20	c	21	a	22	c	23	d	24	a

5.2.2. Fill in the blanks

1. .... is safe moisture content for groundnut seeds

2. .... and .... are the important factors affecting seed storage

F. Many orthodox seeds can be stored for .... periods at ambient temperature provided their moisture content is low

L. The moisture content recommended for vegetable seeds in moisture impervious container is ....

V. Recalcitrant seeds cannot withstand .... besides desiccation.

/. Breeder seeds can be stored for long period under ....

2. If the moisture content of seed is reduced by 1%, the life span of seed is ....

% If the temperature is reduced by 10°F, the life of seed is ....

-. If the relative humidity in percentage and temperature in °F is added to ...., it is considered to be the best place for storage

10. Seed will absorb or desorb moisture depending upon the .... prevailing at the place of storage

11. The seed moisture content will attain an equilibrium with the storage atmosphere irrespective of the initial moisture content is known as ....

12. Optimum temperature for safe seed storage is ....

1F. Of all pulses, ... is least affected by bruchids

1L. Good cold storage for seeds should not exceed .... per cent in relative humidity

1V. If the temperature is sufficiently low during storage, a special type of .... damage will result in loss of viability.

1/. All the storage fungi are completely inactive below .... per cent relative humidity and that there is very little activity below .... per cent relative humidity.

12. During storage there is no insect activity at seed moisture content below .... per cent, but if seed is infected, increased activity may generally be expected upto .... per cent moisture content

1% Harington's thumb rule is applicable for a moisture range of .... per cent

1-. The maintenance of seed moisture content during storage is a function of .... and ....
20. The most important factor influencing seed viability during storage is ....
21. .... are good storers
22. The temperature below .... are considered unsafe for storage insect activity.
2F. The seed started to decline in viability where the seeds on the plant from .... maturity until harvest.
2L. In .... stores also the seeds get deteriorated.
2V. Among the many environmental factors at maturity, ... is the one which accelerates the deterioration of seeds on the mother plant itself .
2/. The seeds having .... moisture content stores for longer period in storage.
22. An increase in oxygen pressure, the seed tends to .... the period of viability.
2% When the RH of the storage environment is below /2%, the .... activity is very low.
2-. For pre-storage treatments .... fungicides are used
F0. Best treatment to keep off bruchid infestation in pulses is ....

♦ Answer Keys ♦

Q.No.	Ans.	Q.No.	Ans.	Q.No.	Ans.
1	%%	2	Moisture content, RH	3	Long
4	/%	5	Low temperature	6	Cold storage conditions
7	Doubled	8	Doubled	9	100
10	Relative Humidity	11	Absorption Isotherm	12	F0°C
13	Horsegram	14	Sixty	15	Freezing injury
16	/2, 2V	17	% 1V	18	V-1L
19	Relative humidity, temperature	20	Moisture content	21	Starchy seeds
22	12 to 22°C	23	Physiological	24	Retail
25	Heavy rainfall	26	% 10%	27	Decrease
28	Fungi and bacteria	29	Non-mercurial	30	Activated clay treatment

5.2.3. State True or False

1. In seed storage, container effect is not felt much in a place which is cool and dry
2. Seed moisture affects the storage life of a seed

F. The insect activity will be high at low seed moisture content

L. Seeds can be safely stored at low RH than at high RH

V. The safe moisture content for the long-term storage of paddy is 1F%

/. Storability of any seed depends on the moisture content and temperature of the storage place

2. The sum of temperature in °C and RH in % should be 100, during storage

% Drying of orthodox seeds to 12% moisture content or below will reduce the viability of seeds

-. Recalcitrant seeds and fruits respire actively and hence require good ventilation

10. Recalcitrant seeds have longer viability

11. Reduction of temperature of seed storage condition will decrease the storability of seeds

12. Mercurial formulations are used for pre-storage treatment of seeds

1F. The best treatment to ward off bruchid infestation is the treatment with activated clay

♦ **Answer Keys** ♦

Q.No.	Ans.	Q.No.	Ans.	Q.No.	Ans.	Q.No.	Ans.	Q.No.	Ans.
1	True	2	True	3	False	4	True	5	False
6	True	7	False	8	False	9	True	10	False
11	False	12	False	13	True				

5.3. VIABILITY NOMOGRAPH

5.3.1. Fill in the blanks

1. Viability equation was first suggested by ....

2. The degree of seed aliveness is referred as ....

F. Father of seed viability is ....

♦ **Answer Keys** ♦

Q.No.	Ans.	Q.No.	Ans.	Q.No.	Ans.
1	Roberts	2	Vigour	3	Dr Roberts

5.3.2. State True or False

1. Loss of viability preceeds vigour

2. Nomographs are used to predict seed vigour

♦ **Answer Keys** ♦

Q.No.	Ans.	Q.No.	Ans.
1	False	2	False

5.4. SEED DETERIORATION

5.4.1. Choose the correct answer

1. Evaluation of membrane damage of seed is assumed by
   (a) Electrical conductivity (b) Protein content
   (c) Moisture content (d) pH
2. Antioxidants scavenges
   (a) Toxins (b) Free radicals
   (c) Phenols (d) Ions
3. Seed deterioration is rapid during
   (a) Physiological maturity until harvest maturity
   (b) Storage in warehouse
   (c) Storage from harvest until processing
   (d) Transit
4. Phospholipids plays a vital role in
   (a) Photosynthesis (b) Cell-permeability
   (c) Transpiration (d) Water absorption
5. Leaching of solutes is considerably greater in
   (a) Non aged seeds (b) Aged seeds
   (c) Fresh seeds (d) Viable seeds
6. Causes of seed deterioration is
   (a) Chromosome aberrations (b) Soil
   (c) Seed type (d) Dormancy
7. The chemicals used for mid storage correction is
   (a) Potassium dihydrogen sulphate (b) Potassium nitrate
   (c) Disodium phosphate (d) Thiourea
8. Which one of the following is the physiological deterioration of seed?
   (a) Loss of weight (b) Loss of germination
   (c) Protein decrease (d) Insect damage
9. Which of the following is the biochemical deterioration of seed?
   (a) Loss of weight (b) Loss of germination
   (c) Protein decrease (d) Insect damage
10. Which one of the following is the biochemical deterioration of seed?
   (a) Loss of weight (b) Free radical production
   (c) Loss of germination (d) Loss of vigour
11. Which one of the following enzyme is increased while seed deterioration is increased?
   (a) Peroxidase (b) Lipases
   (c) Proteinase (d) Amylase

12. Which one the following is the antioxidant?
    (a) Calcium chloride (b) Vitamin F
    (c) Tocopherol (d) Calcium alginate
13. In which reaction, the free radicals are synthesized?
    (a) Lipid peroxidation (b) Rancidity
    (c) Starch degradation (d) Protein degradation
14. In which crop, the rancidity takes place
    (a) Wheat (b) Sorghum
    (c) Groundnut (d) Redgram
15. How the free radical production is controlled?
    (a) Fungicide treatment (b) Insecticide treatment
    (c) Biofertilizer treatment (d) Hydration–dehydration treatment
16. The loss of DNA is referred as
    (a) Physical deterioration (b) Physiological deterioration
    (c) Biochemical deterioration (d) Molecular deterioration

♦ Answer Keys ♦

Q.No.	Ans.	Q.No.	Ans.	Q.No.	Ans.	Q.No.	Ans.
1	a	2	b	3	c	4	b
5	b	6	a	7	c	8	b
9	c	10	b	11	a	12	c
13	a	14	c	15	d	16	d

5.4.2. Fill in the blanks

1. Bio-assay test involves the action of .... from old seeds
2. Desiccation tolerance in seeds is achieved by .... sugars
F. Flint corn types are .... storers
L. Increased peroxidases in seed indicate ....
V. Free radical are quantified by ....
/. Physico-chemical method to control of seed deterioration is ....
2. .... chemical used for hydration and dehydration treatment in seeds
% .... treatments maintain cell wall integrity in stored seeds
-. .... is an irreversible degenerative change in the quality of a seed.
10. The increased leakage of electrolytes in the aged seeds is due to a decreased .... of dead tissues.
11. .... are an important constituent of biological membranes
12. .... refers to the potentiality of seeds to germinate and produce normal seedlings when exposed to a favourable environment.
1F. .... is the central cause of seed deterioration
1L. The loss of weight is the .... deterioration

1V. The lipid peroxidation results in the .... production

1/. The free radicals damage the ....

12. The cell membrane is damaged by ....

1% The damage of cell membrane leads to .... loss

1-. Quenching of .... is done by treating the seeds with antioxidants

20. Quenching of free radicals is done by treating the seeds with ....

21. In most oil seeds, ... indicates the seed deterioration process

22. .... is the free radical scavenging enzyme

2F. The loss of .... preceeds the loss of viability

2L. The loss of vigour .... the loss of viability

2V. The loss of vigour preceeds the loss of ....

2/. The loss of DNA is referred as ....

22. The poor multiplication of chromosomal DNA is associated with ....

♦ Answer Keys ♦

Q.No.	Ans.	Q.No.	Ans.	Q.No.	Ans.
1	Aldehydes	2	Sucrose	3	Good
4	Deterioration	5	Electro Sensitive Resonance	6	Hydration – Dehydration
7	KH_2PO_4	8	Mid storage	9	Seed deterioration
10	Membrane integrity	11	Phospholipids	12	Seed viability
13	Lipid peroxidation	14	Physical	15	Free radical
16	Cell membrane	17	Free radicals	18	Membrane permeability
19	Free radicals	20	Antioxidant	21	Rancidity
22	Catalase/peroxidase	23	Vigour	24	Preceeds
25	Viability	26	Molecular change	27	Molecular change/seed deterioration

5.4.3. State True or False

1. The rate of deterioration in the normal and hard seeds are one and the same

2. Auto – oxidation and lipid peroxidation results in free radical peroxidation

F. Lipid peroxidation results in the formation of minute pores on the seed coat

L. Reduction in germination percentage of seeds is the most accepted and the easiest to observe visible manifestation of seed deterioration.

V. Viable seeds show considerably higher enzyme activities than seeds with reduced viability

/. Phospholipids have been found to decrease as seed deteriorates.

2. Rancidity takes place in castor

% Amylase is the free radical scavenging enzyme

-. Catalase is the free radical scavenging enzyme

10. Hydration and dehydration treatment is the pre storage treatment

11. Soaking and drying is the mid storage treatment

12. Aspirin is an antioxidant

1F. Amoxillin is an antioxidant

1L. Loss of viability preceeds the loss of vigour

1V. Loss of vigour preceeds the loss of viability

1/. During deterioration of seeds, the loss of viability takes place first and than the loss of vigour

♦ Answer Keys ♦

Q.No.	Ans.	Q.No.	Ans.	Q.No.	Ans.	Q.No.	Ans.
1	False	2	True	3	False	4	True
5	True	6	True	7	True	8	False
9	True	10	False	11	True	12	True
13	False	14	False	15	True	16	False

5.5. SEED PACKAGING CONTAINERS AND SEED TREATMENTS

5.5.1. Choose the correct answer

1. For long term storage of seeds, the suitable seed storage container is
   (a) Gada cloth bag (b) Gunny bag
   (c) Polythene bag (d) Paper bag

2. The aluminium foil pouch containers are
   (a) Moist pervious (b) Moist impervious
   (c) Moisture resistant (d) Moisture vapour proof

3. .... polythene bags are generally considered more satisfactory for seed storage.
   (a) High density (b) Medium density
   (c) Low density (d) Very high density

4. If moisture vapour proof containers are used for packaging cauliflower and cabbage, moisture content must not exceed beyond ...., respectively
   (a) L.0 and L.0% (b) V.0 and V.0%
   (c) /.0 and /.0% (d) L.0 and V.0%

5. If any other kind of seed except vegetables and flower is to be stored in moisture vapour proof containers, seed moisture may be kept at
   (a) 2.0% (b) V.0%
   (c) 10.0% (d) /.0%

6. For effective storage, the moisture proof materials like polyethylene and asphalt should have a thickness of
   (a) 10 mm & F mm, respectively (b) V mm & 10 mm, respectively
   (c) F mm & 10 mm, respectively (d) 10 mm & V mm, respectively
7. The suitable storage container for recalcitrant seed is
   (a) Moisture pervious container (b) Moisture impervious container
   (c) Moisture resistant container (d) Moisture vapour proof container
8. For longer storage, the best suited container for orthodox seeds is
   (a) Moisture pervious container (b) Moisture impervious container
   (c) Moisture resistant container (d) Moisture vapour proof container
9. Polyethylene bags of <300 gauge thickness allows
   (a) Entry of water in the form of vapour and liquid
   (b) Entry of water in the form of vapour and not in liquid
   (c) Entry of water in the form of liquid and not in vapour
   (d) Entry of water not in the form of both liquid and vapour
10. Which one of the following packaging material is the moisture vapour proof container?
    (a) Gunny bag
    (b) Cloth bag
    (c) F00 gauge polythene bag
    (d) Aluminium polyethylene laminated pouches
11. Which one of the following packaging material is the moisture vapour proof container?
    (a) Gunny bag (b) Cloth bag
    (c) F00 gauge polythene bag (d) Sealed tins
12. Which one of the following packaging material is the moisture vapour proof container?
    (a) Gunny bag (b) Cloth bag
    (c) F00 gauge polythene bag (d) Gasketed rigid plastic container
13. Which one of the following packaging material is the moisture resistant container?
    (a) Gunny bag (b) Cloth bag
    (c) 200 gauge polythene bag (d) Heat sealed plastic pouches
14. Which one of the following packaging material is the moisture vapour proof container?
    (a) Gunny bag (b) Cloth bag
    (c) F00 gauge polythene bag
    (d) Paper laminated aluminium foil pouches
15. Which one of the following packaging material breeder can use for storing nucleus seed/breeder seed for easy hardening?
    (a) Metal box (b) Aluminium foil
    (c) Glass bottle (d) Polyethylene bag

16. What is the moisture percentage of seed when they are packed in moisture pervious container?
    (a) L% (b) /%
    (c) %% (d) 10%
17. What is the moisture percentage of seed when they are packed in moisture impervious container?
    (a) L% (b) /%
    (c) %% (d) 10%

♦ Answer Keys ♦

Q.No.	Ans.	Q.No.	Ans.	Q.No.	Ans.	Q.No.	Ans.	Q.No.	Ans.	Q.No.	Ans.
1	c	2	c	3	a	4	a	5	b	6	a
7	a	8	d	9	b	10	d	11	d	12	d
13	d	14	d	15	a	16	d	17	c		

5.5.2. Fill in the blanks

1. Example for impermeable container is ....
2. The most effective way of preventing orthodox seeds regaining moisture is to store them ....

F. F00 gauge polyethylene bag serves as a moisture .... container

L. The moisture vapour proof containers used for packing vegetable seeds is ....

V. Different containers used for seed packing are .... and ....

/. Moisture pervious, moisture resistant and .... are the types of containers used for seed storage

2. Any two types of seed storage containers are .... and ....

% When the ambient environment is not conductive for storage, the seeds can be stored in ....

-. .... is a moisture resistant container

10. .... is a moisture vapour proof container

11. F00 gauge polyethylene bag is a ....

12. Seed storage in moisture proof materials is less .... than insulating storage

1F. Seed storage in moisture proof materials is expensive than ....

1L. The seeds that are packed in moisture vapour proof container should have .... moisture

1V. The seeds that are packed in pervious container should have .... moisture

♦ Answer Keys ♦

Q.No.	Ans.	Q.No.	Ans.	Q.No.	Ans.
1	Aluminium foil	2	Moisture impervious container	3	Resistant
4	Aluminium foil	5	Cloth bag, Polyethylene bag, HDP, PAFP	6	Moisture vapour proof
7	Gunny bag Polythene bag	8	Moisture vapour proof	9	Aluminium foil pouches/F00 gauge polyethylene bag
10	200 gauge polyethylene bag	11	Moisture resistant container	12	Expensive
13	Insulating storage	14	%%	15	10%

5.5.3. State True or False

1. Cloth bags are vapour proof containers
2. Thin polyethylene (200 gauge) is as good as a pervious container
F. 200 gauge polyethylene bag is moisture vapour proof container
L. F00 gauge polythene bag is a MVP container
V. Any number of fumigations can be done to disinfect the seed lot
/. Paper laminated aluminium foil pouches are moisture vapour proof container.
2. Heat sealed plastic pouches are moisture resistant containers
% Cloth and gunny bags are moisture pervious container
-. At 10% moisture, the seed can be stored in moisture vapour proof container
10. At %% moisture, the seed can be stored in moisture vapour proof container

♦ Answer Keys ♦

Q.No.	Ans.	Q.No.	Ans.	Q.No.	Ans.	Q.No.	Ans.	Q.No.	Ans.
1	False	2	True	3	True	4	False	5	False
6	True	7	True	8	True	9	False	10	True

5.6. STORAGE STRUCTURE

5.6.1. Choose the correct answer

1. Storage structure must not be a
   (a) Heat trap (b) Vapour trap
   (c) Moisture trap (d) Moisture vapour trap
2. The relative humidity and temperature of germplasm storage building are
   (a) F0% RH and V°C and 10°C (b) 20% RH and F0°C and 2V°C
   (c) %0% RH and V0°C and F°C (d) V0% RH and 2V°C and 0°C

3. The foundation of seed warehouse should be extended above the ground level with the height of
   (a) V0 cm (b) 20 m
   (c) - 0 cm (d) 110 cm
4. Temperature control in seed storage room is achieved by
   (a) Ventilation (b) Deep freezing
   (c) Water spraying (d) Keeping ice cubes
5. In refrigerated storage, how much temperature should be maintained for 60% Relative humidity?
   (a) V2°F (b) /2°F
   (c) 20°F (d) /2°F
6. All the openings in the storage structure should be sealed to control
   (a) Fungi (b) Bacteria
   (c) Rodents (d) Mites
7. How to construct a seed storage structure for providing a cool environment?
   (a) By providing thatched roofing
   (b) By providing a insulation in walls
   (c) By providing a window in walls
   (d) By providing filed roofing
8. For germplasm storage, the seed storage structure should have the temperature range of
   (a) V and 10°C (b) 10 and 1V°C
   (c) V - 10°C (d) 10 - 1V°C
9. For germplasm storage, the seed storage structure should be provided to maintain the RH of
   (a) /0% RH (b) V0% RH
   (c) L0% RH (d) F0% RH
10. What material should be used for construction of foundation of seed storage structure?
    (a) Stone (b) Wood
    (c) Brick (d) Hallow brick
11. How much height the foundation of storage structure from the ground level?
    (a) F0 cm (b) /0 cm
    (c) - 0 cm (d) 120 cm
12. What is the measurement of lip around the building at the 90 cm height extending out while constructing the seed storage godown?
    (a) 20 cm (b) 1V cm
    (c) 10 cm (d) V cm

13. In the seed storage structure, the temperature control can be achieved by
    (a) Refrigeration (b) Windows
    (c) Too many doors (d) No doors

14. What may be temperature difference, if the heat flow of the seed storage structure is twice?
    (a) V°C difference (b) 10°C difference
    (c) 1V°C difference (d) 20°C difference

15. Which one of the following material is a good insulator of heat in seed storage building?
    (a) Wood (b) Glass wool
    (c) Stone (d) Steel

16. Which one of the following material is a fair conductor of heat in seed storage building?
    (a) Wood (b) Glass wool
    (c) Stone (d) Steel

17. Which one of the following material is a poor conductor of heat in seed storage building?
    (a) Wood (b) Glass wool
    (c) Stone (d) Steel

18. Which one of the following material is a conductor of heat in seed storage building?
    (a) Wood (b) Glass wool
    (c) Stone (d) Steel

19. Which one of the following is applicable to steel as a good insulator of heat
    (a) Heat conductor (b) Insect proof
    (c) Bird proof (d) Fungi proof

20. Which one of the following is applicable to glass wool as a good insulator of heat
    (a) Fire proof (b) Insect proof
    (c) Bird proof (d) Fungi proof

21. The common desiccants used for chemical dehumidification in seed storage is
    (a) Activated clay
    (b) Activated alumina
    (c) Magnesium carbonate
    (d) Jalasakthi

22. The silica gel when it turns from blue to pink at about 45% RH, it is reactivated at
    (a) 1VV°C (b) 1/V°C
    (c) 12V°C (d) 1%V°C

23. What is the quantity of silica gel can be mixed with the seed for safer storage?
    (a) 1 kg/10 kg of seed (b) 0.Vkg/10 kg of seed
    (c) 1 kg/Vkg of seed (d) 0.Vkg/Vkg of seed
24. At what height, the seed bags can be stacked in the storage godown
    (a) V-10 feet (b) 10-1Vfeet
    (c) 1V-20 feet (d) 20-2Vfeet
25. Which one of the following material is placed on the floor of storage godown while stacking the seed bags?
    (a) Taurpalene (b) Gunny bag
    (c) Wooden pallets (d) Metal sheet
26. Deep freezing of seeds is only used for
    (a) Long term (b) Short term
    (c) Medium term storage (d) Cryopreservation

♦ Answer Keys ♦

Q.No.	Ans.	Q.No.	Ans.	Q.No.	Ans.	Q.No.	Ans.	Q.No.	Ans.	Q.No.	Ans.
1	a	2	a	3	c	4	a	5	d	6	c
7	b	8	a	9	d	10	a	11	c	12	b
13	a	14	d	15	b	16	a	17	c	18	d
19	a	20	a	21	b	22	c	23	a	24	c
25	c	26	a								

5.6.2. Fill in the blanks

1. Seed warehouse should have no ....
2. At moisture content below 1L per cent, the seed stacks could be .... height.

F. The desiccant which has the ability to extract and retain moisture on its surface by a phenomenon known as ....

L. When the sun is very hot, the seeds should not be stored in ....

V. The seed storage godown should be constructed in a such way that the .... cannot enter and that no serious gain in moisture may not occur.

/. The seed storage structure for germplasm storage, the temperature required is ....

2. The RH of ....% should be maintained in the storage structure for germplasm storage.

% A cracking in the wall and flour will harbour ....

-. Ventilation openings must be .... against the birds, rodents and insects.

10. Ventilation openings must be screened against ....

11. In the seed storage structure, the temperature control can be achieved by ....
12. Insulation of seed storage structure should be done to reduce the flow of .... from the warmer exterior environment
1F. The basic concept of refrigeration is to keep the storage temperature below the usual .... temperature
1L. The chemical dehumidification may be achieved by use of ....
1V. Silica gel can absorb water up to .... of its dry weight
1/. Liquid lithium chloride can be used as a dessicant in storage, but it is a ....
12. The silica gel stored along with seed turns from blue to .... at about LV% RH
1% The silica gel stored along with seed turns from blue to pink at about .... RH
1-. If the temperature difference 20°C, the heat flow is ....
20. For long term storage, the places with low .... should be preferred
21. When the seed bags are stacked at a height of 1V-20 feet, the seed moisture should be below ....
22. .... of seed to prevent direct contact with a damp floor while stacking the seed bags
2F. For long term storage of crop varieties, the RH and temperature recommended for maintenance under godown condition is .... respectively
2L. The only practical method of controlling microflora activity by temperature alone is ....

♦ Answer Keys ♦

Q.No.	Ans.	Q.No.	Ans.	Q.No.	Ans.
1	Windows	2	1V to 20 feet	3	Adsorption
4	Steel bins	5	Rain	6	V and 10°C
7	F0%	8	Insects	9	Screened
10	Birds, rodents and insects	11	Ventilation/insulation/ refrigeration	12	Heat
13	Ambient	14	Desiccants	15	L0%
16	Toxic	17	Pink	18	LV%
19	Twice	20	RH	21	1L%
22	Palleting	23	V0% and V0°C	24	Deep freezing

5.6.3. State True or False

1. The seed storage structure should have no windows
2. The seed storage structure should have only one window and door
F. A cracking in the wall and flour will harbor mold
L. A crack around the electrical provisions will also harbor the insects

V. Ventilation may control the temperature requirement in the seed storage building

/. Providing 2 or F doors may control the temperature in the seed storage building

2. Brick is a good insulator of heat in seed storage godown

% Mineral wool is the good insulator

-. Sand is the good insulator

10. Wood is the fair insulator

11. Aluminium is the conductor of heat

12. Floss wool as a good insulator of heat in the seed store building and it is a rat and fire proof, but it can harbor insects

1F. The alternative method of refrigeration is to storing the seeds dry by dehumidification

1L. Liquid lithium chloride can be served as a desiccant in seed storage maintenance

1V. The silica gel stored along with seed turns from blue to white at about LV% RH

1/. If the temperature difference is 10°C, the heat flow is twice

12. While stacking to the height of 1V-20 feet, the bottom bags get spoiled leading to loss of viability and vigour

1% Good ware house should have too many windows for better preservation

♦ Answer Keys ♦

Q.No.	Ans.	Q.No.	Ans.	Q.No.	Ans.	Q.No.	Ans.	Q.No.	Ans.	Q.No.	Ans.
1	True	2	False	3	False	4	True	5	True	6	False
7	False	8	True	9	False	10	True	11	True	12	True
13	True	14	True	15	False	16	False	17	True	18	False

5.7. GERMPLASM STORAGE

5.7.1. Choose the correct answer

1. Size of base collection of heterogeneous material to store in a gene bank is
   (a) 12,000 seeds (b) 10,000 seeds
   (c) V,000 seeds (d) 1V,000 seeds

2. The largest germplasm storage bank in the world is situated in
   (a) Colorado, USA (b) Tokyo, Japan
   (c) New Delhi, India (d) Moscow, Russia

3. A place of storage in which seeds of rare plants or obsolete varieties are kept in order to prolong their viability is called
   (a) Plant bank (b) Seed bank
   (c) Bureau bank (d) Cryo bank

4. Where kew garden is located

(a) USA (b) Mexico

(c) England (d) Germany

5. Which country is governing Svalbard global seed vault?

(a) England (b) Norway

(c) USA (d) Russia

6. Which institute is responsible for the germplasm collection of paddy?

(a) IRRI (b) CIMMYT

(c) ITTA (d) AURDC

7. Which institute is responsible for germplasm collection of sweet potato?

(a) IRRI (b) CIMMYT

(c) ITTA (d) AVRDC

♦ Answer Keys ♦

Q.No.	Ans.	Q.No.	Ans.	Q.No.	Ans.	Q.No.	Ans.
1	a	2	a	3	b	4	c
5	b	6	a	7	d		

5.7.2. Fill in the blanks

1. Size of base collection of homogenous material to store in a gene bank is ....

2. Maintenace of germplasm under natural environment is called as ....

F. The world famous germplasm collection centre is ....

L. The germplasm storage of seed in a layer of soil that is permanently frozen in very cold region of the world is named as ....

V. Expansion for NBG ....

/. Storing of germplasm seeds in repositories is known as ....

2. Field conservation of accessions is termed as ....

% Expansion for CIMMYT ....

♦ Answer Keys ♦

Q.No.	Ans.	Q.No.	Ans.
1	L000 seeds	2	*In situ* conservation
3	Kew garden	4	Ferma frost
5	Nordic Genetic Resource Centre	6	*Ex situ* maintenance
7	Gene banks	8	International Maize & Wheat Improvement Centre

5.7.3. State True or False

1. The history of Svalbard global seed vault starts in early 1-%0's
2. AVRDC is located at Peru
F. International Institute of Tropical Agricultural is located at Nigeria

♦ Answer Keys ♦

Q.No.	Ans.	Q.No.	Ans.	Q.No.	Ans.
1	True	2	False	3	True

5.8. CRYOPRESERVATION

5.8.1. Choose the correct answer

1. Cryopreservation is used for storage of the following
   (a) Seed (b) Embryo
   (c) Pods (d) Fruits
2. Application of cryogenics to the conservation of plants material, proposed for the first time in the year
   (a) 1-/% (b) 1-20
   (c) 1-22 (d) 1-2L
3. At storing cells at -196°C, the cells are at
   (a) Active (b) Inactive
   (c) Viable (d) Non viable
4. Which activity of the cell is stopped during cryopreservation?
   (a) Metabolic (b) Biochemical
   (c) Enzyme (d) Respiration
5. An example for cryoprotectant is
   (a) Ethylene glycol (b) Calcium chloride
   (c) Manganese sulphate (d) Manganese sulphide

♦ Answer Keys ♦

Q.No.	Ans.	Q.No.	Ans.	Q.No.	Ans.	Q.No.	Ans.	Q.No.	Ans.
1	b	2	a	3	b	4	a	5	a

5.8.2. Fill in the blanks

1. Cryopreservation refers to the storage of cells, tissues and organs at ultra low temperature of ....
2. In cryopreservation, the cells, tissues and organs are stored in liquid nitrogen at .... temperature.
F. For cryogenic storage of seeds, the required temperature is ....

♦ Answer Keys ♦

Q.No.	Ans.	Q.No.	Ans.	Q.No.	Ans.
1	Liquid nitrogen	2	-1-/°C	3	-1-/°C

5.8.3. State True or False

1. In cryopreservation, the containers with seeds are kept immersed in liquid nitrogen

♦ Answer Keys ♦

Q.No.	Ans.
1	True

5.9. SEED STORE MAINTENANCE

5.9.1. Choose the correct answer

1. For fumigating the seed in the store room, the moisture content should be
   (a) Below 10% (b) Below 12%
   (c) Below 1L% (d) Below 1/%
2. Which one of the following is a fumigant?
   (a) Methyl bromide (b) Methyl cyanite
   (c) Methyl dematon (d) Methyl ethyl chloride
3. Which one the following is a fumigant?
   (a) Hydrogen chloride (b) Hydrogen cyanide
   (c) Hydrogen peroxide (d) Hydrogen oxide
4. The exposure period of storage godowns to hydrogen cyanide is
   (a) 12 hrs (b) 2L hrs
   (c) L%hrs (d) F/ hrs

♦ Answer Keys ♦

Q.No.	Ans.	Q.No.	Ans.	Q.No.	Ans.	Q.No.	Ans.
1	b	2	a	3	b	4	b

5.9.2. Fill in the blanks

1. Spilled seeds of storage room should be removed by ....
2. Replace the .... bags immediately in the storage godown.

F. The cleaned floors and walls of seed store room should be sprayed with insecticide such as ....

L. The fumigant hydrogen phosphide is otherwise called as ....

V. The seeds with moisture content greater than ....% be dried before fumigation

/. The seeds with moisture content greater than 1L% should be .... before fumigation

2. For insect free storage, the godowns are sprayed with .... as sanitation spray

♦ Answer Keys ♦

Q.No.	Ans.	Q.No.	Ans.	Q.No.	Ans.	Q.No.	Ans.
1	Floor sweeping	2	Torn/damaged	3	Malathion	4	Phostoxin
5	1L%	6	Dried	7	Malathion		

5.9.3. State True or False

1. Good sanitation in the seed store is necessary to keep off the seed from birds
2. Good sanitation in the seed store is necessary to keep off the seed from insects and rodents

F. The exposure period of fumigation using hydrogen cyanide is F to 2 days

L. The exposure period of fumigation using methyl bromide is 2Lh

V. The exposure period of fumigation using hydrogen phosphide is F to 2 days

/. The cracks in the walls and floors can be repaired frequently to control the harbouring of insects

2. Fumigation is a method used for controlling insects in fields

% Polyethylene bag of more than 200 gauge thickness is moisture pervious container for storage

-. In a seed storage godown, fumigation should be done

♦ Answer Keys ♦

Q.No.	Ans.	Q.No.	Ans.	Q.No.	Ans.	Q.No.	Ans.	Q.No.	Ans.
1	False	2	True	3	False	4	True	5	True
6	True	7	False	8	False	9	True		

5.10. MATCH THE FOLLOWING

1. Question

1.	Ventilation	A.	Heat conductor
2.	Glass wool	B.	2L hours
F.	Stone	C.	Below 12%
L.	Wood	D.	Good insulator
V.	Steel	E.	F tablets for m^F
/.	Aluminium phosphide	F.	Regulate temperature
2.	Methyl bromide exposure period	G.	Poor insulator
%	Optimum seed moisture for fumigation	H.	Fair insulator
-.	Seed store sanitation spray	I.	Safe storage
10.	RH (%) + tempearature (°F) = 100	J.	Malathion

♦ Answer Keys ♦

Q.No.	Ans.	Q.No.	Ans.	Q.No.	Ans.	Q.No.	Ans.	Q.No.	Ans.
1	F	2	D	3	G	4	H	5	A
6	E	7	B	8	C	9	J	10	I

2. Question

1.	Methyl bromide	A.	Blue turns pink colour
2.	Pulse beetle	B.	1V-20 metre height
F.	Harrington thumb rule	C.	Store less than one year
L.	Cryogenic storage	D.	Fumigation
V.	Siliga gel	E.	Rapid mould growth
/.	Microbiotic seed	F.	-1- /°C
2.	Seed stacking	G.	Storage pest
%	CO_2 storage	H.	Decrease of 1% moisture doubles the shelf life
-.	12 to 1L% seed moisture	I.	Heating occurs
10.	1%to 20% seed moisture	J.	Vaccum storage

♦ Answer Keys ♦

Q.No.	Ans.	Q.No.	Ans.	Q.No.	Ans.	Q.No.	Ans.	Q.No.	Ans.
1	D	2	G	3	H	4	F	5	A
6	C	7	B	8	J	9	E	10	I

3. Question

1.	Onion	A.	Moisture resistant container
2.	Starchy seeds	B.	Moisture pervious container
F.	Silicagel	C.	Vapour proof container
L.	Stacking of seeds	D.	Poor storer
V.	Aluminium foil pouch	E.	Decrease seed viability
/.	Cloth bag	F.	Short lived
2.	200 gauge polythene bag	G.	Not susceptible to chilling injury
%	Recalcitrant seed	H.	Wooden pallets
-.	Increase of O_2	I.	Longer storability
10.	Orthodox seeds	J.	Desiccant

♦ Answer Keys ♦

Q.No.	Ans.	Q.No.	Ans.	Q.No.	Ans.	Q.No.	Ans.	Q.No.	Ans.
1	F	2	I	3	J	4	H	5	A
6	B	7	C	8	D	9	E	10	G

4. Question

1.	Loss of vigour	A.	F to V°C
2.	Refrigerated storage	B.	Fumigation
F.	Liquid nitrogen	C.	Mid storage correction
L.	Cold storage	D.	Quenching of free radicals
V.	Gas gutted top glass jar	E.	Dehumidification
/.	Lipid peroxidation	F.	Antioxidant
2.	Hydrogen cyanide	G.	Preceeds viability
%	Hydration dehydration	H.	Free radical production
-.	Halogenation	I.	Cryopreservation
10.	Aspirin	J.	Moisture impervious container

♦ Answer Keys ♦

Q.No.	Ans.	Q.No.	Ans.	Q.No.	Ans.	Q.No.	Ans.	Q.No.	Ans.
1	G	2	E	3	I	4	A	5	J
6	H	7	B	8	C	9	D	10	F

5. Question

1.	Short term storage	A.	Intermediate seeds
2.	Macrobiotic	B.	Viviparous germination
F.	Orthodox seed	C.	Accelerated ageing
L.	Recalcitrant seed	D.	Poor storer
V.	Neem	E.	Good storer
/.	Mangroves seeds	F.	Low temperature storage
2.	Immature seeds	G.	Thumb rule
%	High initial seed quality	H.	Less than a year
-.	Wood stock	I.	More than L years
10.	Harington	J.	Susceptible to chilling injury

♦ Answer Keys ♦

Q.No.	Ans.	Q.No.	Ans.	Q.No.	Ans.	Q.No.	Ans.	Q.No.	Ans.
1	H	2	I	3	F	4	J	5	A
6	B	7	D	8	E	9	C	10	G

6. Question

1.	Phospholipids	A.	Poor storability
2.	Greater leaching of solutes	B.	Maintain seeds at safe moisture level
F.	Seed deterioration	C.	A break down of membrane

L. Mid storage correction — D. -1V to -20°C

V. Intermediate seed — E. Cell permeability

/. High seed oil — F. Loss of membrane permeability

2. < V% seed moisture — G. Aged seeds

% Refrigerated seed storage — H. Disodium phosphate

-. Hermetic storage — I. Chromosome aberrations

10. Free radical damage — J. Neem

♦ Answer Keys ♦

Q.No.	Ans.	Q.No.	Ans.	Q.No.	Ans.	Q.No.	Ans.	Q.No.	Ans.
1	E	2	G	3	I	4	H	5	J
6	A	7	C	8	D	9	B	10	F

7. Question

1. Physical deterioration — A. Mitochondrial DNA damage

2. Physiological deterioration — B. Accessan storage

F. Biochemical deterioration — C. Germination enzymes

L. Genetic deterioration — D. Loss of seed weight

V. Cryopreservation — E. Insect damage of seeds

/. Germplasm storage — F. Mould growth

2. Peroxidase — G. Loss of seed vigour

% Alpha -amylase — H. Cannot store whole seed

-. Webbing of seeds — I. Protein denaturation

10. Charring of seeds — J. Scavenging enzyme

♦ Answer Keys ♦

Q.No.	Ans.	Q.No.	Ans.	Q.No.	Ans.	Q.No.	Ans.	Q.No.	Ans.
1	D	2	G	3	I	4	A	5	H
6	B	7	J	8	C	9	E	10	F

5.11. FIND OUT THE INCORRECT STATEMENT

1. (a) Cloth bags are vapour proof containers

(b) F00 gauge polythene bags are moisture resistant container

(c) 200 gauge polyethylene bag is moisture vapour proof container

2. (a) If the moisture content of seed is reduced by 1%, the life span of seed is doubled

(b) If the temperature is reduced by 10°C the life of seed is doubled

(c) If the relative humidity in percentage and temperature in °F is added to 100, it is considered to be the best place for storage

♦ Answer Keys ♦

Q.No.	Ans.	Q.No.	Ans.
1	a	2	b

5.12. ARRANGE IN ORDER

1. Loss of weight –Vigour loss – Moisture increase – Germination loss

   Answer: Moisture increase –Loss of weight – Vigour loss – Germination loss

2. Mid storage treatment – Prestorage treatment – Seed storage – Sanitation

   Answer: Sanitation – Prestorage treatment – Seed storage – Mid storage treatment

3. Free radical – Lipid peroxidation – Cell membrane damage – Loss of membrane permeability– Quenching of free radical – Prolong the storage – Anti oxidant

   Answer: Lipid peroxidation – Free radical – Cell membrane damage – Loss of membrane permeability – Anti oxidant – Quenching of free radical – Prolong the storage

4. -196°C – Embryo – Seed – Put in the liquid nitrogen – Cryopreservation

   Answer: Seed – Embryo – Put in the liquid nitrogen – -1-/°C – Cryopreservation

5.13. DIFFERENTIATE THE FOLLOWING

1. Differentiate between orthodox seeds and recalcitrant seed with example

Orthodox seed	Recalcitrant seed
Can be dried to /-10% moisture depending upon the species	Cannot be dried below F0-FV% moisture and should be maintained at this moisture
No decline in germination, if dried	A decline in germination, if dried below this moisture
Tolerant to desiccation (V-2% moisture) and low temperatures (0-VoC)	Intolerant to desiccation (V-2% moisture) and low temperature (12-1V°C)
Examples. Cereals, Pulses, Oilseeds etc.	Example. Rubber, Coffee, Jack fruit, Mango, Citrus, Jamun
Prevalent in arid and semi arid environments	Prevalent in warm humid climates
Can be stored for several years	Can be stored for few days to few months

Orthodox seed	Recalcitrant seed
Small to medium size seed often with hard seed coat	Usually medium to large size and heavy seeds
Dormancy often occurs	Dormancy absent or weak
Not metabolically active when shed	Metabolically active when shed
Accumulation of dry weight ceases before maturation	Accumulation of dry weight upto the time of seed dispersal

2. Differentiate between mesobiotic seed and macrobiotic seed with examples

Mesobiotic seed	Macrobiotic seed
Seeds can be stored upto F-1V years	Can be stored for more than 1V years to 100 years
Eg. Barley, Flax, Tall fescue, Carrot	Eg. Trifolium, Lupine seeds, Acacia, Hard seeded legumes

3. Differentiate between short term storage and long term storage

Short term storage	Long term storage
Storage of seeds from harvest to next planting (/-%months)	Storage of germplasm, nucleus seeds, samples tested in seed testing laboratory for regulatory purpose for V-20 years
Seeds must be dried to 12% moisture content for starchy seeds and less than - % for oil seeds	Seeds must be dried to %% moisture content for starchy seeds and V% for oil seeds
Can be stored under ambient conditions of cool and dry climate	Storage conditions is 2V% RH at F0°C (or) LV% RH at 20°C or less and refrigerated storage (V-10°C)
Moisture pervious containers or moisture proof containers can be used Eg. Cloth bag, Polythene bag of F00 gauge thickness	Moisture vapour proof containers can be used Eg. Polythene bag of 200 gauge thickness

4. Differentiate between moisture pervious container and vapour proof container with examples

Moisture pervious container	Vapour proof container
These containers allow entry of water in the form of vapour and liquid	These containers will not allow entry of moisture in the form of liquid or vapour
The seeds in these containers will attain seed equilibrium moisture	These are used for long term storage even in hot humid conditions, if the

Moisture pervious container	Vapour proof container
with surrounding atmosphere	seeds are sealed at optimum moisture content
This type of container used for low value and high volume seeds	This is used for high value and low volume seeds
Used only for short term storage only	Used for long term storage
Example. Cloth bags, gunny bags, Paper bags etc.	Eg. Polyethylenc bags of > 200 gauge thickness, Poly laminated aluminium foil pouches

5. **Differentiate between moisture resistant container and moisture vapour proof container with examples**

Moisture resistant container	Moisture vapour proof container
These allow entry of water in the form of vapour and not in liquid	These containers will not allow entry of moisture in the form of liquid or vapour
The seeds in the containers can't be carried over for long period in hot humid conditions	These are used for long term storage even in humid conditions, if the seeds are sealed at optimum moisturc content
Eg. Polythene bags < F00 gauge thickness, Polylined gunny bags or urea bags	Eg. Polythene bags > 200 gauge thickness, Aluminium foil pouch, rigid plastics etc.

6. **Differentiate between ex situ conservation and in situ conservation**

***Ex situ* conservation**	***In situ* conservation**
The conservation outside their habitats by perpetuating sample populations	The conservation of genetic through their maintenance with in natural or even human made ecosystems in which occur
Conservation in artificially created environment	Conservation in the natural environment itself
Includes zoos, botanical garden, gene banks, aquarium, cryopreservation etc.,	Includes national parks, sanctuaries, biosphere reserves etc.,

7. **Differentiate between natural ageing and artificial ageing**

Natural ageing	Artificial ageing
Storing in ambient conditions of cool and dry climate	Subjecting the seeds to adverse conditions to predict storability of seeds

Natural ageing	Artificial ageing
Prediction of storability is longer process	Prediction of storability very fast and quick (within weeks time)
Temperature and humidity can't be controlled	Exposing the seeds L0±1°C temperature and - V±V% R.H

8. Differentiate between heat insulator and heat conductor with examples

Heat insulator	Heat conductor
To reduce the flow of heat from the warmer exterior, through the walls, roof and floor of the storage, to the cooler air and seeds in storage	It conducts the heat, so the temperature of seed store room can not be controlled
Eg. Glass wool, Mineral wool, Saw dust	Eg. Steel, Aluminium, Copper
Seed can be stored for longer period	Can't be stored for longer period
Dry and cool climate is maintained so less insect activity and mould growth and maintain high germination and vigour	Uncontrolled temperature so more insect activity, mould growth and loss of germination and vigour

9. Differentiate between internal factors of seed storage and external factors of seed storage

Internal factors of seed storage	External factors of seed storage
The physical condition and physiological state of seeds greatly influence their life span	Storage environment will affect the longevity of seeds
Types of seed, initial seed quality, provenance and seed moisture content will affect the viability of seeds	Relative humidity and temperature will affect the life span of seeds
Higher the initial quality of seed, longer the duration of storage	High relative humidity and high temperature will drastically reduce the seed storage period
A one per cent decrease in moisture content nearly doubles storage life of seeds	A 10oF decrease in temperature, doubles the storage life of seeds

10. Differentiate between physiological deterioration and biochemical deterioration of seeds

Physiological deterioration	Biochemical deterioration
Physiological deterioration of seeds observable in their lower	Biochemical deterioration of seeds observable by loss of enzyme activity,

Physiological deterioration	Biochemical deterioration
performance during germination, delayed seedling emergence, slower rate of seedling growth and development, decreased germination and loss of field emergence	reduced respiration, increase in seed leachates, increase in free fatty acids, lipid peroxidation, loss of membrane permeability etc

11. Differentiate between short lived seeds and long lived seeds with examples

Short lived seeds	Long lived seeds
Remain viable for less than one year	Seeds are having viability for longer periods of more than 1Vyears
Eg. Onion, Soybean, Groundnut	Eg. Lotus, Lupine

12. Differentiate between conditioned storage and sealed storage

Conditioned storage	Sealed storage
Conditioned storage involves placing the seeds in a dry and cool environment for extended periods	Packing of seeds in moisture resistant or hermetically sealed containers for storage Eg. 200 gauge polyethylene bag
For longer storage 10% seed moisture content at 20oC or less is recommended	Purpose of such containers is to maintain the seeds at safe storage moisture level of / -%%

13. Differentiate between cryopreservation and germplasm storage

Cryopreservation	Germplasm storage
Conserve only plant cells, tissues and organs	Conserve seeds, whole plants and vegetative propagules
Ex situ conservation	Both ex situ and in situ conservation
Stored in liquid nitrogen at -1- /°C	Can be stored in natural environment and controlled condition
Highly valuable materials like endangered, rare and recalcitrant, poor storer seeds	All type of seeds
Expensive	Less expensive

Unit 6

SEED HEALTH

6.1. SEED PATHOLOGY

6.1.1. Choose the correct answer

1. Which one of the following bacteria have direct systemic infection via vascular system
   (a) *Cladosporium* spp. (b) *Ustilago* spp.
   (c) *Verticillium* spp. (d) *Xanthomonas* spp.
2. Which one of the disease causing pathogen enters into seed through pollen tube?
   (a) Whip tail (b) Grain smut
   (c) Loose smut (d) Wilt
3. Designated disease for wheat is
   (a) Kernel bunt (b) Ergot
   (c) Black rot (d) Kernel smut
4. Designated disease for sorghum is
   (a) Kernel bunt (b) Ergot
   (c) Black rot (d) Kernel smut
5. Designated disease for rice is
   (a) Kernel bunt (b) Ergot
   (c) Black rot (d) Kernel smut
6. Methods employed for the detection of virus associated with seed is
   (a) NaOH test (b) Blotter paper method
   (c) Agar plate method (d) ELISA/PCR
7. Method employed for the detection of bacterial pathogen associated with seed is
   (a) NaOH test (b) Blotter paper method
   (c) PCR (d) ELISA
8. Passport data of seed is
   (a) Seed health status
   (b) Germination
   (c) Purity
   (d) Complete information about the seed

9. NSC was started during
   (a) 1983 (b) 1973
   (c) 1963 (d) 1953

10. The pioneer organization involved in seed production and quality control is
   (a) NSC (b) ISTA
   (c) CSTL (d) SSFC

11. National nematode collection of India is located at
   (a) Anand (b) Madurai
   (c) New Delhi (d) Ranchi

12. Test for pathogen infection in seed is known as
   (a) Health test (b) DUS test
   (c) Patent test (d) Vigour test

13. Blotter test is recommended by ISTA for analyzing
   (a) Seed germination (b) Seed vigour
   (c) Seed health (d) Mechanical damage

14. Blotter method is suitable for detection of
   (a) Fungi (b) Bacteria
   (c) Virus (d) Nematode

15. Thc toxin produced by *Aspergillus flavus* is
   (a) Aflatoxin (b) Ochratoxin
   (c) Fumerosin (d) Zeatin

16. The concentration of 2,4-D used for 2,4-D blotter test
   (a) 100 ppm (b) 1000 ppm
   (c) 0.2% (d) 2%

17. Chemical used for surface sterilization of seeds before testing seed health on agar plate
   (a) NaOH (b) NaCl
   (c) NaHCl (d) NaOCl

18. Double diffusion test is used for the detection of seed borne
   (a) Fungi (b) Bacteria
   (c) Virus (d) Nematode

19. PCNB media is most useful media to detect
   (a) *Aspergillus* (b) *Phythium*
   (c) *Penicillium* (d) *Fusarium* spp.

20. Double diffusion test is used for the detection of
   (a) Seed borne bacteria (b) Seed borne nematodes
   (c) Seed borne viruses (d) Seed borne insects

21. Kernel bunt of paddy can be identified by soaking the seeds in
   (a) KOH (b) NaOH
   (c) Phenol (d) GA_3

22. Designated disease of paddy
    (a) Black leg (b) Leaf spot
    (c) Black bunt (d) Blast
23. The presence of fungal fruiting bodies in a seed lot is
    (a) Contamination (b) Infection
    (c) Infestation (d) Impurity
24. Presence of pathogen inside the seed coat, seed tissue is called as
    (a) Contamination (b) Infection
    (c) Infestation (d) Admixture
25. Wheat variety which is susceptible to loose smut is
    (a) Kalyansona (b) Arjun
    (c) Sonalika (d) Varun
26. The designated disease present all over India except Tamil Nadu is
    (a) Bunt of wheat (b) Smut of wheat
    (c) False smut of rice (d) Bunt of paddy
27. Designated disease of wheat
    (a) Bunt and smut (b) Blast
    (c) Rust (d) Blight
28. Designated disease of onion
    (a) Smut (b) Bulb rot
    (c) Silver shoot (d) *Aspergillus* rot
29. Black rot of cabbage is caused by
    (a) Fungi (b) Bacteria
    (c) Virus (d) Nematode
30. Brown spot of rice is caused by
    (a) *Fusarium* (b) Bacteria
    (c) *Alternaria* (d) *Macrophomina*
31. Designated disease of chilli is
    (a) Leaf spot (b) Fruit rot
    (c) Downey mildew (d) Powdery mildew
32. Designated disease of sesamum is
    (a) Powdery mildew (b) Leaf spot
    (c) Downey mildew (d) Leaf blight
33. Isolation distance of wheat in the presence of loose smut in certified seed is
    (a) 3 m (b) 5 m
    (c) 140 m (d) 180 m
34. Maximum permissible limit of loose smut infected wheat plants in a seed production field of foundation seed is
    (a) 0.25% (b) 0.50%
    (c) 0.10% (d) 0.20%

35. Maximum permissible limit for sunflower downy mildew in certified seed production plot is
    (a) 0.05% (b) 0.25%
    (c) 0.50% (d) 0.10%

36. A mycotoxin produced by *Aspergillus flavus* strains when growing on groundnut and cereals is called
    (a) Mycotoxin (b) Aflatoxin
    (c) Cytotoxin (d) Endotoxin

37. Fungi and bacteria usually enters through
    (a) Stomata (b) Wounds
    (c) Hydathodes (d) Insect puncture

38. If the pathogen is present on the seed surface and affecting its quality, then that seed is called as
    (a) Undersize seed (b) Infected seed
    (c) Infested seed (d) Inert matter

39. Loose smut is an
    (a) Internally seed borne (b) Externally seed borne
    (c) Insect borne (d) Air borne

40. Solar treatment is used for
    (a) Stem rust (b) Loose smut
    (c) Powdery mildew (d) Blast

41. Maximum permissible limit for leaf spot of tomato in foundation class is
    (a) 0.25% (b) 0.05%
    (c) 0.01% (d) 0.5%

42. Fusarium wilt in *Cicer arietinum* is spread through
    (a) Seed (b) Soil
    (c) Wind (d) Water

43. The casual organism of downey mildew in maize is present in
    (a) Endosperm (b) Scutellum
    (c) Upper part of the plumule (d) Radicle

44. An application of chemicals or fungicides to cover and protect seeds from parasites, pathogens or adverse environmental conditions that cause damage to the seed or seedling is called
    (a) Seed divider (b) Seed dressing
    (c) Seed extraction (d) Seed cleaning

45. Death of germinated seed or young seedlings in the nursery resulting from attach by certain soil living fungi is defined as
    (a) Rotting (b) Damping – off
    (c) Glotting (d) Fermentation

46. Fungi that produce distinct mycelium or spore mass on the surface of their hosts are called as
(a) Necrosis (b) Mould
(c) Masting (d) Mycorrhiza

47. Grain smut is a designated disease in
(a) Wheat (b) Sesame
(c) Mustard (d) Sorghum

48. *Tilletia indica* is important seed borne pathogen of
(a) Wheat (b) Barley
(c) Oats (d) Rice

49. The common fungicide used for pre-storage seed treatment is
(a) Bavistin (b) Dithane M 45
(c) Endosulphan (d) Monocrotophos

50. As per the schedule of plant quarantine regulation 1981, the quarantine pest of mango is
(a) Seed weevil (b) Mango slug
(b) Red ant (d) Tea mosquito bug

51. A certificate from a recognized plant quarantine service that indicates that a sample is substantially free from disease or pests is called as
(a) DUS certificate (b) Certification
(c) Quarantine (d) Phytosanitary certificate

52. The prophylactic measure that is used to prevent the entry of new diseases, insects and weeds from other countries is called
(a) Truthful certificate (b) Certification
(c) Quarantine (d) Phytosanitary certificate

53. The headquarters of plant quarantine is located at
(a) Hyderabad (b) Faridabad
(c) New Delhi (d) Calcutta

54. .... is a seed borne viral disease
(a) Cabbage yellow mosaic (b) Tobacco mosaic
(c) Common bean mosaic (d) Bhendi mosaic

55. Which is the designated disease of cumbu seed production?
(a) Grain smut (b) Ergot
(c) Downy mildew (d) Bunt

56. *Aspergillus niger* is a
(a) Seed fungi (b) Seed bacteria
(c) Storage fungi (d) Field fungi

57. Aflatoxins are produced by the fungi
(a) *Aspergillus* spp. (b) *Penicillium notatum*
(c) *Macrophomina* spp. (d) *Claviceps purpurea*

58. Seed treatment will not be useful for the control of
(a) Blast (b) Downy mildew
(c) Root rot (d) Long smut

59. If the seed transmission has to take place, the plants must be infected
(a) Before bud formation (b) After bud formation
(c) During bud formation (d) During fertilization

60. .... disease is transmitted through pollen grains
(a) Cabbage yellow (b) Chrysanthemum stunt
(c) Cocoa swollen shoot (d) Squash ring spot

61. Which is the designated disease of paddy seed production?
(a) Blast (b) Brown spot
(c) Foot rot (d) Sheath rot

62. Which of the following is a storage fungi?
(a) *Alternaria* (b) *Aspergillus*
(c) *Colletotrichum* (d) *Helminthosporium*

63. In blotter paper technique in order to inhibit the germination of seeds, the blotters are moistened with
(a) 0.2% sodium salt of 2,4-dichlorophenoxy acetic acid
(b) 0.2% sodium hypochlorite
(c) 0.2% acetic acid
(d) 0.2% sodium hypochlorite and acetic acid

64. .... method is the most practical procedure for determining the fungi in the seed sample
(a) Seed washing technique (b) Blotter paper technique
(c) Examining the growing plants (d) Agar plate technique

65. The post harvest losses of grains in our country as estimated by Government of India is
(a) 10% (b) 9%
(c) 8% (d) 9.5%

66. Deterioration due to fungi in storage can be detected by
(a) Mould growth (b) Formation of lumps
(c) Musty odour (d) Insect activity

67. To prevent .... infection, seeds are treated with groundnut oil
(a) *Trigoderma granarium* (b) *Tribolium confusum*
(c) *Stogobium panicium* (d) *Callosobruchus maculatus*

68. .... inhibits the seed germination in blotter paper technique
(a) PEG (b) 2,4-D
(c) $HgCl_2$ (d) K_2HPO_4

69. Which of the following disease is not tested by seed washing technique
    (a) Downy mildew of soybean (b) Crown gall of coriander
    (c) Loose smut of wheat (d) Short smut of sorghum
70. In blotter method, seeds are incubated at 20 ± 2°C usually for 7 days in alternating cycles of
    (a) 12 hr light and 12 hr darkness (b) 10 hr light and 10 hr darkness
    (c) 24 hr light and 24 hr darkness (d) 15 hr light and 15 hr darkness
71. The method specially used to detect loose smut of wheat and barley is
    (a) Seedling symptom test (b) Blotter method
    (c) NaOH seed soak method (d) Embryo count method
72. Which among the following is not a designated disease of pearl millef
    (a) Grain smut (b) Green ear
    (c) Ergot (d) Leaf blight
73. The storage bacteria require atleast .... per cent relative humidity for its growth
    (a) 75 (b) 80
    (c) 90 (d) 70
74. In storage certain organisms can grow at temperature as low as .... and others at temperatures as high as ....
    (a) -15°C, 100°C (b) -10°C, 100°C
    (c) -8°C, 80°C (d) -15°C, 90°C
75. The practical method of controlling microflora activity is by
    (a) Increasing temperature (b) Deep freezing
    (c) Increasing the relative humidity above 75 per cent
    (d) Increasing the relative humidity above 90 per cent

♦ Answer Keys ♦

Q.No.	Ans.	Q.No.	Ans.	Q.No.	Ans.	Q.No.	Ans.	Q.No.	Ans.	Q.No.	Ans.
1	d	2	c	3	a	4	b	5	d	6	d
7	d	8	d	9	c	10	a	11	c	12	a
13	c	14	a	15	a	16	c	17	d	18	d
19	c	20	d	21	b	22	c	23	a	24	a
25	c	26	d	27	a	28	a	29	b	30	a
31	b	32	c	33	d	34	b	35	c	36	b
37	a	38	c	39	a	40	b	41	d	42	a
43	b	44	b	45	b	46	d	47	a	48	a
49	a	50	d	51	d	52	c	53	c	54	d
55	b	56	c	57	a	58	a	59	d	60	a
61	b	62	b	63	c	64	b	65	d	66	b
67	d	68	b	69	c	70	a	71	d	72	d
73	c	74	c	75	b						

6.1.2. Fill in the blanks

1. The penetration through ovary wall is by .... fungi in onion.
2. Seed health represents that seed should be free from .... infestation and .... infection
3. Seed health test is recommended in seed testing as per ....
4. Washing test is used to determine ....
5. 2,4-D is used in blotter paper method to inhibit ....
6. The accurate and less time consuming serological technique used to detect pathogen is ....
7. The toxin produced by fungi is called as ....
8. UV light will induce .... of fungi during incubation period
9. Most fungi infecting seeds require .... humidity for germination and growth
10. An example for externally seed borne pathogen is ....
11. Designated disease in bhendi is ....
12. .... is the designated disease of bajra
13. Ergot is a designated disease of ....
14. Presence of pathogen inside the seed is called as ....
15. The pathogen carried passively on surface of seed is called as .... seed borne pathogen
16. The pathogen imbedded in seed tissue is called as .... seed borne pathogen
17. A common fungicide used for seed treatment is ....
18. Black eyed seeds in chilli is caused by ....
19. .... is the pioneer organization involved in seed production and quality control
20. A legal system for restriction of spread of pests through seed material by import or export of seed is called ....
21. The phytosanitary certificate in India is issued by ....
22. In sodium hydroxide test, the infected seed gives .... appearance
23. Deterioration of stored grain containing fats and oils is accompanied by increase in ....
24. The International Plant Protection Convention was established in ....
25. The designated disease of wheat is ....
26. The designated disease of brinjal is ....
27. The designated disease of potato is ....
28. The NBPGR is located at ....
29. Determination of 1000 seed weight is called as ....
30. All the species of *Aspergillus* produce ....

31. .... is a mycotoxicosis caused by seed borne pathogen of lupin.
32. Seed borne diseases are common and severe in .... family.
33. Establishment of a pathogen inside or outside of the seed is called as ....
34. Transfere of the seed borne pathogen and establishment of infection in plants from the seed is called as ....
35. Black mould is ....

♦ Answer Keys ♦

Q.No.	Ans.	Q.No.	Ans.	Q.No.	Ans.
1	*Botrytis* spp.	2	Insect, Disease	3	Seed Bill, 2004
4	Pathogens	5	Sprouting of seeds	6	Polymerase Chain Reaction
7	Mycotoxin	8	Sporulation	9	High
10	Alternaria	11	Yellow mosaic virus	12	Green ear
13	Pearl millet	14	Seed infection	15	Externally
16	Internally	17	Bavistin	18	Fusarium
19	NSC	20	Plant quarantine	21	The Director, NBPGR
22	Black	23	Fatty acids	24	1951
25	Loose smut	26	Little leaf	27	Brown rot
28	New Delhi	29	Weight determination	30	Aflatoxin
31	Myotroph	32	Leguminaceae	33	Seed – borne
34	Seed transmitted	35	*Aspergillus niger*		

6.1.3. State True or False

1. Presence of designated disease affected plants in seed production plot affects the genetic purity of seed crop
2. Seed health is not mandatory in seed testing as per the Seed Act, 1966
3. ELISA is a simple test for genetic purity assessment
4. Ochratocin is produced by *Aspergillus niger*
5. Growing on test is used for the detection of seed borne bacteria
6. Dome test is used for the detection of virus
7. The best and easiest method of detecting seed borne fungi is blotter test.
8. Ergot is the designated disease of bajra seed crop
9. Blast of paddy is a designated disease
10. Fusarium is an internally seed borne pathogen
11. *Aspergillus spp* are externally seed borne pathogens
12. Bavistin is a systemic fungicide
13. Wheat leaf rust is a designated disease

♦ Answer Keys ♦

Q.No.	Ans.	Q.No.	Ans.	Q.No.	Ans.	Q.No.	Ans.	Q.No.	Ans.
1	False	2	True	3	False	4	False	5	True
6	False	7	True	8	True	9	True	10	False
11	True	12	True	13	False				

6.2. SEED ENTOMOLOGY

6.2.1. Choose the correct answer

1. Important stored grain pest for pulse is
   (a) Rice moth (b) Grain moth
   (c) Pulse beetle (d) Khapra beetle
2. Important stored grain pest for rice is
   (a) Rice moth (b) Pulse beetle
   (c) Khapra beetle (d) Red flour beetle
3. An example for internal feeder is
   (a) Red flour beetle (b) Saw toothed beetle
   (c) Rice weevil (d) Cigarette beetle
4. An example for external feeder is
   (a) Rice weevil (b) Granery weevil
   (c) Seed beetles (d) Red flour beetle
5. An example for secondary feeder is
   (a) Rice weevil (b) Red flour beetle
   (c) Seed beetles (d) Common grain mite
6. The symptom of pulse beetle attack is
   (a) Circular holes (b) Irregular holes
   (c) Damaged seed (d) Webbing
7. Musty odour and dust is the symptoms of
   (a) Red flour beetle (b) Rice weevil
   (c) Pulse beetle (d) Mites
8. In maize, the symptom of khapra beetle is
   (a) Circular holes (b) Seed reduced to husk
   (c) Webbing (d) Damaged seed
9. Irregular holes is the symptom of
   (a) Red flour beetle (b) Pulse beetle
   (c) Lesser grain borer (d) Mites
10. The common insect infesting the pulse seed in storage is
    (a) *Callasobruchus* spp. (b) *Rhyzopertha* spp.
    (c) *Tribolium* spp. (d) Anguis moth
11. For grain cum seed storage of pulses, seeds are treated with
    (a) Activated clay (b) Dithane M45
    (c) Vitavax (d) Bavistin

12. Activated clay is mainly used for
    (a) Improving germination (b) Against storage insects
    (c) Against storage pathogens (d) For reducing moisture content
13. Fumigation is very effective only in
    (a) Air-tight storage (b) Vapour proof storage
    (c) Gas-tight storage (d) Moisture proof storage
14. Fumigation period, generally followed in storage godowns is
    (a) 1-3 days (b) 2-3 days
    (c) 3-5 days (d) 5-7 days
15. Infestation of legume seeds by bruchids results in complete loss of viability within .... months of storage
    (a) 3-5 (b) 5-8
    (c) 4-6 (d) 2-4
16. Which among the following is the fastest fumigant
    (a) Ethylene bromide (b) Malathion 50 EC
    (c) Fenitrothion 50 EC (d) Aluminium phosphide
17. For fumigation, aluminium phosphide is applied at the rate of ....
    (a) 5 tablets of 5 grams each per tonne of grain for 10 days
    (b) 21 tablets of 3 grams each per tonne of grain for 10 days
    (c) 3 tablets of 3 grams each per tonne of grain for 5 days
    (d) 21 tablets of 3 grams each for 28 cu. metres for 5 days
18. The activity of .... in storage causes complete loss of seed
    (a) Rodents (b) Birds
    (c) Fungi (d) Insects
19. The optimum temperature for insect activity of most important storage insects ranges from
    (a) 20 to 30°C (b) 17 to 22°C
    (c) 25 to 35°C (d) 28 to 38°C

♦ Answer Keys ♦

Q.No.	Ans.	Q.No.	Ans.	Q.No.	Ans.	Q.No.	Ans.	Q.No.	Ans.
1	c	2	a	3	c	4	d	5	d
6	a	7	d	8	b	9	c	10	a
11	a	12	b	13	c	14	d	15	d
16	a	17	c	18	a	19	d		

6.2.2. Fill in the blanks

1. .... is secondary pest of storage
2. .... is an internal feeder
3. .... is an external feeder

4. .... is one of the stored grain pest and only larvae causes damage
5. Webbing of rice seeds is characteristics symptom of ....
6. *Sitophilus oryzae, Sitotroga cerealella* and .... are stored grain pests which infestation starts from the field itself and carried to the godowns
7. .... is a major stored grain pest of groundnut
8. For the long term safe storage of seeds, the moisture content of the seeds should be ....
9. Chlorpyrifos methyl (Reldan) is effective against stored product insects except ....
10. .... is a stored product insect which is more sensitive/ susceptible to spinosad
11. .... is a first case of insecticide resistance stored grain pests in India
12. .... is a only stored grain pest of groundnut which attacks both kernels and pods
13. The droppings are .... shaped in Root rat, Norway rat and Mouse
14. *Chlorpyrifos methyl* (Reldan), *Cyfluthrin* (Tempo) and *Pirimiphos methyl* (Actellic) are used for the management of .... pests
15. Pusa bin is used for the control of ....
16. One aluminium phosphide tablet (three grams) release one gram of .... gas
17. Hidden infestation can be detected by ...., ninhydrin method and KOH method
18. .... is a acute rodenticides (single dose and quick acting)
19. Warfarin and bromodiolone are .... (multi dose and slow acting)
20. Slow acting poison, anticoagulant refers to .... poison
21. Red squill, strychnine are rodenticide of .... origin
22. Methyl bromide, sulfuryl floride (Profume) and phosphine are the .... used for the management of stored grain pests

♦ Answer Keys ♦

Q.No.	Ans.	Q.No.	Ans.	Q.No.	Ans.
1	Long headed flour beetle	2	Rice weevil	3	Khapra beetle
4	Khapra beetle	5	Rice moth	6	*Callosobruchus* spp
7	*Caryedon serratus*	8	7 to 8%	9	*Rhyzopertha dominica*
10	*Rhyzopertha dominica*	11	*Tribolium castaneum*	12	*Caryedon serratus*
13	Spindle	14	Stored grain	15	*Trogoderma*
16	Phosphine	17	Staining method	18	Zinc phosphide

Q.No.	Ans.	Q.No.	Ans.	Q.No.	Ans.
19	Chronic rodenticides	20	Bormodiolone	21	Botanical
22	Fumigant				

6.2.3. State True or False

1. Grubs of callosobruchus cause damage to seed
2. Larval and adult stages are more susceptible than egg and pupae due to phosphine fumigation
3. Phosphine fumigation has more reaction with most commodities
4. Phosphine fumigation does not affect the germination and milling qualities
5. Methyl bromide fumigation is effective for all insect stages and requires less exposure period
6. Methyl bromide is not widely used for plant quarantine and food processing facilities
7. Methyl bromide destroys stratospheric ozone layer

♦ Answer Keys ♦

Q.No.	Ans.	Q.No.	Ans.	Q.No.	Ans.	Q.No.	Ans.
1	False	2	True	3	False	4	True
5	True	6	False	7	True		

6.3. SEED BORNE NEMATODE

6.3.1. Choose the correct answer

1. What is the minimum seed certification standard for nematode infection in sweet potato certified seed crop?
   (a) 2% (b) 6%
   (c) 8% (d) 10%
2. In rice, which of the following is the seed borne nematode?
   (a) Black ram (b) Silver leaf
   (c) White tip (d) Green ear
3. How to control the infection of white tip nematode in rice?
   (a) Hot water treatment at 35°C (b) Hot water treatment at 45°C
   (c) Hot water treatment at 55°C (d) Hot water treatment at 65°C
4. How to control the ufra nematode (*Ditylenchus angustus*)
   (a) Carbofuran spray (b) Carbendazim spray
   (c) Endosulfan spray (d) Atrazine spray
5. The infection of nematode leads to modification of floral structure into a seed gall by
   (a) *Anguina* spp. (b) *Ditylenchus* spp.
   (c) *Aphelenchoides* spp. (d) *Heterodera* spp.

6. The infection of nematode leads to discolouraton of testa in groundnut by
   (a) *Anguina* spp. (b) *Ditylenchus* spp.
   (c) *Aphelenchoides* spp. (d) *Heterodera* spp.
7. Paddy kernels infected by .... is symptomless
   (a) *Anguina* spp. (b) *Ditylenchus* spp.
   (c) *Aphelenchoides* spp. (d) *Heterodera* spp.
8. Which one of the following is the testa nematode of groundnut
   (a) *Aphelenchoides besseyi* (b) *Phelenchoides arachidis*
   (c) *Ditylenchus destructor* (d) *Ditylenchus angustas*
9. Which one of the following is the white tip nematode of rice?
   (a) *Aphelenchoides besseyi* (b) *Aphelenchoides arachidis*
   (c) *Ditylenchus destructor* (d) *Ditylenchus angustas*
10. Which one of the following is the seed borne nematode of chrysanthemum ?
    (a) *Aphelenchoides ritzema-bosi* (b) *A. besseyi*
    (c) *A. arachidis* (d) *Anguina* spp

♦ Answer Keys ♦

Q.No.	Ans.	Q.No.	Ans.	Q.No.	Ans.	Q.No.	Ans.	Q.No.	Ans.
1	d	2	c	3	c	4	a	5	a
6	b	7	c	8	b	9	a	10	a

6.3.2. Fill in the blanks

1. Seed borne nematodes may be .... or may occur as seed infestation.
2. The ufra nematode is a serious pest of .... rice in Assam.
3. Endoparasitic nematodes species are found abundantly in .... soil.
4. Seed borne nematode *Anguina tritici* has developed special adaptive mechanisms, such as ability to withstand .... for their survival and dispersal.
5. .... of groundnut has developed special adaptive mechanisms, such as ability to withstand desiccation, for their survival and dispersal.

♦ Answer Keys ♦

Q.No.	Ans.	Q.No.	Ans.	Q.No.	Ans.
1	Internal	2	Deep water	3	Rhizosphere
4	Desiccation	5	*Aphelenchoides arachidis*		

6.4. MATCH THE FOLLOWING

1. Question

	Question		
1.	Karnal bunt	A.	Detection of nematode
2.	Kernel bunt	B.	Detection of virus
3.	Black rot	C.	Insect damage
4.	Orobanche	D.	Rice
5.	Cuscuta	E.	Detection of bacteria
6.	NaOH seed soaking method	F.	Wheat
7.	Soil washing technique	G.	Sunflower
8.	ELISA test	H.	Detection of fungal pathogen
9.	PCR	I.	Sweet potato
10.	Visual examination	J.	Lucerne

♦ Answer Keys ♦

Q.No.	Ans.	Q.No.	Ans.	Q.No.	Ans.	Q.No.	Ans.	Q.No.	Ans.
1	F	2	D	3	I	4	G	5	J
6	H	7	A	8	E	9	B	10	C

2. Question

	Question		
1.	Disinfestation	A.	Damping off
2.	Disinfection	B.	Seed borne fungi
3.	*Rhizopus*	C.	Seed borne nematode
4.	*Pythium*	D.	Bruchids
5.	*Phytopthora*	E.	External pathogen
6.	*Fusarium*	F.	Red flour beetle
7.	*Xanthomonas*	G.	Internal pathogen
8.	Whiptail	H.	Storage fungi
9.	Internal feeder	I.	Field fungi
10.	External feeder	J.	Seed borne bacteria

♦ Answer Keys ♦

Q.No.	Ans.	Q.No.	Ans.	Q.No.	Ans.	Q.No.	Ans.	Q.No.	Ans.
1	E	2	G	3	H	4	A	5	I
6	B	7	J	8	C	9	D	10	F

3. Question

	Question		
1.	Seed health	A.	Eats embryo only
2.	Storage entomology	B.	Control of seed borne pathogen

3. Seed pathology — C. Vector transmitted
4. Transmission of fungi — D. Study of both pests and diseases
5. Bruchids — E. Seed borne virus
6. Red flour beetle — F. Study of pests only
7. Malathion — G. Study of seed borne diseases
8. Vitavax — H. Through micropyle
9. Bunchy top of banana — I. Damages whole seed
10. Yellow vein mosaic — J. Control of storage insects

♦ Answer Keys ♦

Q.No.	Ans.	Q.No.	Ans.	Q.No.	Ans.	Q.No.	Ans.	Q.No.	Ans.
1	D	2	F	3	G	4	H	5	A
6	I	7	J	8	B	9	E	10	C

4. Question

1. *Collasobruchus maculatus* — A. Khapra beetle
2. *Tribolium castaneum* — B. Ergot of sorghum
3. *Sitophilus oryzae* — C. Seed borne nematode
4. *Oryzaephilus surinamensis* — D. Loose smut of wheat
5. *Rhizopertha dominica* — E. Seed borne bacteria
6. *Trogoderma granarium* — F. Bruchids
7. *Ditylenchus* spp — G. Irregular holes
8. *Xanthomonas* spp — H. Red flour beetles
9. *Claviceps purpurea* — I. Circular holes
10. *Ustilago segetum* — J. Saw toothed grain beetle

♦ Answer Keys ♦

Q.No.	Ans.	Q.No.	Ans.	Q.No.	Ans.	Q.No.	Ans.	Q.No.	Ans.
1	F	2	H	3	I	4	J	5	G
6	A	7	C	8	E	9	B	10	D

6.5. ARRANGE IN ORDER

1. Insect damage – Moisture increase – Webbing or charing – Complete loss

   Answer: Moisture increase – Insect damage – webbing or charing – Complete loss

2. Field fungi - Bavistin – *Ustilago tritici* – Storate fungi – *Aspergillus niger*

Answer: Field fungi – *Ustilago tritici* – Bavistin – Storage fungi – *Aspergillus niger*

3. Prestorage treatment – Preharvest sanitation spray – Endosulphan – Imidacloprid – Fumigation – Malathion – Seed store sanitation spray – Aluminium phosphide

   Answer: Preharvest sanitation spray – Endosulphan – Prestorage treatment – Imidacloprid – Seed store sanitation spray – Malathion – Fumigation – Aluminium phosphide

4. Blotter method – Washing technique – Dry seed examination – Seedling test – NaOH test

   Answer: Dry seed examination – Washing test – Blotter method – Seedling test – NaOH test

6.6. DIFFERENTIATE THE FOLLOWING

1. Differentiate between post entry quarantine and pre entry quarantine

Post entry quarantine	Pre entry quarantine
Post entry quarantine refers to the inspection of some plants for planting or for propagation, where pests inclu-ding virus that may be difficult to detect during the inspection at airports and seaports may lay dormant	Prior to commencing pre entry quar-antine, the plants will be inspected by QDPIF inspectors for signs of diseases and pests
Plants subjected to post entry quarantine shall be inspected at 2 national quarantine sites, but because all the imported volumes cannot be accommodated	Plants are grown in the pre entry quarantine at the QDPIF plant quarantine facility.
Eg. Potato tubers and sweet potato tubers: 10 kg per inspection unit	

2. Differentiate between external seed borne pathogen and internal seed borne pathogen with examples

External seed borne pathogen	Internal seed borne pathogen
Pathogen externally available on the seed coat surface	Pathogen deeply seated in endosperm and embryo
Pathogen may be present as spores, oospores and chlamydospores	Pathogen establish within the seed with definite relationship with seeds
Eg. Grain smut of sorghum, Head smut of sorghum, Early blight of tomato and chilli	Eg. Loose smut of wheat, Anthracnose of chilli, Cowpea, French bean and Indian bean, Kernl bmut of rice

3. Differentiate between external feeders and internal feeders with examples

External feeders	Internal feeders
External feeders are so called because the egg is laid outside the whole grains	Internal feeders are so called because the female deposits a single egg inside a whole kernel of grain
Larvae feeds on the grain or product from the outside	The egg hatching the larvae feeding and the pupation all occur within that grain kernel
These insects are often found in homes	These insects are primarily a problem in fields or in grain storage and processing facilities
Eg. Cigarette beetle, red flour beetle, saw toothed beetle etc.	Eg. Rice weevil, Lesser & larger grain borer, *Angoumois* grain moth etc.

4. Differentiate between protective seed treatment and fumigation with examples

Protective seed treatment	Fumigation
Seed treatment refers to the application of fungicide, insecticide or a combination of both to seeds so as to disinfect them from seed borne or soil borne pathogenic organisms and storage insects	Fumigation is a chemical substance with at specified temperature and pressure can exist in gaseous form in sufficient concentration to be lethal to the pests.
Seed treatment is done to prevent spread of plant diseases, protects seed from seed rot and seeding blights, improves germination	Fumigation with fumigants is widely practiced for curative or preventive action, as they are cost effective, efficient and easy to use against targets pests
Eg. Bavistin, Imidacloprid	Eg. Aluminium phosphide, Methyl bromide, Ethyldibromide and EDCT

5. Differentiate between seed disinfestation and seed disinfection with examples

Seed disinfestation	Seed disinfection
Killing of spores, myclelia or propagules of micro organisms on seed surface	Elimination of pathogen that has penetrated into living cells of seed (Eg. Smut or bunt)
Non systemic fungicide can be used	Systemic fungicide can be used
Eg. Dithane M45	Eg. Carbendazim

6. Differentiate between loose smut and head smut

Loose smut	Head smut
Loose smut is caused by *Ustilago tritici*	Head smut is caused by *Sphacelotheca reiliana*
It is internally seed borne disease of wheat	Externally seed borne disease of sorghum
Most obvious symptom is seen at heading time as the diseased heads emerge from the boot	Yield loss due to the disease is variable and directly dependent upon the incidence of disease
The kernels and glumes have been converted to masses of black spores	Ears and tassels may be replaced with smut sori filled with teliospores. Infected plants may develop bizarre symptoms on either the ear or the tassel
Wheat plants are infected only during flowering by wind blown spores from nearby smutted heads	Ears of infected plants may be recognized by their tear drop shape and lack of silks where the husks are filled with the teliospores
Seed treatment is very effective and inexpensive way to control loose smut. Seed disinfectant fungicide with systemic action is necessary to eliminate the fungus in the embryo of the seed	Seed treatment with non systemic fungicides in the areas with a high incidence of the disease. Planting earlier to avoid the optimum temperature for teliospore germination
Eg. Carbendazim	Eg. Dithane M45

7. Differentiate between field fungi and storage fungi

Field fungi	Storage fungi
Field fungi are *Alternaria, Cladosporium*	Common storage fungi are *Apergillus* and *Penicillium*
Field fungi invade the seeds before harvest while the crop is still in the field	Stored fungi are fungi which invade grains or seeds during storage
Field fungi may affect the appearance and quality of seed and grain	Storage fungi are usually not present to any serious extent before harvest
Damages caused by field fungi can be detected by routine inspection	Detected by blotter test or agar plate test
Most field fungi are more prevalent when rainfall is above normal during grain fill and harvest	Development of storage fungi in stored grain is influenced by the moisture content, temperature and RH

Field fungi	Storage fungi
Invasion by field fungi may be more severe is the crop which has already been damaged by insects, birds etc.	

8. Differentiate between biocontrol agents and botanicals

Biocontrol agents	Botanicals
Biocontrol agents are used as seed treatment that protect seeds and seedlings from various pathogens	Botanical seed treatments means the treating seeds with plant products
Biocontrols are free of side effects	Botanicals provide 100% protection to the environment
They are safe to handle or use	Adaptable even under water stress conditions
Biocontrol agents occurs naturally	Ecofriendly and user friendly
Biocontrol agents are cost effective	Usage of the botanicals is large and handling is difficult
Contain plant growth promoting substances	
Eg. *Trichoderma, Pseudomonas*	Eg. Pungam, Neem, Arappu, Notchi

9. Differentiate between Trichoderma viride and Pseudomonas fluorescens

Trichoderma viride	Pseudomonas fluorescens
Trichoderma viride is the potential antagonistic fungus which prevents the crops from diseases *viz.*, root rot, brown rot, damping off, charcoal rot and other soil borne diseases	*Pseudomonas fluorescens* is a bacteria and have a strong oxidising power that helps them to break down environmental pollutants and provide useful enzymes and oxygen for plant growth
Trichoderma viride is able to suppress more than 60 species of pathogens (*Pythium, Botrytis, Phoma, Sclerotinia, Fusarium, Alternaria* & others)	It improves soil quality with subsequent uses.
It is suitable for sugarcane, pulses, oilseeds, cotton, vegetables, banana, coconut, oil palm, chilli	It controls the rice blast and sheath blight of paddy
The active components of biopesticides made on the base of this fungus antagonistic are their	It has the capability to destroy the cell wall of the fungal pathogens and annihilate them

Trichoderma viride	Pseudomonas fluorescens
spores, mycelium and products of metabolism	
Trichoderma viride utilizes the protoplasm as a source of food and multiply its spores	It should not be mixed with the antibacterial agents and inorganic fertilizers
In the process of development, *Trichoderma viride* develops a variety of antibiotics (gliotoxin, viridine, tricho dermin and others). These destroy the cell walls of phytopathogenic fungi and produce biologically active substances	It secrets hydrogen cyanide and antibiotics such as pycocyanin and phenazine which inhibit the growth of disease causing pathogens

10. Differentiate between biological control and chemical control with examples

Biological control	Chemical control
Control of plant pest and diseases by using plant products, botanicals, biocontrol agents (microorganism)	Control of plant pest and disease by using chemicals
Eg. Botanicals: Neem leaf powder, Neem oil, Neem seed kernel extract, Nochi leaf powder, Turmeric and Vasmbu rhizome powder Biological agents: *Trichoderma viride, Pseudomonas fluorescens, Trichogramma* spp.	Eg. Carbendazim, Mancozeb, Imidacloprid, Dithane M45
Ecofriendly	Not ecofriendly
Free of side effects and safe to use	Hazardous to health
Slow in action	Quick action
Not available easily	Easily available

11. Differentiate between ELISA test and PCR test

ELISA test	PCR test
ELISA – Enzymes Linked Immuno Sorbant Assay	PCR- Polymerase Chain Reaction
This is highly sensitive method used for detection of seed borne virus and bacteria	This is used for varietal identification and detection of virus associated with seed
Based on antigen-antibody interaction	Based on DNA present in the pathogen

12. Differentiate between agar plate test and blotter paper method

Agar plate test	Blotter paper method
Those fungi which are not easily detectable in the blotter method can be easily identified	Widely used method for detecting fungus associated with seed
In this method, the seeds are placed on the surface of 2% malt extract sterilized agar	Seeds are placed on moistened blotters, filter paper or on other paper substrate
Eg. *Xanthomonas, Fusarium oxysporum, Pseudomonas*	Eg. *Alternaria brassicola, Ascochyta rabiei*

13. Differentiate between seed borne fungi and seed borne bacteria

Seed borne fungi	Seed borne bacteria
Detected by dry seed examination, washing techniques, NaOH seed soaking method, blotter paper method	Detected by dry seed examination, seed washing techniques and ELISA method
Eg. Loose smut of wheat, downey mildew in bajra	Eg. Black rot of cabbage

14. Differentiate between circular holes and irregular holes due to seed damages by insects

Circular holes	Irregular holes
Circular holes are made by pulse beetle, rice beetle, rice weevil, angoumois grain moth	Irregular holes are made by lesser grain borer, cigarette beetle
Eg. Pulses, Rice, Wheat, Maize, Sorghum	Eg. Spices

15. Differentiate between fruit blight and leaf blight

Fruit blight	Leaf blight
Circular gray to brown spot with light coloured centre usually observed on leaves, yellow and pale spots developed on fruits	The leaf spot appears as circular brown concentric rings crowded with black pycnidia
Seed borne pathogen	Seed and soil borne pathogen
Eg. *Phomopsis vexans* in brinjal	Eg. *Ascochyta phaseolorum* in cowpea
Control by foliar application of Dithane M45	Seed treatment with carbendazim

16. **Differentiate between primary feeder and secondary feeder with example**

Primary feeder	Secondary feeder
Insect feed the seeds either by internal or external	Those are the insects or mites living on the seed which are already damaged by other insects
Eg. Beetles, weevils	Eg. Mites
Damage the seeds by making circular, irregular holes; seeds are reduced to husk and then webbing	Musty odour and dust

Unit 7

SEED INDUSTRY DEVELOPMENT AND MARKETING

7.1. SEED INDUSTRY

7.1.1. Choose the correct answer

1. Year of starting of National Seed Association
   (a) 2005 (b) 2006
   (c) 2007 (d) 2008
2. The program upgraded as Directorate of Seed Research
   (a) National Seed Programme (b) National Seed Policy
   (c) Central Seed Policy (d) State Seed Policy
3. The headquarters of Directorate of Seed Research
   (a) New Delhi (b) Mau
   (c) Chennai (d) Hyderabad
4. The year of starting of Federation of International Seed Trade
   (a) 1924 (b) 1930
   (c) 1935 (d) 1940
5. National Seed Corporation was established during
   (a) 1963 (b) 1960
   (c) 1961 (d) 1975
6. At present, the private sector share of seed industry is
   (a) 70% (b) 90%
   (c) 100% (d) 50%
7. High value low volume seeds are
   (a) Cereals (b) Pulses
   (c) Vegetables (d) Oilseeds
8. In India, NSC started its function during
   (a) 1963 (b) 1965
   (c) 1947 (d) 1962
9. National organization involved in seed production is
   (a) NSC (b) ISTA
   (c) ISST (d) ICAR

10. IIVR is located in
    (a) New Delhi (b) Varanasi
    (c) Hyderabad (d) Bangalore
11. CPRI is located in
    (a) Wellington (b) Shimla
    (c) Kashmir (d) Ooty
12. Indian domestic seed industry stands globally at
    (a) 7th position (b) 5th position
    (c) 8th position (d) 4th position
13. The restriction for import of foreign germplasm was removed by
    (a) Seed Act, 1966
    (b) Seed Control Order, 1983
    (c) New Seed Development Policy, 1988
    (d) Seed Bill, 2004
14. Establishment of separate organization within agriculture department to deal with seed distribution and testing was recommended by
    (a) Agriculture Production Team 1959
    (b) Famine Enquiry Commission, 1944
    (c) Food Grain Policy Committee, 1944
    (d) Royal Commission on Agriculture, 1925
15. All India seed growers, merchants and nursery men's association was started in
    (a) 1946 (b) 1939
    (c) 1941 (d) 1952
16. All India Co-ordinated Maize Improvement Programme was started in
    (a) I Five year plan (b) II Five year plan
    (c) III Five year plan (d) IV Five year plan
17. First Indo-American Agricultural Production Team was headed by
    (a) Dr. Randhawa (b) Dr. Sherman E.Johnson
    (c) A.S. Carter (d) Sir John Russell
18. First All India Co-ordinated Crop Improvement Project started in
    (a) Sorghum (b) Bajra
    (c) Maize (d) Paddy
19. High yielding varieties programme was launched in
    (a) 1964 (b) 1961
    (c) 1966 (d) 1969
20. Seed Review Team was constituted by Govt. of India in
    (a) 1968 (b) 1970
    (c) 1966 (d) 1971

21. Tarai Development Corporation Ltd. established in
    (a) 1965 (b) 1969
    (c) 1971 (d) 1973
22. Indian Society of Seed Technology was formed in
    (a) 1969 (b) 1972
    (c) 1975 (d) 1971
23. During the first phase of National Seed Programme, State Seed Corporation was established in
    (a) Karnataka (b) Punjab
    (c) Rajasthan (d) Bihar
24. Seed is declared as an essential commodity under
    (a) Seed Act, 1966 (b) New Seed Policy, 1988
    (c) Seed Control Order, 1983 (d) Seed Bill, 2004
25. Systematic large scale seed certification in our country began with the formation of
    (a) State Seed Corporation (b) National Seed Corporation
    (c) State Farm Corporation (d) Agriculture Departments
26. Activities of state seed certification agencies is co-ordianated by
    (a) NSC (b) Central Seed Committee
    (c) Central Seed Certification Board (d) SFCI
27. Seed law enforcement is performed by
    (a) NSC (b) Seed Certification Agency
    (c) State Seed Corporation (d) Agriculture Department
28. At present, number of State Seed Corporations in India are
    (a) 17 (b) 13
    (c) 16 (d) 15
29. Salient feature of National Seed Bill, 2004 is
    (a) Promotion of domestic private sector seed industry
    (b) Seed distribution and marketing
    (c) Varietal development
    (d) Compulsory registration of varieties
30. Currently, number of state seed certification agency in the country are
    (a) 20 (b) 22
    (c) 18 (d) 23
31. International Seed Network Initiative was started during
    (a) June 1999 (b) June 2000
    (c) June 1998 (d) June 1996
32. The largest producer of vegetable seeds in the world
    (a) India (b) China
    (c) USA (d) Germany

33. Under All India Co-ordianted Crop Improvement Programme, first hybrid was released in
(a) Sorghum (b) Maize
(c) Pearl millet (d) Cotton

34. The seed year is
(a) 1961 (b) 1975
(c) 1980 (d) 1981

35. First AICCIP was introduced in
(a) Maize (b) Paddy
(c) Groundnut (d) Redgram

36. In which year, All India Co-ordinated Maize Improvement Project was initiated?
(a) 1955 (b) 1956
(c) 1957 (d) 1958

37. In which year, All India Co-ordinated Sorghum Improvement Project was initiated?
(a) 1957 (b) 1958
(c) 1959 (d) 1960

38. In which year, the first SAU was started?
(a) 1957 (b) 1958
(c) 1959 (d) 1960

39. Where the first SAU was started?
(a) Coimbatore (b) Pantnagar
(c) New Delhi (d) Punjab

40. Syngenta belongs to which country?
(a) USA (b) UK
(c) Germany (d) Switzerland

41. Bayer Crop Science belongs to which country
(a) USA (b) UK
(c) Germany (d) Switzerland

42. In which year, seed review team was started?
(a) 1958 (b) 1968
(c) 1978 (d) 1988

43. National commission of Agriculture submitted a report to the Government of India during the year
(a) 1951 (b) 1961
(c) 1971 (d) 1981

44. Where Bharwale foundation is located?
(a) Jalna (b) Hydrabad
(c) Bangalore (d) Coimbatore

45. Where Indo American hybrid seed company is located?
(a) Jalna (b) Hydrabad
(c) Bangalore (d) Coimbatore

46. Where Rasi seed is located?
(a) Hydrabad (b) Bangalore
(c) Coimbatore (d) Attur

47. Where the Mahyco-Monsanto seed company is located?
(a) Hydrabad (b) Bangalore
(c) Jalna (d) Mumbai

48. Seed multiplication ratio for wheat is
(a) 1:20 (b) 1:40
(c) 1:80 (d) 1:100

49. Seed multiplication ratio for paddy is
(a) 1:20 (b) 1:40
(c) 1:80 (d) 1:100

50. Seed multiplication ratio for groundnut is
(a) 1:40 (b) 1:20
(c) 1:10 (d) 1:8

51. Seed multiplication ratio for redgram is
(a) 1:40 (b) 1:50
(c) 1:80 (d) 1:100

52. Seed multiplication ratio for safflower and castor is
(a) 1:40 (b) 1:50
(c) 1:60 (d) 1:70

53. Seed multiplication ratio for tomato is
(a) 1:100 (b) 1:200
(c) 1:300 (d) 1:400

54. When TARAI Development Corporation was established?
(a) 1959 (b) 1969
(c) 1979 (d) 1989

55. Where was the first seed production programme for temperate vegetables started in India
(a) Quetta (b) Bangalore
(c) Ranibennur (d) Jalna

♦ Answer Keys ♦

Q.No.	Ans.	Q.No.	Ans.	Q.No.	Ans.	Q.No.	Ans.	Q.No.	Ans.	Q.No.	Ans.
1	c	2	a	3	b	4	a	5	a	6	a
7	c	8	a	9	a	10	b	11	b	12	b
13	c	14	15	a	16	b	17	b	18	c	15
19	c	20	a	21	b	22	d	23	b	24	c
25	b	26	c	27	b	28	d	29	d	30	b

Q.No.	Ans.	Q.No.	Ans.	Q.No.	Ans.	Q.No.	Ans.	Q.No.	Ans.	Q.No.	Ans.
31	a	32	c	33	c	34	a	35	a	36	c
37	d	38	c	39	b	40	d	41	c	42	b
43	c	44	b	45	c	46	d	47	a	48	a
49	c	50	d	51	d	52	c	53	d	54	b
55	a										

7.1.2. Fill in the blanks

1. In India, first SAU was started at ....
2. The ratio at which the seed multiplies is called ....
3. Seed multiplication ratio is worked out by ....
4. SRR is ....
5. Establishment of .... will ensure seed supply during natural calamities
6. The seed multiplication ratio in varieties is based on .... and ....
7. The All India Seed Growers, Merchants, Nurserymen's Association was established in ....
8. Tremendous progress made in development of high yielding varieties was possible in India due to the establishment of ....
9. A National Organization involved in seed production, processing and seed supply is ....
10. A successful co-operative venture in seed production and distribution of quality seeds in our country is ....
11. .... is the first company to commercially grow and market the Bt cotton in India
12. The expansion of OECD is ....
13. Seed multiplication ratio for cole crops are ....
14. Seed multiplication ratio for potato is ....
15. Expand NPDP ....
16. Expand IPRD ....
17. Expand TRIPS ....
18. In which year, the first sorghum hybrid was released in India ....
19. The SFCI stands for ....
20. .... is the first temperate vegetable seed production place in India.
21. Indian total seed industry is worth about .... in 2010-11
22. Indian Agriculture seed industry is worth .... approximately in 2010-11
23. Indian vegetable seed industry is worth .... in 2010-11

24. The contribution of public sector is .... per cent in Indian seed industry
25. The contribution of private sector is .... per cent in Indian seed industry
26. The contribution of unorganized sector is .... per cent in Indian seed industry

♦ Answer Keys ♦

Q.No.	Ans.	Q.No.	Ans.
1	Pantnagar	2	Seed multiplication ratio
3	Average seed yield/Seed rate used	4	Seed replacement rRate
5	Gene bank	6	Seed rate and yield
7	1946	8	AICRP
9	NSC	10	TDC
11	MAHYCO	12	Organization for Economic Co-operation and Development
13	1:433	14	1:4
15	National Pulse Development Programme	16	Integrated Programme for Rice Development
17	Trade Related Intellectual Property Rights and Services	18	1964
19	State Farm Corporation of India	20	Quetta (Kashmir valley)
21	7500-8000 crores	22	6000 crores
23	1500 crores	24	30
25	57	26	13

7.1.3. State True or False

1. State seed farms are involved in the production of nucleus seed
2. Seed demand forecasting is great help in production of needed seeds with minimum cost and maximum production efficiency
3. The first private temperate vegetable seed companies developed in Quetta and Kashmir valley
4. TDC started in India to assist certified seed production
5. 1965 is considered as the seed year
6. Seed village concept is the best approach to produce genetically and physically pure seeds
7. Compact area approach solves the isolation problem
8. Potato is example for very low seed multiplication ratio
9. Cotton seed multiplication ratio is 1:50
10. The first temperate vegetable seed production programme was started by private entrepreneurs

♦ Answer Keys ♦

Q.No.	Ans.	Q.No.	Ans.	Q.No.	Ans.	Q.No.	Ans.	Q.No.	Ans.
1	False	2	True	3	True	4	True	5	False
6	True	7	True	8	True	9	True	10	True

7.2. SEED MARKETING

7.2.1. Fill in the blanks

1. Seed marketing depends on .... and supply of seeds.
2. The seed demand is the multiplication of total area in acres and ....
3. The percentage of farm saved seed is ....
4. The percentage of quality seed supply to the farmer is ....
5. Seed distribution system by State Agricultural Department is by ....
6. Seed distribution system by State Agricultural University is by ....
7. Seed distribution system by private seed industry is by ....
8. The main sales promotion activity is promoted by ....
9. Supply and demand of seed decide the ....
10. The price of seed is .... than the grain

♦ Answer Keys ♦

Q.No.	Ans.	Q.No.	Ans.
1	Demand	2	Seed rate
3	75 to 80%	4	20 to 25%
5	Single window system	6	Multiple window system
7	Multiple window system	8	Publicity/advertisement
9	Seed price	10	Twice

7.2.2. State True or False

1. The seed demand is worked out based on total cultivated acreage multiplied by seed yield.
2. The total seed requirement of a particular crop is the relationship between the seed rate and total cultivated area.
3. Based on the demand and supply of the seed, producer can fix the seed price.
4. The co-operative society's markets their seed through single window system.
5. The state government markets their seed through single window system.
6. The local input traders can influence the farmer better than seed producer for sale of seed.
7. The overhead charges are not included in the seed price

8. The seed growers by fixing seed price may consider production cost only
9. The seed price includes transport cost also
10. Participating in kissan melas is also important medium of seed sales promotional activities.

♦ Answer Keys ♦

Q.No.	Ans.	Q.No.	Ans.	Q.No.	Ans.	Q.No.	Ans.	Q.No.	Ans.
1	False	2	True	3	True	4	False	5	True
6	True	7	False	8	False	9	True	10	True

7.3. MATCH THE FOLLOWING

1. Question

1.	Phyto sanitary certificate	A.	Attur
2.	Seed control order	B.	Jalna
3.	NBPGR	C.	Public seed industry
4.	Rasi seeds	D.	1961
5.	Nuziveedu seeds	E.	Import of plant genetic material
6.	MAHYCO seeds	F.	Wheat
7.	Low volume high value seed	G.	1951-56
8.	Low value high volume seeds	H.	1983
9.	First maize hybrid released	I.	Private seed industry
10.	First five year plan	J.	Hydrabad

♦ Answer Keys ♦

Q.No.	Ans.	Q.No.	Ans.	Q.No.	Ans.	Q.No.	Ans.	Q.No.	Ans.
1	F	2	H	3	E	4	A	5	J
6	B	7	I	8	C	9	D	10	G

2. Question

1.	Tarai Development Corporation	A.	1:50
2.	NSC	B.	30 to 40%
3.	Du-pont	C.	Pre-entry quarantine
4.	Indo-American seeds	D.	20 to 25%
5.	SMR for linseed	E.	1:16
6.	SMR for cucumber	F.	Uttar Pradesh
7.	SRR for self pollinated crop	G.	USA
8.	SRR for cross pollinated crop	H.	New Delhi
9.	Sea port	I.	Bangalore
10.	SMR of soybean	J.	1:200

♦ Answer Keys ♦

Q.No.	Ans.	Q.No.	Ans.	Q.No.	Ans.	Q.No.	Ans.	Q.No.	Ans.
1	F	2	H	3	G	4	I	5	A
6	J	7	D	8	B	9	C	10	E

3. Question

1.	New seed policy	A.	1957
2.	Monsanto	B.	Pant Nagar
3.	MAHYCO	C.	1966
4.	All India co-ordinated maize improvement project	D.	1994
5.	All India co-ordinated sorghum and pearl millet improvement project	E.	New Delhi
6.	First SAU	F.	1988
7.	First sorghum hybrid released	G.	Bt brinjal
8.	High yielding variety programme	H.	1959
9.	GATT agreement	I.	Bt cotton
10.	ISST	J.	1964

♦ Answer Keys ♦

Q.No.	Ans.	Q.No.	Ans.	Q.No.	Ans.	Q.No.	Ans.	Q.No.	Ans.
1	F	2	I	3	G	4	A	5	H
6	B	7	J	8	C	9	D	10	E

4. Question

1.	Monsanto	A.	India
2.	Bayer CropScience	B.	Multiple window system
3.	Syngenta	C.	Single window system
4.	MAHYCO seeds	D.	Demand and supply
5.	Seed review team	E.	USA
6.	National Commission on Agriculture	F.	New Delhi
7.	Seed sale by State Agricultural Department	G.	Germany
8.	Seed sale by Quasi Government Organisation	H.	1968
9.	Seed pricing	I.	Switzerland
10.	Indian Seed Association	J.	1971

♦ Answer Keys ♦

Q.No.	Ans.	Q.No.	Ans.	Q.No.	Ans.	Q.No.	Ans.	Q.No.	Ans.
1	E	2	G	3	I	4	A	5	H
6	J	7	C	8	B	9	D	10	F

7.4. ARRANGE IN ORDER

1. Small seed industry – 40% - Public seed industry – 20% - Private large seed industry 40%

   Answer: Public seed industry – 40% - Private large seed industry 40% - Small seed industry – 20%

2. Phyto sanitary certificate – Post entry quarantaine – NBPGR – Pre entry quarantaine – Import of seed

   Answer: NBPGR – Pre entry quarantaine – Post entry quarantaine – Phyto sanitary certificate – Import of seed

3. The sequence of planning for seed production

   Capital investment – Execution of plan – Marketing – Demand forecasting

   Answer: Demand forecasting – Capital investment – Execution of plan – Marketing

4. The most important problem affects the seed marketing

   Lack of organization – Marketing location - Kind of seed - Purchase price

   Answer: Lack of organization - Kind of seed - Purchase price - Marketing location

7.5. DIFFERENTIATE THE FOLLOWING

1. Differentiate between public sector seed companies and private sector seed companies with examples

Public sector seed companies	Private sector seed companies
Public sector involved in the seed industry on a national scale began at the beginning of the green revolution with the establishment of the National Seed Corporation in 1963	Easing of government regulations in the late 1980's spurned enormous development within the seed industry by attracting several foreign seed companies to India
Eg. NSC, SSC, SFCI etc.	Eg. MAHYCO, Rasi seeds, Monsanto, Du-pont, Bayer CropScience, Syngenta etc.
Stated with the responsibility of promoting seed industry development from production through processing, storage and marketing and establishing a system of quality control	They identified potential crops for hybridization and started research and development activities. There are now undergoing a transition followed the Indian governments focus on biotechnology research, as a means of increasing agricultural production and also

Public sector seed companies	Private sector seed companies
	driven by trends in domestic and world seed market
Role is mostly confined to certified seeds of high volume, low value segment of high yielding varieties of cereals, pulses and cotton with a limited presence in the high value hybrid sec-tors of cotton, cereals and vegetables	Private seed firms undertake seed production of high value low volume seeds especially hybrids and vegetables
Public sector companies lag behind in research and they are mostly dependent in public research institutions, under the region of ICAR and SAU's for breeding seed requirements	There is no significant government intervention in the pricing of these hybrids and the Indian seed regulations permit.
Involves is production of breeder seed and foundation seed as well as certified seed	Produce only certified seed and truthful labelled seeds

2. Differentiate between seed day and kissan mela

Seed day	Kissan mela
Organized to impart awareness on quality seed production	Farmers through kissan melas to adopt latest agricultural practices
Ministry of Agriculture, GOI is funding this activity to promote seed village concept among seed producers	State/Central Agricultural Universities provided financial assistance to organize kissan melas
To popularise the high yielding varieties and new hybrids for adoption by farmers	It disseminates the technological advances and encourage the farmers to adopt new technologies in agriculture & allied areas for boosting agricultural production and their export
The importance of field standards and seed standards are being disseminated	It also disseminate latest technological knowledge of agriculture and allied sectors

3. Differentiate between single window system and multiple window system of seed marketing with examples

Single window system	Multiple window system
The seed producing organization supplies the seed to the farmer through independent retail outlet	The seed producing organization sup-plies seeds to the farmer through national distributors, wholesale sellers or regional distributor who, in

Single window system	Multiple window system
	turn, supply sub-distributors or the retail outlet
The seed supplier relies on the retailer for contact with consumer	The supplier having no direct contact with the consumer
Retail networks require strong service and support from the supplier	Products being strongly promoted in order to create demand
Good administrative control must be provided by the sales management	Supplying seed to the distributors in sufficient time to achieve timely availability at the retail level
The suppliers distribution system must be well organized and responsive	Management ensuring that there is a good system of monitoring sales and obtaining feedback from the consumer
Product quality at the retail level must be monitored for deterioration and adulteration and a return system should be considered	The distributor being interested only in the strongest selling lines
Although the product may be well promoted, the supplier relies on the retailer to make the final sale	If the seed supply is of good quality, the sale is very effective
Eg. Department of Agriculture, Co-operative societies	Eg. Private seed companies

4. Differentiate between government seed companies and semi government seed companies

Government seed companies	Semi government seed companies
Public sector corporations have an independent management or financial structure. As such, they have some financial autonomy but their operational strategies and approach to pricing are usually determined by official policy rather than by market forces	The main function of a semi govern-ment seed companies is to act on behalf of its member farmers in provi-ding them with inputs and in the marketing of their crops
Management is frequently expected to operate amid conflicting social and commercial objectives, while not loosing money	Individual co-operatives can be members of a single centralized state organization or independent, if owned directly by the farmer themselves
Government seed company may be integrated plant breeding, production and marketing	Co-operative groups are well placed to become involved in seed production on an independent or

Government seed companies	Semi government seed companies
organizations or simply act as the seed supplier with breeding being part of government financed research	contract basis and can act very effectively as marketing agents
Distribution may take place through government supply points, direct from seed plants or through Government seed sale outlet	The co-operative sells seed at prices which are agreed with government and surplus funds are distributed to members
Eg. State Agricultural Departments, NSC, SSC	Eg. Co-operatives

5. Differentiate between low volume seed and high volume seed

Low volume seed	High volume seed
Lesser quantity of seeds (less seed rate) are required for sowing	Higher quantity of seeds are required for sowing (higher seed rate)
Seeds may be smaller in size	Seeds may be larger in size
Value and profit is very high	Value and profit is low
Mostly produced and marketed by private seed companies	Mostly produced and marketed by public sector companies
Eg. Vegetable seeds	Eg. Cereals, Pulses

6. Differentiate between domestic seed company and international seed company

Domestic seed company	International seed company
Companies are situated in India	The companies are situated all over the world
Concentrate on cereals and vegetables (both high value and low value crops)	Concentrate mainly vegetables and flower crops (high value crops)
Concentrate both *Bt* and non *Bt* technologies	Concentrate on *Bt* technology in a large scale
Eg. Rasi seeds, MAHYCO, Nuziveedu seeds etc.	Eg. Monsanto, Syngenta, Du-pont Bayer CropScience etc.

7.6. IMPORTANT EVENTS IN INDIAN SEED INDUSTRY

1928 – Royal Commission on Agriculture submitted report citing that facilities for increasing the supply of breeder seed were inadequate and that varietal purity of seeds was not maintained.

1957 – All-India Co-ordinated Maize Improvement Project began.

1960	–	Similar projects on Sorghum and Pearlmillet began.
1961	–	Maize hybrids released.
1963	–	National Seeds Corporation established.
1964	–	First pearlmillet hybrid released; 250t of dwarf wheat was imported from CIMMYT, Mexico.
1966	–	Seed Act passed (came into force in Oct. 1969).
1963-69	–	Certified Seed Production increased from 360 ha to 35,000 ha.
1968	–	Seed Review team report was submitted; Seed rules formed.
1969	–	Tarai Development Corporation was established with World Bank assistance primarily to produce certified wheat seed; Concept of compact area development for seed production initiated.
1971	–	National Commission on Agriculture submitted its report; Government decided to re-organise seed production in India; ISST was formed.
1972	–	ISST organized first All India Seed Seminar
1975	–	Project Report on NSP was submitted.
1981	–	First workshop on Seed Technology under NSP was held.
1987	–	An Expert Group on Seed was constituted in october; The report was submitted in June 1989.
1988	–	A New Seed Policy to encourage import of seed and planting materials was announced in September.
1905	–	IARI
1929	–	ICAR
1952	–	Grow more seed enquiry committee (in First 5 year Plan 1951-56)
1956	–	Seed health testing was established in IARI.
1961	–	Seed Multiplication Team Review and World Seed Year; First STL (CSTL) was established
1969	–	SFCI established
1970	–	STL was started in Tamil Nadu
1971	–	First cotton hybrid (C.T. Patel); MSCS
1972	–	TNAU – Seed Pathology was started.
1973	–	First Seed Technology course in India at TNAU.
1976	–	NSP I
1981	–	NSP II
1983	–	Seed Control Order (under Essential Commodity Act, 1956)
1990	–	NSP III
1994	–	GOI signed – GATT approval; ISTA Rules were revised.

7.7. ABBREVIATIONS

1921 –	ESTA	European Seed Testing Association (Denmark)
1924 –	ISTA	International Seed Testing Association (Switzerland/Norway)
1971 –	ISST	Indian Society of Seed Technology, New Delhi
1963 –	NSC	National Seed Corporation, New Delhi (Central Seed Corporation)
1969 –	TDC	Tarai Development Corporation (World Bank Assistance)
1929 –	ICAR	Indian Council for Agricultural Research, New Delhi
–	ICRISAT	International Crop Research Institute for Semi arid and Tropics, Hydrabad
1969 –	AOSCA	Association of Official Seed Certifying Agency, U.S.A
1919 –	ICIA	International Crop Improvement Association, USA
1969 –	SFCI	State Farm Corporation of India, New Delhi.
1908 –	AOSA	Association of Official Seed Analyst
1922 –	SCST	Society of Commercial Seed Technologist
1922 –	CSAAC	Commercial Seed Analyst Association of Canada
1966 –	ISA	Indian Seed Act – Passed
1968 –	ISR	Indian Seed Rule
1975-76 –	NSP	National Seed Programme

ARIC	–	Agricultural Research Information Centre
AICBIP	–	All India Co-ordinated Bajra Improvement Programme
AICCIP	–	All India Co-ordinated Crop Improvement Programme
AICMIP	–	All India Co-ordinated Maize Improvement Programme
AICRIP	–	All India Co-ordinated Rice Improvement Programme
AICSIP	–	All India Co-ordinated Sorghum Improvement Programme
AICVIP	–	All India Co-ordinated Vegetable Improvement Programme
BP	–	Between paper
BSMT	–	Breeder Seed Monitoring Team
CHAs	–	Chemical Hybridizing Agents
CIAT	–	Centro International du Agriculture Tropic, Columbia
CIP	–	Centro International du Papa, Peru
CSC	–	Central Seed Committee
CSCB	–	Central Seed Certification Board

CSH	–	Co-ordinated Sorghum Hybrid
CVPO	–	Certification on the basis of Varietal Purity Only
CVRC	–	Central Variety Release Committee
DDT	–	Dunkel Draft Text
DUS	–	Distinctness, Uniformity, Stability
EET	–	Embryo Excised Test
ELISA	–	Enzyme Linked Imuno Sorbent Assay
EMC	–	Equilibrium Moisture Content
EPPO	–	European Plant Protection Organisation
FIST	–	International Seed Trade Federation
GA	–	Gibberellic Acid
GADA	–	Glutamic Acid Decarboxylase Activity Test
GATT	–	General Agreement on Tariff and Trade
GOT	–	Grow Out Test
IEF	–	Isoelectric Focussing
IITA	–	International Institute of Tropical Agriculture, Nigeria
IPPC	–	International Plant Protection Convention
IPR	–	Intellectual Property Right
MH	–	Maleic Hydrazide
MSCS	–	Minimum Seed Certification Standard
NPSD	–	New Policy on Seed Development
ODV	–	Other Distinguishable Variety
OEEC	–	Organisation for European Economic Co-operation
OGL	–	Open General License
OT	–	Off Type
PAGE	–	Poly Acrylamide Gel Electrophoresis
PBR	–	Plant Breeder's Right
PEO	–	Programme Evaluation Organisation
PEQ	–	Post Entry Quarantine
PIBA	–	Pot Immo Binding Assay
PLS	–	Pure Live Seed
PPA	–	Plant Protection Advisor
PSC	–	Phyto Sanitary Certificate
PVT	–	Primary Variety Trial
RCA	–	Royal Commission on Agriculture
RFLP	–	Restricted Fragment Length Polymorphism
RST	–	Registered Seed Technologist

RVS	–	Real Value Seed
SCA	–	Seed Certification Agency
SCP	–	Seed Certification Procedure
SCR	–	Seed Certification Regulation
SDS – PAGE	–	Sodium Dodicyl Sulphate – PAGE
SRT	–	Seed Review Team
SVRC	–	State Variety Release Committee
TP	–	Top of paper
TRIMS	–	Trade Related Investment Measures
TRIPS	–	Trade Related Intellectual Property Rights
TRP	–	Trade Related Policies
TS	–	Top of sand
TTT	–	Topographical Tetrazolium Test
TZ	–	Topographical Tetrazolium, 2,3,5–Triphenyl Tetrazolium Chloride
UPOV	–	International Union for Protection of new varieties of plant
URT	–	Uniform Registered Trial
UVT	–	Uniform Variety Trial
VCU	–	Value of Cultivation and Use
WTO	–	World Trade Organisation

7.8. BOOKS AND JOURNALS

7.8.1. Journals

India

1. Seed Research	-	ISST	-	Half yearly
2. Seed Tech News	-	ISST	-	Quarterly
3. Seeds and Farms	-	NSC	-	Monthly
4. Plant Variety Journal	-	PPV & FRA	-	Monthly

International

1. Seed Sci. & Technol. (1973) – Previously it was as Proceedings of ISTA till 1972, ISTA, Monthly.
2. Seed Science Research – Monthly.
3. Plant Varieties and Seeds, UK – Monthly.
4. Journal of Seed Technology – AOSA – Monthly.
5. Seed Pathology News – Monthly, Danish Institute of Seed Pathology.
6. Seed World – Monthly.

7. Seed Man's Digest – Monthly.
8. ISTA News Bulletin – Quarterly.
9. Advances in Research and Technology of Seed – Review articles – edited by ISTA.
10. Seed Technology Journal, USA
11. Seed Research, Iran, quarterly
12. Journal of Seed Science, Brasil
13. Journal of Applied Seed Production, Denmark
14. Research Journal of Seed Science, USA
15. International Journal of Advanced seed and Crop Technology, India
16. Seed Technology, USA
17. Seed Abstract, CAB, UK

7.8.2. Books

International Books

1. Seed Processing (1970) – B.R. Gregg, A.G. Law, S.S. Virdi and J.S. Bat.
2. Seed Storage and Packaging (1970) – J.F. Harrington and J.E. Douglas.
3. Germination of Seeds (1982) – H.M. Mayer and A.P. Mayber.
4. Principles and Practices of Seed Storage (1979), - Justice and Bass
5. Seed Biology (Vol. 1 & 2) – T.T. Kozlowski.
6. Seeds (1961) – USDA year book.
7. An introduction to Seed Technology (1979) – J.R. Thompson.
8. Seed Technology in the tropics – by ISTA (H.F. Chin, I.H. Enoch and R.M. Raja).
9. Principles of Seed Science & Technology. 4th Eddition – L.O. Copeland and M. McDonald.
10. Physiology and Biochemistry of seed in relation to seed germination (Vol.-I/II) – J.D. Bewley and M. Black.
11. Viability of Seed – E.H. Roberts, 1972
12. Vegetable Seed Production – Raymond, A.T. George.
13. Physiological basis of Seed Germination – K.E. Ovcharov.
14. The Physiology and Biochemistry of seed dormancy and germination – A.A. Khan.
15. Seed Pathology (Vol. I/II) – Dr. Paul Neergard.
16. Handbook of Seed Science and Technology - A.S. Basra, 2006
17. Handbook of Seed Physiology: application to agriculture – R.L. Beneeh – Amold and R.A. Sanches, 2004
18. Seed Technology and its Biological Basis – M. Black and D. Bewley, 2000
19. Seed Technology in the Tropics - F. Ader *et al.*, 1997
20. Vegetable Seed Technology

21. Advances in Research and Technology of Seeds Part 1 to 11 – J.R. Thompson
22. Seed Development : OMICS Technologies Forward Improvement of Seed Quality and Crop Yield : OMICS in Seed Biology - G.K. Agrawal and R. Rakwal, 2012
23. Planting Seeds - Kathryn Clay, 2016
24. Seed Inoculation, Coating and Precision Pelleting : Science, Technology and Practical Applications - G.M. Bennett and J. Lloyd, 2015
25. The Seed Production Technology of Xinjiang Vegetables and Melons – Jia Liltui, 2012
26. Seed Science and Technology Lab. Manual - McDonald, 1992
27. Seed conditioning, Volume 2: Technology .... Parts A&B – B. Gregg and G. Billups, 2010
28. Seeds: Their Conservation Principles and Practices – Veena Gupta, 2008
29. Seeds: Proceedings of Technical Conference on Improved Seed Production – FAO, 1983
30. Seed Identification Manual – A.C. Martin, 2010
31. The Encyclopaedia of Seeds : Science, Technology and Uses – M. Black and D.D Bewley, 2006
32. Seed Testing : Principles and Practices - S.G. Elias and L.O. Copeland, 2012
33. Synthetic Seed: A Challenging Technology in Plant Propagation, Transportation and Conservation - Bidhan Roy, 2010
34. Role of Seed Certification and Seed Industry – M.B. McDonald, 1986
35. Seeds: Ecology, Biogeography and Evolution of Dormancy and Germination - C.C Baskin and J.M. Baskin, 2014
36. Seeds: The Ultimate Guide to Growing Successfully from Seed - J. McVicar and M. Majerus, 2003
37. Seeds: Biology, Development and Ecology - S.W. Adkins and S.C. Navie, 2007
38. Seeds Handbook : Processiong and Storage - B.B. Desai, 2004
39. Principles of Seed Pathology - V.K. Agarwal and J.B. Sinclair, 1996
40. Seed Conditioning, Volume 1 : Management : A Practical Advanced – A level guide - B.R. Gregg and G.L. Billups, 2009
41. Seed Science and Technology - G. Merkebu, 2013

Indian Books

1. Recalcitrant Seeds : Causes & Effects - K. Sivasubramaniam
2. Seed Physiology of Crops - Prakash
3. A Handbook of Seed Certification - Mishra
4. A Handbook of Seed Processing & Marketing - Gaur
5. A Handbook of Seed Testing - Renugadevi

6. Advances in Seed Science & Technology (Vol. 1) : Recent Trends in Seed Technology & Management - K. Vanangamudi *et al.*, 2006
7. Advances in Seed Science & Technology (Vol. 2) : Quality Seed Production in Vegetable Crops - K. Vanangamudi *et al.*, 2006
8. Advances in Seed Science & Technology (Vol. 3) : Forest Tree Seed Technology & Management - K. Vanangamudi *et al.*, 2006
9. Advances in Seed Science & Technology (Vol. 4) : Forest Tree Seed Production - K. Vanangamudi *et al.*, 2007
10. Advances in Seed Science & Technology (Vol. 5) : Quality Seed Production in Spices, Plantation & Medicinal Crops - K. Vanangamudi *et al.*, 2008
11. Advances in Seed Science & Technology (Vol. 6) : Fruit Seed Production – K. Vanangamudi *et al.*, 2011
12. Advances in Seed Science & Technology (Vol. 7) : Flower Seed Production -K. Vanangamudi *et al.*, 2012
13. Agro's Dictionary of Seed Science & Technology - Ghosh
14. Principles of Post Harvest Seed Physiology & Technology - R.K. Maiti, 2006
15. Seed borne Diseases of Field Crops & their Management - Bhale
16. Plant Breeding & Hybrid Seed Production - Krishnan
17. Seed Production Techniques of Major Crops - Verma
18. Seed Technology & Seed Pathology - Suri
19. Seed : A Global Perspective - Kalloo
20. Seed Physiology - K. Vanangamudi *et al.*, 2006
21. X - Radiography of Seeds - K. Vanangamudi *et al.*, 2001
22. Textbook of Seed Science & Technology - Prakash
23. Agri Facts : Seed Science & Technology Model Papers & Dictionary Useful for JRF, ARS, NET, SAU, PhD, IFFCO, NFL - Prasad
24. Agrobiology : Essays on Problems of Genetics, Plant Breeding & Seed Growing -Lysenko
25. Breeding Biotechnology & Seed Production of Field Crops - Roy
26. Encyclopaedia of Seed Technology - Barua
27. Forest Tree Improvement & Seed Technology - Dutta
28. Fundamentals of Seed Production & Testing - Udai
29. Weed Seed Biology - K. Vanangamudi *et al.*, 2013
30. Handbook of Seed Industry : Prospects & its Costing - Ramdeo
31. Marketing of Seeds - Sharma
32. Organic Seed : Traditional Varieties & Technologies - K. Vanangamudi *et al.*, 2011
33. Principles & Practices of Seed Storage - Justice
34. Principles of Seed Pathology (2nd Edn) - Agarwal
35. Principles of Seed Science & Technology (4th Edn) - Copeland

36. FAO : Quality Declared Seed
37. Quarantine for Seed - Mathur
38. Recent Trends in Seed Technology - Mallick
39. FAO : Restoring Farmers Seed Systems in Disaster Situations
40. Seed Economics - Karanth
41. Seed Identification Manual - Martin
42. Seed Legislation - Sharma
43. Seed Legislation in India - K. Ramamoorthy *et al.*
44. Seed Pathology - Singh
45. Seed Pathology - Karanth
46. Seed Production of Agricultural Crops - Kelly
47. Seed Production of Field Crops - Tiwari
48. Seed Production Technology of Vegetables - Singh and B.S. Asati, 2013
49. Seed Quality - Singh
50. Seed Quality Enhancement : Principles & Practices - K. Vanangamudi *et al.*, 2010
51. Seed Science & Seed Technology - Shagufta
52. Seed Science & Technology - Gumam Singh, 2006
53. Seed Technology - Karanth
54. Seed Technology & Quality Control - P.C. Trivedi, 2010
55. Seed Technology of Horticultural Crops - Geetharani
56. Seed Testing - Singh
57. Seeds Handbook : Biology, Production, Processing & Storage (2nd Edn) - Desai
58. Seeds : Their Preservation & Longevity - Barton
59. Synthetic Seeds for Commercial Crop Production - Kumar
60. A Textbook of Practical Approaches in Seed Science & Technology - Chakraborty
61. A Textbook of Seed Science & Technology - Padmavathi
62. Vegetable Hybrid Seed : Production & Management- K. Vanangamudi *et al., 2012*
63. Vegetable Seedling Production Manual - David
64. The Encyclopaedia of Seeds : Science, Technology & Uses - Michael
65. Saving Seeds : The Economics of Conserving Crop Genetic Resources *ex situ* in the Future Harvest Centres - Bonwoo
66. Securing the Harvest : Biotechnology Breeding & Seed Systems for African Crops - Devries
67. Seed Fate : Predation, Dispersal & Seedling Establishment - Forget
68. Seed Trade in Rural Markets : Implications for Crop Diversity & Agricultural Development - Lipper
69. Seeds of Concern : The Genetic Manipulation of Plants - Murray

70. Testing Methods for Seed Transmitted Viruses : Principles & Protocols - Albrechtsen
71. The Tiny Seed - Eric Carie, 2000
72. Seed Technology Second Edition - R.L. Agrawal, 2014
73. Seed Technology : 2nd Revised Edition D. Kare, 2014
74. Objective Seed Science and Technology - K. Vanangamudi, 2017
75. Seed Science and Technology - S.V. Sai Prasad *et al.*, 2014
76. Principles of Seed Technology - Phundan Singh, 2013
77. Question Bank : Seed Science and Technology - K. Vanangamudi *et al.*, 2014
78. Objective Seed Science and Technology - P.R. Renganayagi *et al.*, 2008
79. Seed Science and Technology - Subir Sen and Nabinananda Ghosh, 2014
80. Practical Handbook of Seed Quality Testing Technology - Nabinananda Ghosh, 2012
81. Seed Potato Technology - H. Zandstra and P.C. Strulk, 1999
82. Seed Science & Technology - K. Raja *et al.*, 2012
83. Seed Science and Technology (An illustrated text book) - K. Vanangamudi, 2014
84. Seed Technology - O.P. Verma, 2011
85. Question Bank on Seed Science and Technology - G.V. Basavaraju, 2014
86. Objective Seed Technology - D. Prabha and Y.K. Negi, 2013
87. Vegetable Seed Production Technology - D.K. Singh and Y.K. Singh, 2006
88. Seed Technology - Apoorva Karanath, 2013
89. Forest Improvement and Seed Technology - M. Dutta, 2010
90. Seed Technology: A Practical Approach - J.K Sharma, 2012
91. Principles of Seed Technology - B.L. Jana, 2015
92. Recent Trends in Seed Technology - A.K. Chakraporty, 2013
93. Vegetable Seed Production and Hybrid Technology - P. Hazra and M.G. Som, 2009
94. Seed Technology and Seed Pathology - Shalini Suri, 2011
95. Seed Technology in the Tropics - D. Khare and M. Bhale, 2007
96. Experimental Seed Science and Technology - R Umarani *et al.*, 2008
97. Seed Science, Seed Technology and Seed Pathology - A. K. Choudhary, 2004
98. Advances in Seed Science and Technology - P.S. Shukla and B. Prasad, 2015
99. Principles of Seed Technology - Sukhdev Ransing and Aniket Kalhapure, 2013
100. Seed Technology and Seed Pathology - O.P. Chaubey and R.P.A. Sharma, 2014

101. Seed Technology and Quality Control - O.P Khedar and D.S. Chaudhary, 2012
102. Seed Technology, Plant Growth and Cropping System - P.K. Tyagi, 2015
103. Seed Technology and Seed Pathology - B. Kumar, 2012
104. Seed Technology and Hybrid Seed Production - D.K. Das, 2014
105. Seed Technology and Pathology - Ashok Kumar, 2014
106. Seed Science and Technology - B.D. Singh, 1999
107. Seed Science and Seed Technology - Shagufta, 2012
108. Encyclopaedia of Seed Technology - Sid Barua and Murali Krishnan, 2012
109. Principles of Seed Technology - P.K. Agrawal, 2010
110. Principles of Seed Technology - Sanjeev Kumar, 2014
111. Dictionary of Plant Breeding and Seed Technology - L.L Somani, 2010
112. Seed Production Techniques of Major Crops - O.P. Verma, 2013
113. Handbook of Seed Industry : Prospects and its Costing - Avinash Ramdeo, 2011
114. Seed Production of Field Crops - S.S. Mondal, 2009
115. Seed and Nursery Technology of Forest Trees - D.G.W. Edwards and S.C. Naithani, 1999
116. Techniques of Seed Science and Technology - P.K. Agarwal and M. Dadlani
117. Medicinal Plants and Seeds - K. Vanangamudi *et. al.*, 2014
118. Seeds and Agro's Dictionary of Plant Sciences - K. Vanangamudi *et. al.*, 2007
119. Agro's Dictionary of Plant Sciences, 2nd Edition - K. Vanangamudi *et. al.*, 2013
120. Text Book on Principles of Seed Production and Quality Control - M. Bhaskaran, *et. al.*, 2005
121. An Introduction to Tree Seed Technology - R. Umarani and K. Vanangamudi, 2004
122. Seed Hardening, Pellelting and Coating : Principles and Practices - K. Vanangamudi *et. al.*, 2006
123. Hybrid Seed Production of Agronomic Crops - K. Vanangamudi and A. Vijayakumar, 2014
124. Seed Physiology and Biochemistry Vol. 1: Seed Development and Maturation - K. Vanangamudi and Malika Vanangamudi, 2015
125. Recent Techniques and Participatory Approaches on Quality Seed Production - K. Vanangamudi *et. al.*, 2001
126. A Handbook of Seed Science and Technology - K. Vanangamudi *et. al.*, 2016
127. MCQ's in Plant Breeding, Biotechnology and Seed Science - K. Vanangamudi *et. al.*

7.9. INTERNATIONAL STL

1. ISTL – Thrandt/Saxony – Dr. Frederick Nobbe.
2. ISTA – Zurich, Switzerland.
3. Seed Technological Lab, Mississipi State University.
4. STL, Dept. of Primary Industry, Australia.
5. Govt. Seed Testing Station, Wageningen, Netherlands.
6. Manitoba STL, Canada.
7. New Zealand STL and Seed Technology Centre.
8. Danish Govt. Institute of Seed Pathology for developing countries, Copenhagen, Denmark.
9. Official Seed Testing Centre, Cambridge, U.K.

7.10. SCIENTISTS

1.	DNA finger printing	–	Jeffreys (1985)
2.	Phenol test	–	Waltz (1965)
3.	TTZ	–	Moore (1961)
4.	GADA test	–	Grab (1964)
5.	Peroxidase	–	Maxwell (1950)
6.	Sugar leachate	–	Takayanagi and Murakami (1969)
7.	EC	–	Mathew and Bradnock (1967)
8.	AA (Accelerated ageing)	–	Delouche (1960)
9.	Cold test	–	Isley (1950)
10.	Brick Gravel Test	–	Hiltner (1911)
11.	Seed size	–	Willey *et al.* (1966)
12.	Field emergence	–	Heydecker (1962)
13.	Speed of germination	–	Lawrence (1963)
14.	Vigour index	–	Abdul-Baki and Anderson (1973)

Unit 8

PROTECTION OF PLANT VARIETIES

8.1. CHOOSE THE CORRECT ANSWER

1. Chapter VII of PPV&FRA 2001 deals with
   (a) Finance, Accounts and Audit (b) Farmers right
   (c) Compulsory licence (d) Miscellaneous
2. The duration of protection of varieties of rice and maize under PPV&FRA, 2001 is
   (a) 15 years (b) 18 years
   (c) 6 years (d) 9 years
3. The gene fund shall be utilized for
   (a) Contribution (b) Compensation
   (c) Commercial protection of varieties (d) Collection of varieties
4. Duration of protection of extant varieties is
   (a) 15 years from the date of notification
   (b) 15 years from the date of DUS test
   (c) 9 years initially and renewal
   (d) 8 years initially and renewal
5. The periodicity of PVJ of India is
   (a) Quarterly (b) Half yearly
   (c) Annual (d) Monthly
6. The Nodal agency for implementing the PPV&FR act is
   (a) ICAR
   (b) Ministry of Agriculture, GOI
   (c) Ministry of Agriculture, State Govt.
   (d) SAUs
7. Compulsory licence will be granted after the expiry of
   (a) 3 years from the date of issue of certificate
   (b) 4 years from the date of issue of certificate
   (c) 5 years from the date of issue of certificate
   (d) 2 years from the date of issue of certificate
8. Protection of plant varieties under PPV&FRA, 2001 is
   (a) Mandatory (b) Voluntary
   (c) Based on request (d) Based on demand of seeds

9. The act that provides for protection of farmers and plant varieties is
   (a) Protection of Plant Varieties and Farmers' Rights Act
   (b) Essential Commodities Act
   (c) Seed Control Order
   (d) Environmental Protection Act

10. As on 30th November 2012, who is the chairman of PPV&FR Authority
    (a) Dr. N. Nagarajan (b) Dr. D.K. Mishra
    (c) Dr. P.L. Gautam (d) Dr. R.S. Paroda

11. The journal published by PPV&FR Authority is
    (a) Plant Variety Journal of India
    (b) Plant Variety Registration Journal
    (c) Protection of Plant Varieties Journal
    (d) Registration of Plant Varieties Journal

12. Chapter VI under PPV&FR Act, 2001 deals with
    (a) Compulsory license
    (b) Surrender/revocation of certificate
    (c) Appellate Tribunal
    (d) Farmes' Rights

13. How many members are present in PPV&FR Authority
    (a) 12 (b) 15
    (c) 18 (d) 21

14. Period of protection given for trees and vines crops under PPV&FR Act, 2001
    (a) 12 years (b) 15 years
    (c) 18 years (d) 21 years

15. The Head Quarter of PPV&FR Authority located at
    (a) New Delhi (b) Chennai
    (c) Bangalore (d) Mumbai

16. Number of Agro-biodiversity hot spots identified by PPV&FR Authority
    (a) 22 (b) 36
    (c) 18 (d) 42

17. As per PPV&FR Act, 2001, a variety is not sold/commercialized within India
    (a) Not earlier than one year from the date of application for registration
    (b) Not earlier than two year from the date of application for registration
    (c) Earlier than one year from the date of application for registration
    (d) Earlier than two year from the date of application for registration

18. Researchers Right under PPV&FR Act, 2001 is given in
    (a) Section 30 (b) Section 45
    (c) Section 39 (d) Section 47

19. As on 30th November, 2012, number of crops approved for registration under PPV&FR Act, 2001 is
   (a) 52 (b) 57
   (c) 62 (d) 67
20. Which one of the following award instituted under National Gene Fund of PPV&FR Authority
   (a) Plant Genetic Resources Conservation Award
   (b) Best Farmer/Community in Genome Conservation
   (c) Plant Genome Senior Community Award
   (d) Genetic Resources Conservation Award
21. Periodicity of publication of plant variety journal of India by PPV&FR Authority is
   (a) Bi-monthly (b) Monthly
   (c) Quarterly (d) Half yearly
22. The total validity period of certificate of registration issued to a crop species other than trees and vines is
   (a) 15 years (b) 12 years
   (c) 21 years (d) 18 tears
23. The provision for compulsory licence is given in
   (a) Chapter V (b) Chapter VI
   (c) Chapter VII (d) Chapter VIII
24. In PPV&FR Act 2001, provision of National Gene Fund is given under
   (a) Section 32 (b) Section 39
   (c) Section 47 (d) Section 45
25. As per PPV&FR Act 2001, initial validity of certificate of registration issued to tree and vine crops is
   (a) 5 years (b) 6 years
   (c) 7 years (d) 9 years
26. The right granted by a Government to an inventor to exclude others from imitating, manufacturing, using or selling the invention in question for commercial use during the specified period is called
   (a) Certificate (b) Copyright
   (c) Patent (d) License
27. The rights granted by the Government to a plant breeder to exclude others from producing or commercializing the propagating material of that variety for a minimum period of 15-20 years is known
   (a) Plant Breeders Rights (b) Patent Right
   (c) Copy Right (d) License Right

28. A variety that is released and granted a certificate of plant variety protection under the legal status of given country is referred to as
   (a) Protected variety (b) Extant variety
   (c) Notified variety (d) Obsolute variety
29. As per PPV&FR Act 2001, time limit to file opposition against an application is
   (a) 1 month from the date of registration
   (b) 1 month from the date of advertisement
   (c) 3 months from the date of registration
   (d) 3 months from the date of advertisement
30. COYU (The Combined-Over-Years Uniformity Criterion) has been proposed by
   (a) GOI (b) ISTA
   (c) UPOV (d) OECD
31. The UPOV convention was adopted in PARIS during
   (a) 1999 (b) 1963
   (c) 1961 (d) 1972
32. The secretariat of UPOV is located at
   (a) Pairs (b) Geneva
   (c) Rome (d) Washington D.C
33. As on 31st October 2012, the status of India in UPOV is
   (a) Member (b) Observer
   (c) Officiating member (d) Permanent member.
34. WTO is located at
   (a) Paris (b) New Delhi
   (c) Geneva (d) Bangalore
35. The duration of protection of varieties of rice and maize under PPV&FR act, 2001 is
   (a) 15 years (b) 18 years
   (c) 6 years (d) 9 years
36. The gene fund shall be utilized for
   (a) Contribution (b) Compensation
   (c) Commercial protection of varieties
   (d) Collection of varieties
37. The recommended quantity of seed of variety and hybrid for conducting DUS testing in rice is
   (a) 3000 gm (b) 2000 gm
   (c) 2500 gm (d) 1500 gm
38. In quantitative characteristics, the whole range of variation of a character is divided into
   (a) Nine states (b) Eight states
   (c) Seven states (d) Three states

39. The most used experimental design in DUS trial is
(a) Randomized incomplete block design
(b) Randomized complete block design
(c) Completely randomized design
(d) Split plot design

40. DUS conducting test fee for vegetables is
(a) Rs. 30,000 (b) Rs. 20,000
(c) Rs.15,000 (d) Rs. 39,000

41. Use of experimental design in DUS testing is largely determined by
(a) Number and nature of varieties (b) Test locations
(c) Period of testing (d) Number of growing cycle

42. The intensity of phenol colour reaction depend on the extent of
(a) Dehydrogenase activity (b) Tyrosinase activity
(c) Alpha amylase (d) Peroxidase activity

43. Newly identified characteristics in DUS testing are called
(a) Additional characteristics (b) Special characteristics
(c) Standard characteristics (d) Supporting characteristics

44. For cross pollinated species, the assessment of uniformity is based on
(a) Number of off-types (b) Variation in comparable variety
(c) Population standard (d) Relevant characteristics of variety

45. The tolerance limit of offtypes for establishment of uniformity in self pollinated crop for DUS testing is
(a) 0.5% (b) 1.0%
(c) 1.5% (d) 2.0%

46. In cereals, how many variants in 150 ears in row progency of the variety is considered as unstable
(a) 4 variants (b) 6 variants
(c) 8 variants (d) 10 variants

47. Which one of the following is the member of PPV & FR Authority?
(a) DDG (Edn) (b) DDG (Crop science)
(c) DDG (Horticulture) (d) DDG (Extension)

48. How many varieties/crops are registered in PPV & FR act as on 2012?
(a) 37 (b) 47
(c) 57 (d) 67

49. PPV&FR Authority is located in which place in New Delhi?
(a) NASC complex (b) IARI
(c) ICAR (d) Krishi Anusandhan Bhavan

50. Plant variety journal of India is published by
    (a) PPV & FRA (b) PBR
    (c) ICAR (d) IARI

51. DUS test centre for *Eucalyptus* is located at
    (a) ICFRE, Dehradun (b) IFGTB, Coimbatore
    (c) IARI, New Delhi (d) CICR, Bhopal

52. Where UPOV is located?
    (a) New York (b) Paris
    (c) London (d) Geneva

53. Which act deals with Essentially Derived Varieties?
    (a) UPOV 1951 act (b) UPOV 1961 act
    (c) UPOV 1971 act (d) UPOV 1981 act

54. Which section of UPOV 1961 act deals with assessment of essentially derived varieties?
    (a) Section I (b) Section II
    (c) Section III (d) Section IV

55. Which of the following is not a measure of biodiversity?
    (a) Economic diversity (b) Ecosystem diversity
    (c) Genetic diversity (d) Species diversity

56. Which of the following conditions does not define the fundamental niche of a species
    (a) Humidity (b) Predators
    (c) Temperature (d) Salinity

57. Among the following, whose responsibility is to inform the buyer, if the variety is protected?
    (a) Seed dealer (b) Grower
    (c) Farmer (d) Seed conditioner

58. In which year, the first GMOs was produced?
    (a) 1952 (b) 1962
    (c) 1972 (d) 1982

59. In which crop, the first GMO plant was developed?
    (a) Tapioca (b) Turmeric
    (c) Tobacco (d) Tamarind

60. In tobacco, the first GMO crop, what type of resistance has been developed in the first GMO variety
    (a) Insect resistance (b) Herbicide resistance
    (c) Disease resistance (d) Herbicide tolerance

61. Which one of the following crop has many Bt hybrids on commercial scale
    (a) Rice (b) Redgram
    (c) Cotton (d) Sorghum

62. Which one of the following crop has most genetically modified plants by research?
(a) *Lactuca sativa* (b) *Zea mays*
(c) *Glycine max* (d) *Arabidopsis thaliana*

63. Which seed company has the largest share of the GMO crops planted globally
(a) Dupont (b) Pionner Hi-Breed
(c) Monsanto (d) Bayer CropScience

64. In which crop, GMO plant has been developed to improve fuel product
(a) Neem (b) Pungam
(c) Jatropha (d) Paradise tree

65. Which country is growing GMO crops in a larger area
(a) USA (b) India
(c) Australia (d) Europe

66. In USA during 2009-10, the GMO crop that is grown in higher percentage is
(a) Cotton (b) Maize
(c) Soybean (d) Sugarbeet

67. In golden rice, the genetic transformation is done to increase the content of
(a) Vitamin A (b) Vitamin B
(c) Vitamin C (d) Vitamin D

♦ Answer Keys ♦

Q.No.	Ans.	Q.No.	Ans.	Q.No.	Ans.	Q.No.	Ans.	Q.No.	Ans.	Q.No.	Ans.
1	c	2	a	3	b	4	a	5	d	6	b
7	a	8	b	9	a	10	c	11	a	12	d
13	b	14	c	15	a	16	a	17	c	18	a
19	b	20	c	21	b	22	a	23	c	24	d
25	d	26	c	27	a	28	a	29	d	30	c
31	c	32	b	33	b	34	c	35	a	36	a
37	a	38	a	39	b	40	b	41	a	42	b
43	a	44	b	45	b	46	c	47	b	48	c
49	a	50	a	51	b	52	d	53	b	54	b
55	a	56	b	57	a	58	d	59	c	60	a
61	c	62	d	63	c	64	c	65	a	66	d
67	a										

8.2. FILL IN THE BLANKS

1. Abbreviation of PPV&FRA is Protection of Plant Varieties and Farmers' Right ....
2. The Protection of Plant Varieties and Farmers' Right Act (PPV& FR Act) was implemented during ....
3. The provision for National Gene Fund was given under section .... of PPV&FR Act, 2001
4. Objections against an application should be filled within .... months from the date of advertisement
5. Initial validity period of certificate of registration issued by the PPV&FR Authority for tree and vine crops is ....
6. Section 39 of PPV&FR Act, 2001 deals with ....
7. Chapter VII of PPV&FR Act, 2001 deals with ....
8. The office of PPV&FR Authority is situated at ....
9. The total period of protection for trees and vine crops is ....
10. The journal published by PPV&FR Authority is ....
11. World Trade Organization located at ....
12. A variety needs to be tested for .... independent growing cycles for DUS test to obtain certificate of registration
13. .... system of DUS testing is in operation in Canada
14. Plant Variety Right (PVR) is also known as ....
15. PVR is the right granted to ....
16. UPOV is located at ....
17. UPOV expands for ....
18. The criteria for new varieties to be protected for Distinctness, ... and Stability.
19. The first UPOV act was drafted in ....
20. If distinctness, conformity and .... from an initial variety are not fulfilled, there is no essential ....
21. The proof of predominant derivation is tested by ....
22. Section .... in the PPV and FR Act 2001 deals with essentially derived varieties.
23. The other name for the Biodiversity Act is ....
24. Biological Diversity Act came into existence from ....
25. Abbreviate BMC ....
26. Amendments in the patent act came into existence in ....
27. Protection of Plant Varieties and Farmer's Right Act (PPV&FRA) came into existence from ....
28. Abbreviate PPV&FR ....
29. TRIPS came into existence during ....
30. Abbreviation for TRIPS ....

31. The variety which been traditionally cultivated and evolved by the farmers in their fields is called as ....
32. Any other variety which is in public domain is called as ....
33. New seed policy came in to existence from ....
34. Act which was enacted in India to protect the new plant varieties is ....
35. Term of plant protection for trees and vines is about .... from registration of the variety.
36. Term of plant protection for extant variety is about .... from registration of the variety.
37. Genetically modified crops are plants in which .... has been modified using ....
38. The first GMOs was produced in 1982, using an antibiotic resistant .... plant
39. In USA, GMO tomato was approved for sale during the year ....
40. In Europe, GMO tobacco was approved for sale during the year ....
41. Mansanto Bt cotton is resistant for ....
42. Most GMOs are generated by the ....
43. .... plants have gene inserted into them that are derived from another species
44. .... plants are made using genes found within the same species or closely related one.
45. The GMO rice for increased nutritive value is called as ....
46. The GMO plants are developed in sugar beet for resistance to ....
47. Expand VCU ....
48. Expansion of WIPO ....
49. Expansion of PCT ....
50. Expansion of EPC ....
51. Expansion of WTO ....
52. Expansion of BDA ....
53. Expansion of PVP ....
54. Expansion of PPV & FR ....
55. Expansion of PPV & FRA ....
56. Expansion of IPR ....

♦ Answer Keys ♦

Q.No.	Ans.	Q.No.	Ans.
1	Authority	2	2001
3	45	4	Three

Q.No.	Ans.	Q.No.	Ans.
5	9 years	6	Farmers' Right
7	Compulsory license	8	Delhi
9	18 years	10	Plant Variety Journal of India
11	Geneva, Switzerland	12	Two
13	Breeder	14	Plant Breeders Right (PBR)
15	Breeder	16	Geneva
17	International Union for the Protection of New varieties of Plants	18	Uniformity
19	1961	20	Pre-dominant derivation/deviation
21	Molecular characteristics/ Combining ability/ Phenotype characters/ Breeding records	22	2(i)
23	Biological Diversity Act	24	2002
25	Biological Management Committee	26	1970
27	2001	28	Protection of Plant Varieties and Farmer's Right
29	1994	30	Trade Related Intellectual Property Rights Agreement
31	Farmer's variety	32	Extant variety
33	2002	34	PPV&FR
35	18 years	36	15 years
37	DNA, Genetic Engineering	38	Tobacco
39	1994	40	1995
41	Herbicide	42	Biolistic method (Particle or Agro bacterium mediated transformation)
43	Transgenic	44	Cisgenic
45	Golden rice	46	Glyphosate
47	Value for cultivation and use	48	World Intellectual Property Organization
49	Paris Convention Treaty	50	European Patent Convention
51	World Trade Organization	52	Biodiversity Act
53	Plant Variety Protection	54	Protection of Plant Varieties and Farmers Right Act
55	Protection of Plant Varieties and Farmers Right Authority	56	Protection of Intellectual Property Rights

8.3. STATE TRUE OR FALSE

1. PPV&FR Act does not define the term distinct
2. Compulsory licensing can be issued only after 3 years from the date of issue of certificate of registration
3. A person can apply for registration under PPV&FR Act, 2001 for maximum of two varieties in a year

4. A variety of foreign origin cannot be registered under PPV&FR Act, 2001
5. Certificate of registration issued by the PPV&FR Authority can be revoked at any time
6. Any one can apply for variety protection under PPV&FR Act, 2001
7. PPV&FR Authority has all the powers of a civil matters related to variety registration
8. Farmers right given under PPV&FR Act, 2001 is unique in the world
9. PPV&FR Act, 2001 replaced the Seed Act, 1966
10. Vegetatively propagated crops/tree crops cannot be registered under PPV&FR Act, 2001
11. PPV&FR Act does not define the term distinctness.
12. Transgenic/GM crops are not eligible for registration for making decision on distinctness
13. National test guidelines for DUS have been developed for 45 crops
14. For DUS testing, seed should have germination higher than IMSCS of the crop
15. In DUS trial, only one treatment factor is available
16. India is a founder member of UPOV
17. In DUS testing, completely randomized design is recommended, if the number of test varieties is large
18. The variety of common knowledge does not include protected varieties
19. Truly qualitative characteristics are influenced by the environment
20. For making decision on distinctness, choice of proper characteristics is necessary
21. PCR based marker are second generation DNA profiling technique
22. A rice plant developed for stress resistance is called as Golden rice
23. The GM *Arabidopsis* are produced to bioremediate and clean up TNT and RDS explosive contaminants from the soil
24. GM corn is threat to the environment
25. In India, *Bt* maize is the fruit and only GM crop commercially cultivated
26. The largest share of GM crops planted globally are from the seed created by Europe firm Monsanto
27. The first tomato GM plant has developed in USA for longer shelf life
28. Bt cotton is resistance for herbicide
29. GM plant of canola/rapeseed is resistant for herbicide
30. GM plant of tomato is to retard softening of fruit after harvest
31. GM plant of sweet pepper is resistant to herbicide
32. VCU test of newly evolved variety is compared with existing cultivars for 2 to 3 years

◆ **Answer Keys** ◆

Q.No.	Ans.	Q.No.	Ans.	Q.No.	Ans.	Q.No.	Ans.	Q.No.	Ans.q	Q.No.	Ans.
1	True	2	True	3	False	4	False	5	True	6	True
7	True	8	True	9	False	10	False	11	False	12	True
13	False	14	True	15	True	16	False	17	False	18	False
19	False	20	True	21	True	22	False	23	True	24	False
25	False	26	False	27	True	28	False	29	True	30	True
31	False	32	True								

8.4. MATCH THE FOLLOWING

1. Question

1.	UPOV Act	A.	1970
2.	Plant variety registry	B.	Rome
3.	GATT	C.	Indian variety
4.	WTO	D.	1930
5.	Indian patents act	E.	1991
6.	FAO	F.	Geneva
7.	Basumati	G.	Registration of plant variety
8.	Exmati	H.	1994
9.	USA plant patent act	I.	Genetic engineering
10.	DNA technology	J.	US trade mark registry

◆ **Answer Keys** ◆

Q.No.	Ans.	Q.No.	Ans	Q.No.	Ans	Q.No.	Ans	Q.No.	Ans
1	E	2	G	3	H	4	F	5	A
6	B	7	C	8	J	9	D	10	I

2. Question

1.	Distinctness of crops	A.	2001
2.	VCU test	B.	Traditional variety
3.	Duration of protection of varieties of trees and vines	C.	A system of their own
4.	Duration of protection of varieties of other crops	D.	Two years
5.	Compulsory license	E.	Notified variety
6.	PPV & FR Act	F.	2 to 3 years
7.	*Suigeneris*	G.	Initial variety
8.	Extant variety	H.	18 years

9.	Farmers variety	I.	15 years
10.	Essentially derived varieties	J.	PPV & FR registered varieties are not available to the public

♦ Answer Keys ♦

Q.No.	Ans.	Q.No.	Ans.	Q.No.	Ans.	Q.No.	Ans.	Q.No.	Ans.
1	D	2	F	3	H	4	I	5	J
6	A	7	C	8	E	9	B	10	G

8.5. FIND OUT THE INCORRECT STATEMENT

1. a. Farmers right given under PPV & FR Act 2001 is unique in the world

b. Compulsory licensing can be issued only after 3 years from date of issue of certificate of registration

c. A variety of foreign origin cannot be registered under PPV & FR Act 2001

♦ Answer Keys ♦

Q.No.	Ans.
1	c

8.6. DIFFERENTIATE THE FOLLOWING

1. Differentiate between breeder's right and breeder's exemptions

Breeder's right	Breeder's exemptions
Breeders right has been specially created as intellectual property right in plant breeding	Breeders exemption allows plant breeders to freely use protected varieties for the development of new varieties
It protects one variety and the genetic material and the breeding procedures of this variety is not protected for further breeding	Breeder exemption allows breeders to use protected varieties, source variety of plants or for other experimental purpose
It provides sufficient protection for breeders to recover their investments on a new variety	Breeders exemptions leaves other bre-eders the opportunity to use the protected variety to develop new varieties
Breeder's right also includes a farmer's privilege and gives farmers under certain condition the right to reuse seeds of protected varieties	Farmer's privilege allows farmers to save and reuse harvested materials such as seeds for the next season

2. Differentiate between intellectual property and intellectual property rights

Intellectual property	Intellectual property rights
An idea, a concept, a design, a process, etc., which can ultimately be translated into a useful product	The right of an individual to desire benefits from his intellectual property, and to exclude others from doing so
The IP has to be developed into a product of economic value	IPR may be protected in several ways, the main forms of protection being trade secret, patent, plant breeder's right and copy right.

3. Differentiate between farmer's right and farmer's privilege

Farmer's right	Farmer's privilege
Farmers can save, use, reuse, exchange, share or sell the farm product, including seed (except for branded seed) of a protected variety	A PBR system allows a farmer to use a part of the material produced on his farm, from a protected variety, for planting his own fields without any obligation to the PBR title holder, this is called farmer's privilege
Reward for those farmers who are engaged in the conservation of genetic resources of land races and wild relatives of economic plants and their improvement through selection and preservation provided that the materials so selected and preserved have been used as donors of genes in the registered varieties	Farmers privilege allows a farmer to use his own produce, a seed (= propagule), but does not allow him to exchange seeds with other farmer

4. Differentiate between novelty and stability

Novelty	Stability
Variety must not have been exploited commercially for more than one year before granting of PBR protection	Variety must be stable in appearance and its claimed characteristic over generations under the specified environment

5. Differentiate between farmer's variety and extant variety

Farmer's variety	Extant variety
Farmer's variety is one that was evolved by farmers/farming communities over several years and has proven special features compared to other materials (or) a	A variety which is notified under Seed Act 1966 or a variety about which there is common knowledge or a farmer's variety or any other which is public domain, is

Farmer's variety	Extant variety
variety which is wild relative or land race of a variety about which farmers possess common knowledge	considered as an extant variety. Registration of extant varieties will be done within a specified period and subject to their meeting the criteria of distinctness, uniformity and stability Duration of the protection is 15 years from the date of notification of that variety

6. Differentiate between Essentially Derived Variety and Genetically Modified Variety

Essentially Derived Variety	Genetically Modified Variety
An essentially derived variety is a variety predominantly derived from the initial variety, which retains the expression of the essential characteristics from the genotype or combination of genotypes of the initial variety	Genetically modified crops or variety or plants, in which the DNA has been modified using genetic engineering techniques, to resist pests, weeds and other agents causing harm to plants and to improve the growth of these plants including nutritional quality to assist in farmers efficiency
For example, a variety produced by mutation or transfer through back cross of a single gene will be regarded as an essentially derived variety	*Bt.* cotton is resistant to boll worm (pest resistance)
Essentially derived varieties developed from protected variety are subjected to the plant breeders right protection granted to initial variety	GM variety does not cover under PPV & FR Act
The breeders of essentially derived variety must obtain authorization from the breeder of the initial variety	Not eligible for registration in National Variety Registry

7. Differentiate between breeder's right and farmer's right

Breeder's right	Farmer's right
Breeder's right extend to seeds and/ or propagating material of the protected variety to production, selling, marketing, distribution, export and import of seed	Farmer's right entitled to save, use, resow, exchange, share or sell his farm produce including seed in the same manner as he was entitled earlier
However, if the breeders variety is essentially derived from a farmers	Full disclosure of the expected performance of the seeds or planting

Breeder's right	Farmer's right
variety, the breeder cannot give any authorization without the consent of the farmers or communities from whose varieties the protected variety is derived	material by the plant breeder. Where these fail to perform in the manner claimed by the breeder, the farmer may claim compensation from the plant breeder
	Reward the farmers who is engaged in the conservation and preservation of genetic resources of land races and wild relatives of economic plants and this improvement through selection and presentation

8. Differentiate between distinctness and uniformity

Distinctness	Uniformity
New variety must be distinguishable from other varieties by one or more identifiable morphological, physiological and other characteristics	Variety must be uniform in appearance under the specified environment of its adaptation

9. Differentiate between DUS test and VCU test

DUS test	VCU test
DUS stands for distinctness, uniformity and stability test	VCU stands for value for cultivation and use test
This is a criteria on the basis of which the plant breeder's rights will be granted to a variety by the authority	Objective of VCU test is for assess the differences of productivity, biological characteristics, quality of the obtained produce, chemical and technological characteristics from the standard variety, resistance to pests and diseases as well as other commercially important characteristics
DUS test used as main criteria for deciding the novelty of a variety	To assess the growth suitability of the varieties under varying agro climatic conditions

Part 2

Advances in Seed Science and Technology : Seed Physiology and Biochemistry

Unit 1

SEED DEVELOPMENT AND MATURATION

1.1. THE CELL AND ITS ORGANELLES

1.1.1. Choose the correct answer

1. Which of the following organelles would only be found within a cell that was both eukaryotic and autotrophic?
   (a) Mitochondria (b) Ribosomes
   (c) Rough endoplasmic reticulum (d) Chloroplast
2. After being formed by the ribosomes located on the endoplasmic reticulum, what is the next organelle to which a protein could be transported?
   (a) Mitochondria (b) Smooth endoplasmic reticulum
   (c) Golgi apparatus (d) Nucleus
3. Which is not true about the cell theory?
   (a) Its various parts were described by Schleiden, Schwann, and Virchow.
   (b) It states that all organisms are composed of cells.
   (c) It states that all cells come from preexisting cells.
   (d) It states that bacteria and other small organisms can arise spontaneously.
4. Which is not true of eukaryotic cells?
   (a) A true nucleus contains the chromosomes.
   (b) Eukaryotic cells contain membrane-bounded compartments.
   (c) They contain ribosomes that are smaller than those of prokaryotic cells.
   (d) They all contain mitochondria.
5. Cells that do not produce cell walls are
   (a) Animal cells. (b) Plant cells.
   (c) Bacteria. (d) Prokaryotic cells.
6. The Greek root word that means "true kernel" is
   (a) Prokaryote. (b) Eukaryote.
   (c) Nucleolus. (d) Nucleoid.

7. A eukaryotic organelle that can easily be seen with the light microscope is the
   (a) Endoplasmic reticulum. (b) Nucleus.
   (c) Ribosome. (d) Polyribosome.
8. Which is a true statement about ribosomes?
   (a) Ribosomes contain DNA and protein
   (b) Ribosomes are active in carbohydrate synthesis
   (c) Ribosomal subunits leave the nucleus after being formed by the nucleolus
   (d) Polyribosomes are the subunits of ribosomes
9. Which of these is not part of the endomembrane system of the cell?
   (a) Mitochondria (b) Endoplasmic reticulum
   (c) Lysosomes (d) Golgi complex
10. The Greek root word that means "net" and "within" is
    (a) Prokaryote (b) Eukaryote
    (c) Cytoplasmic (d) Endoplasmic reticulum
11. Which is not true concerning the golgi complex?
    (a) It consists of a stack of saccules.
    (b) Golgi complexes may have an inner face and an outer face.
    (c) Molecules are modified within the lumen of golgi saccules.
    (d) Lysosomes are vesicles that bind to enter the golgi saccules.
12. Lysosomes are produced by the
    (a) Vacuoles. (b) Nucleus.
    (c) Mitochondria. (d) Golgi apparatus.
13. Membrane-bounded vesicles that contain enzymes for oxidizing small organic molecules with the formation of hydrogen peroxide are
    (a) Vesicles. (b) Glyoxisomes.
    (c) Lysosomes. (d) Peroxisomes.
14. Large membranous sacs that are more prevalent in plant cells and some protozoa than in animal cells are called
    (a) Vacuoles. (b) Vesicles.
    (c) Glyoxisomes. (d) Lysosomes.
15. Which is not a characteristic of chloroplasts?
    (a) A chloroplast is surrounded by a double membrane.
    (b) Chlorophyll absorbs solar energy that is used to form chemical bonds.
    (c) A membrane inside a chloroplast is called a thylakoid.
    (d) Chlorophylls are found in the fluid stroma of the chloroplast.
16. Which is not a characteristic of mitochondria?
    (a) A mitochondrion has two membranes.
    (b) Mitochondria are the site of cellular respiration.
    (c) Mitochondria are found in prokaryotic and eukaryotic cells.
    (d) Mitochondria contain DNA and ribosomes.

17. Which of the following is not true concerning cytoskeletal elements?
    (a) Cilia are small extensions of membrane-surrounded microtubules.
    (b) Microtubules are made up of a globular protein called tubulin.
    (c) Intermediate filaments compose flagella.
    (d) Flagella have a 9 + 2 pattern of microtubule structure.
18. Prokaryotes are characterized by all of the following structures except
    (a) A nucleoid. (b) Inclusion bodies.
    (c) Mitochondria. (d) Mesosomes.
19. All of the following are parts of plant cells except
    (a) Centrioles. (b) Mitochondria.
    (c) Chloroplasts. (d) Cell walls.
20. Mitochondria are also called as
    (a) Suicidal bags (b) Digestive bags
    (c) Protein synthesizers (d) Power house of the cell
21. ATP stands for
    (a) Adenosine triphosphate (b) Adenosine diphosphate
    (c) Advanced tactical plae (d) Adenosine terphosphorus
22. Centrosomes are present in ....
    (a) Animal cells only
    (b) Plant cells only
    (c) In both plant cells and animal cells
    (d) In neither plant cells nor animal cells
23. Vacuoles (in plant cells) ....
    (a) Produce energy for the cell
    (b) Synthesize proteins
    (c) Defend the cell
    (d) Provide turgidity and rigidity to plant cells
24. Organelle not present in animal cells is ....
    (a) Plastid (b) Centrosome
    (c) Nucleus (d) Mitochondria
25. The secretory organelle of the cell is also called as
    (a) Ribosomes (b) Golgi apparatus
    (c) Peroxisomes (d) Centrosomes
26. The 'Scavengers' or 'Digestive bags' of a cell are ....
    (a) Chromosomes (b) Centrosomes
    (c) Lysosomes (d) Ribosomes
27. The site of oxidation in a cell is the ....
    (a) Mitochondrion (b) Endoplasmic reticulum
    (c) Golgi apparatus (d) Ribosome

28. Lysosomes are reservoirs of ....
   (a) Fat (b) Secretory glycoproteins
   (c) RNA (d) Hydrolytic enzymes
29. The network of endoplasmic reticulum is present in the
   (a) Nucleus (b) Chromosome
   (c) Cytoplasm (d) Golgi apparatus
30. Centrioles are associated with ....
   (a) Digestion (b) Respiration
   (c) Reproduction (d) Cell division
31. Double membrane is absent in ....
   (a) Lysosomes (b) Chloroplast
   (c) Nucleus (d) Mitochondria
32. Organisms lacking nucleus and membrane bound organelle are ....
   (a) Diploids (b) Prokaryotes
   (c) Haploids (d) Eukaryotes
33. The organelle of the endomembrane system associated with the sorting of lipids and proteins for various cellular functions are
   (a) Rough endoplasmic reticulum. (b) Lysosomes.
   (c) Vesicles. (d) Golgi apparatus.
34. Which of the following gives rise to both lysosomes and vesicles?
   (a) Rough endoplasmic reticulum (b) Mitochondria
   (c) Golgi apparatus (d) Ribosomes
35. Of the following, which is not associated with the mitochondria?
   (a) ATP production (b) Cristae
   (c) Stroma (d) Matrix
36. Which organelles are most directly involved in transporting materials out of the cell?
   (a) Nucleus and ribosomes
   (b) Chloroplast and mitochondria
   (c) Cell membrane and cell wall
   (d) Golgi apparatus and cell membrane
37. Which organelle is the site of cellular respiration in both animal and plant cells?
   (a) Nucleus (b) Chloroplasts
   (c) Mitochondria (d) Vacuole
38. Which statement best describes the cell membrane in a typical plant cell? The membrane
   (a) Selectively controls what enters and exits the cell.
   (b) Is composed of protein and carbohydrates only.
   (c) Has the same permeability to all substances moving in and out the cell.
   (d) Is a composed of two protein layers with lipids floating inside.

39. Which organelle is responsible for photosynthesis?
   (a) Nucleus (b) Chloroplasts
   (c) Mitochondria (d) Vacuole
40. Which of the following is present in a typical plant cell, but not in an animal cell?
   (a) Mitochondria (b) Cell wall
   (c) Ribosome (d) Golgi apparatus
41. Which organelle functions in the storage of water and biomolecules?
   (a) Vacuole (b) Cell Wall
   (c) Endoplasmic reticulum (d) Chloroplast
42. Plants are unable to move to a new location in response to changes in their environment. Which of the following organelle plays a role in maintaining homeostasis in a plant?
   (a) Cell Wall (b) Chloroplast
   (c) Vacuole (d) Nucleus
43. What is the net energy production of glycolysis
   (a) 2 ATP (b) 4 ATP
   (c) 6 ATP (d) 8 ATP
44. In which one of the following, the Kreb cycle occurs
   (a) Mitochondria (b) Cytoplasm
   (c) Ribosome (d) Plastids
45. Kreb cycle also called
   (a) TCA cycle (b) EMP pathway
   (c) Electron transport chain (d) Pentose phosphate pathway
46. Glycolysis also called
   (a) TCA cycle (b) EMP pathway
   (c) Electron transport chain (d) Pentose phosphate pathway
47. The word cell comes from
   (a) Latin (b) Greek
   (c) French (d) English
48. The basic structural and functional unit of all known living organisms
   (a) Cell (b) Lysosome
   (c) Ribosome (d) Plastids
49. Out come of electron transport chain reaction
   (a) 32 ATP (b) 36 ATP
   (c) 34 ATP (d) 38 ATP
50. Cytoplasm and plastids is site of
   (a) Protein synthesis (b) Lipid synthesis
   (c) Starch synthesis (d) Hormone synthesis

51. Word cell coined by

(a) Robert Hooke (b) T.H. Morgan

(c) R. Brown (d) Schleiden and Schwann

52. Formulated the cell theory

(a) R. Bonding (b) M. Hussey

(c) W. Fleming (d) Schleiden and Schwan

♦ Answer Keys ♦

Q.No.	Ans.	Q.No.	Ans.	Q.No.	Ans.	Q.No.	Ans.	Q.No.	Ans.	Q.No.	Ans.
1	d	2	c	3	d	4	c	5	a	6	b
7	b	8	c	9	a	10	d	11	d	12	d
13	d	14	a	15	d	16	c	17	c	18	c
19	a	20	d	21	a	22	a	23	d	24	a
25	b	26	c	27	a	28	d	29	c	30	d
31	d	32	b	33	d	34	c	35	c	36	d
37	c	38	a	39	b	40	b	41	a	42	c
43	a	44	a	45	a	46	b	47	a	48	a
49	a	50	c	51	a	52	d				

1.1.2. Choose the correct answer and explain why?

1. Which structure includes all of the others?

(a) Nucleolus (b) Nucleus

(c) Chromosomes (d) Genes ♦ **Ans:** b

Explanation: The nucleus is the large dark spot in cells. Inside are the chromosomes (pieces of DNA), which contain genes (pieces of chromosomes which correspond to our traits), as well as nucleoli (which are the sites of ribosome synthesis).

2. Most cell membranes are composed principally of

(a) DNA and ATP (b) Proteins and lipids

(c) Chitin and starch (d) Nucleotides and amino acids

♦ **Ans:** b

Explanation: Singer proposed the fluid mosaic model of the cell membrane in which the cell membrane consists of a double lipid layer with proteins embedded in it to act as name tags or to aid in movement of larger molecules across the membrane. DNA is the hereditary information stored in the nucleus and ATP is the energy capturing molecule found throughout the cell. Chitin and starch are large carbohydrates and nucleotides make up DNA, while amino acids are the tiniest proteins.

3. Which structure is usually present only in animal cells?

(a) Vacuole (b) Cell wall

(c) Nucleus (d) Centriole ♦ **Ans:** d

Explanation: Animal cells contain centrioles, cylindrical organelles that direct the movement of the chromosomes during mitosis. Plant cells tend to have bigger vacuoles and they also have rigid cell walls. Both have nuclei to control the cell.

4. **To enter or leave a cell, substances must pass through**
   (a) A microtubule. (b) The golgi apparatus.
   (c) A ribosome. (d) The plasma membrane. ♦ **Ans:** d

   Explanation: The plasma membrane defines the boundary of the cell.

5. **Bacterial cell are prokaryotic; in comparison to a typical eukaryotic cell they would**
   (a) Be smaller.
   (b) Have a smaller nucleus.
   (c) Lack a plasma membrane.
   (d) Have fewer internal membranous compartments. ♦ **Ans:** b

   Explanation: Most prokaryotic cells are much smaller than typical eukaryotic cells. The smallest known cells are bacteria called mycoplasmas.

6. **The maximum size of a cell is limited by**
   (a) Its need for enough surface area for exchange with its environment.
   (b) The number of organelles that can be packed inside.
   (c) The materials needed to build it.
   (d) The amount of flexibility it needs to be able to move. ♦ **Ans:** a

   Explanation: Since volume increases at a faster rate than surface area, large cells have proportionately less plasma membrane surface area.

7. **Which of the following correctly matches an organelle with its function?**
   (a) Mitochondrion _____ Photosynthesis
   (b) Nucleus _____ Cellular respiration
   (c) Ribosome _____ Manufacture of lipids
   (d) Central vacuole _____ Storage ♦ **Ans:** d

   Explanation: But remember, plant cells have a central vacuole, while most animal cells have several smaller vacuole located throughout the cytoplasm

8. **Of the following organelles, which group is involved in manufacturing substances needed by the cell?**
   (a) Lysosome, vacuole, ribosome (b) Ribosome, rough ER, smooth ER
   (c) Vacuole, rough ER, smooth ER (d) Smooth ER, ribosome, vacuole
   (e) Rough ER, lysosome, vacuole ♦ **Ans:** b

   Explanation: Each structure is capable of synthesis.

9. **Which of the following clues would tell you whether a cell is prokaryotic or eukaryotic?**
   (a) The presence or absence of a rigid cell wall
   (b) Whether or not the cell is partitioned by internal membranes
   (c) The presence or absence of ribosomes
   (d) Whether or not the cell carries out cellular metabolism
   (e) Whether or not the cell contains DNA ♦ **Ans:** b

   Explanation: Prokaryotic cells lack any internal membranous compartmentalization.

1.1.3. State True or False

1. The word cell comes from the Greek
2. The word cell was coined by Robert Hooke
3. Chlorophyll gives plants green color to the plants.
4. Jan Evangelista Purkynje was first observed small "granules" in plant cells
5. Matthias Jakob Schleiden & Theodor Schwann was first developed cell theory
6. Centrioles are absent in animal cell
7. In plant cell, lysosomes occur in cytoplasm
8. In animal cell, lysosomes occur in cytoplasm
9. Lysosome is called power house of cell
10. Lysosome is called suicidal bag
11. Plant cells have chloroplasts, because they make their own food
12. Animal cells have chloroplasts
13. Plastids are present in plant cell
14. Plastids are absent in animal cell
15. Cell membrane acts as a boundary layer to contain the cytoplasm
16. Cell membrane provides protection from physical injury
17. Cell wall provides protection from physical injury
18. Plastids are the site of photosynthesis
19. Leucoplasts are used for storage of starch
20. Ribosome is site of protein synthesis
21. Mitochondria is power house of cell
22. Glycolysis occurs in mitochondria
23. Glycolysis is the reaction for anaerobic respiration
24. Kreb's cycle occur in cytoplasm
25. Glycolysis is also called citric acid cycle
26. Electron Transport Chain occurs in cytoplasm

♦ Answer Keys ♦

Q.No.	Ans.	Q.No.	Ans.	Q.No.	Ans.	Q.No.	Ans.	Q.No.	Ans.	Q.No.	Ans.
1	False	2	True	3	True	4	True	5	True	6	False
7	False	8	True	9	False	10	True	11	True	12	False
13	True	14	True	15	True	16	False	17	True	18	True
19	True	20	True	21	True	22	False	23	False	24	False
25	False	26	False								

1.1.4. Fill in the blanks

1. .... composed of cellulose, hemicelluloses, pectins, and proteins.
2. .... site of malate storage in CAM plants.

3. .... site of Krebs Cycle reactions.
4. .... organized in stacks of grana.
5. .... site of RuBP carboxylation.
6. .... takes up most of a cell's volume, accounting for the turgidity of the cell.
7. .... site of photosystems I and 11.
8. .... during photorespiration, site of catalase activity (destruction of peroxide).
9. .... assembled by rosette protein complexes.
10. .... primary site of DNA transcription.
11. .... site of sucrose synthesis.
12. .... site of phosphoenolpyruvate carboxylation in C-4 plants.
13. .... site of CO_2 evolution.
14. .... enzyme broth whose pH rises at sunrise.
15. .... convert sunlight to chemical energy.
16. .... and .... help to support the plant cell and help it to maintain its shape.
17. .... store food or pigments.
18. The .... converts food into energy. It is found in both plant cells and animal cells.

♦ Answer Keys ♦

Q.No.	Ans.	Q.No.	Ans.	Q.No.	Ans.
1	Cell wall	2	Vacuole	3	Mitochondrion
4	Thylakoid membrane	5	Chloroplast stroma	6	Vacuole
7	Thylakoid	8	Peroxisome	9	Cell wall
10	Nucleus	11	Cytoplasm	12	Cytoplasm
13	Mitochondrion	14	Chloroplast stroma	15	Chloroplasts
16	Cell wall and vacuole	17	Plastids	18	Mitochondrion

1.1.5. Match the following

1. Match each organelle to its function/description

S.No.	Organelles		Functions
1	Cytoplasm	a	Structure that organizes motion of chromosomes.
2.	Lysosome	b	Stack of membranes that packages chemicals.
3.	Mitochondria	c	Membrane that protects the nucleus
4.	Centriole	d	Membrane that surrounds and protects the cell.
5.	Endoplasmic reticulum	e	Sac filled with digestive chemicals.

S.No.	Organelles		Functions
6.	Vacuole	f	Structures that converts nutrients to energy.
7.	Plasma membrane	g	Passage ways where chemicals are made.
8.	Nucleus	h	Jelly-like substance within the plasma membrane.
9.	Ribosome	i	Structure that manufactures ribosomes.
10.	Nuclear envelope	j	Structure that contains DNA and directs the cell.
11.	Golgi apparatus	k	Package created by the golgi apparatus.
12.	Vesicle	l	Small structure that synthesizes proteins.
13.	Nucleolus	m	Sac that stores water, nutrients, or waste products.

♦ Answer Keys ♦

Q.No.	Ans.	Q.No.	Ans.	Q.No.	Ans.	Q.No.	Ans.	Q.No.	Ans.	Q.No.	Ans.
1	h	2	e	3	f	4	a	5	g	6	m
7	d	8	j	9	l	10	c	11	b	12	k
13	i										

2. Match each organelle to its function/description

S.No.	Organelles		Functions
1	Ribosomes	a	Uses energy from sunlight to make rich energy food
2	Endoplasmic reticulum	b	Stack of membranes in which enzymes attach carbohydrates and lipids to proteins
3	Golgi apparatus	c	Use energy from food to make high energy compounds
4	Lysosome	d	An internal membrane system in which components of cell membrane and some proteins are constructed.
5	Vacuole	e	Sack like structure that stores materials
6	Chloroplast	f	Small particle of RNA and protein that provides protein following instructions from nucleus
7	Mitochondrion	g	Filled with enzymes used to breakdown food into particles that can be used

♦ Answer Keys ♦

Q.No.	Ans.	Q.No.	Ans.	Q.No.	Ans.	Q.No.	Ans.	Q.No.	Ans.	Q.No.	Ans.
1	f	2	d	3	b	4	g	5	e	6	a
7	c										

1.2. EMBRYOGENY AND STORAGE TISSUE FORMATION

1.2.1. Choose the correct answer

1. The parts of the female pistil are
   (a) Anther and filament (b) Ovary, style and stigma
   (c) Ovary and anther (d) Corolla and calyx
2. The process of transferring pollen from one flower to another is
   (a) Fertilization (b) Pollination
   (c) Nectar production (d) Protandry
3. Flowers with only male reproductive organs are
   (a) Hermaphrodite flowers (b) Pistillate flowers
   (c) Staminate flowers (d) Bisexual flowers
4. Monocotyledons have petals in multiples of
   (a) Three (b) Four
   (c) Two (d) five
5. The cell formed through fertilization is called as
   (a) Gamete (b) Sperm cell
   (c) Zygote (d) Egg cell
6. The two chromosomes that are members of the same pair are referred to as
   (a) Tetrad (b) Homologous
   (c) Haploid (d) Diploid
7. Plants have a/an .... life cycle.
   (a) Asexual-only (b) Sexual-only
   (c) Alternation of generations (d) Diplontic
8. Which of these best describes the gametophyte in the alternation of generations life cycle?
   (a) The diploid generation
   (b) Generation that produces the gametes
   (c) Generation that produces the spores
   (d) Generation that has vascular tissue
9. Which of these best describes the sporophyte generation?
   (a) The haploid generation
   (b) Generation that produces the gametes
   (c) Generation that produces spores
   (d) Generation that lacks vascular tissue
10. What is the function of antheridia?
   (a) To produce eggs
   (b) To produce sperm
   (c) To carry out asexual reproduction
   (d) To absorb minerals and water

11. In the fern, the gametophyte ....
(a) Is a tiny heart-shaped structure (b) Is dominant
(c) Produces the spores (d) Is used as a human food

12. Seed plants use meiosis to produce ....
(a) Pollen and seeds (b) Spores and pollen
(c) Seeds and fruit (d) Microspores and megaspores

13. In seed plants, a microspore develops into ....
(a) A female gametophyte
(b) An egg
(c) A pollen grain
(d) A sporophyte

14. In seed plants, a megaspore develops into ....
(a) A pollen grain (b) An egg-bearing female gametophyte
(c) A male gametophyte (d) Endosperm

15. In seed plants, after fertilization the zygote becomes ....
(a) A seed
(b) An embryo that is enclosed with reserve food and the seed coat
(c) A fruit
(d) An ovule

16. Angiosperms are the .... plants.
(a) "Naked seed" (b) Non-vascular
(c) Flowering (d) Most primitive or earliest

17. Flowering plants include many families, but all are grouped into what two angiosperm classes?
(a) Monocotyledons and dicotyledons (b) Cycads and conifers
(c) Mosses and ferns (d) Sexual and asexual

18. Endosperm is ....
(a) The ovary and sometimes the adjacent part that develops into a fruit
(b) The embryo (c) Food for the developing plant
(d) A combination of the embryo, enclosed food, and a protective seed coat

19. In flowering plants, which structure becomes the seed?
(a) The ovule (b) The pollen grain
(c) The pistil (d) The anther

20. The part of a flower that becomes the fruit is ....
(a) The stamen (b) The pistil
(c) The ovule (d) The ovary

21. Which structure is part of a stamen?
(a) Stigma (b) Ovary
(c) Anther (d) Style

22. The process of double fertilization is unique to
(a) Angiosperms (b) Cycads
(c) Gymnosperms (d) Ginkos

23. In plant life cycle which of the following sequences is correct?
   (a) Sporophyte, mitosis, spores, gametophyte
   (b) Spores, meiosis, gametophyte, mitosis
   (c) Gametophyte, meiosis, gametes, zygote
   (d) Zygote, sporophyte, meiosis, spores
24. The structure traditionally called pistil is more precisely called
   (a) Style (b) Ovary
   (c) Pollen tube (d) Carpel
25. In the flower, the two pollen sacs are in the ....
   (a) Style (b) Filament
   (c) Anthers (d) Sepal
26. Which of the following takes part in the double fertilization and becomes an endosperm?
   (a) Egg cell (b) Synergid cells
   (c) Antipodal cells (d) Polar nuclei
27. The cell formed through fertilization is called as
   (a) Gamete (b) Sperm cell
   (c) Zygote (d) Egg cell
28. The seed coat of angiosperms forms from the
   (a) Ovary (b) Carpel
   (c) Integuments (d) Peduncle
29. The first cell division of the plant zygote is typically
   (a) Longitudinal (b) Diagonal
   (c) Radial (d) Transverse
30. Plant development includes all of the following except
   (a) Seed formation
   (b) Tissue formation
   (c) Formation of embryo and suspensor
   (d) Deteriorating cotyledons during germination
31. Which of these structures is unique to the seed of a monocot?
   (a) Cotyledon (b) Endosperm
   (c) Coleoptyle (d) Seed coat
32. What is typically the result of double fertilization in angiosperms?
   (a) The endosperm develops into a diploid nutrient tissue
   (b) A triploid zygote is formed
   (c) Both a diploid embryo and triploid endosperm are formed
   (d) The fertilized antipodal cells develop into the seed coat
33. The products of meiosis in plants are always which of the following?
   (a) Spores (b) Eggs
   (c) Sperm (d) Seeds

34. Which of the following part is composed of one or more carpels
(a) Pistil (b) Ovary
(c) Stigma (d) anther

35. The development of the female gametophyte (embryo sac) from the megaspore mother cell
(a) Megasporogenesis (b) Megagametogenesis
(c) Double fertilization (d) Ovule primordial

36. The scar on the ovule made where the funiculus detaches at maturity is known as
(a) Hilum (b) Funiculus
(c) Scar (d) Micropyle

37. The part of seed opposite of the micropyle is termed as
(a) Chalaza (b) Hilum
(c) Scar (d) Funiculus

38. Which stage of the embryo develops into two-lobed form
(a) Globular (b) Heart
(c) Torpedo (d) Mature

♦ Answer Keys ♦

Q.No.	Ans.	Q.No.	Ans.	Q.No.	Ans.	Q.No.	Ans.	Q.No.	Ans.	Q.No.	Ans.
1	b	2	b	3	c	4	a	5	c	6	b
7	c	8	b	9	c	10	b	11	a	12	d
13	c	14	b	15	b	16	c	17	a	18	c
19	a	20	d	21	c	22	a	23	d	24	d
25	c	26	d	27	c	28	c	29	d	30	d
31	b	32	c	33	a	34	a	35	b	36	a
37	a	38	b								

1.2.2. State True or False

1. All conifers have spores in cones.
2. The fruit must be produced before the flower.
3. Examples of flowering plants are grasses, rose bushes and eucalyptus.
4. Only insects can pollinate flowers.
5. A spore is the same as a seed.
6. Double fertilization occurs in angiosperms.
7. In the flower, the ovules are in the ovary at the base of the carpel
8. The protective layer covering the ovule of seed plants is the nucellus.
9. The cotyledons become the first leaves of the plant.
10. Pistil is composed of one or more stamens
11. In angiosperms, the seeds originate from meristematic tissue of the ovary wall called ovule primordial

12. The development of the female gametophyte (embryo sac) from the megaspore mother cell is called megasporogenesis
13. The developing ovule is commonly attached to the placenta by the hilum
14. The period of flower development when the stigma is ready to receive the pollen is known as anthesis
15. Chalaza is at the upper end of the embryo sac.
16. Suspensor acts as an "umbilical cord", providing nutrients to the embryo
17. Triple fusion occurs in orchidaceae
18. In Cellular endosperm, each nuclear division is unaccompanied by cell wall formation
19. The helobial endosperm is intermediate between the nuclear and cellular types

Answer Keys

Q.No.	Ans.	Q.No.	Ans.	Q.No.	Ans.	Q.No.	Ans.	Q.No.	Ans.	Q.No.	Ans.
1	False	2	False	3	True	4	False	5	False	6	True
7	True	8	False	9	False	10	False	11	True	12	False
13	False	14	True	15	False	16	True	17	True	18	False
19	True	20									

1.3. ASSIMILATE TRANSLOCATION

1.3.1. Choose the correct answer

1. The pressure flow model of phloem transport states that
   (a) Water flow brings sucrose from a source to a sink
   (b) Phloem sap always flows from the leaves to the root
   (c) Phloem sap always flows from root to the leaves
   (d) Xylem helps in water movement
2. The transport of sugar and other organic molecules within a plant is called as
   (a) Assimilation (b) Transpirational pull
   (c) Translocation (d) Guttation
3. The most abundant solute in phloem sap is
   (a) Starch (b) Glucose
   (c) Potassium ion (d) Sucrose
4. Active loading of sugar into a sieve tube is probably driven by
   (a) Water flow (b) Proton pump
   (c) Solar energy (d) Gravity
5. Mass-flow hypothesis for phloem sap was first put forward and explained by
   (a) Hales (b) Blackman
   (c) Munch (d) Jolly and Dixon

6. The smallest amount of pressure needed to stop fluid from moving by osmosis is referred to as the
   (a) Turgor pressure (b) Water potential
   (c) Pressure potential (d) Solute potential
7. Turgor pressure is also referred to as
   (a) Solute potential (b) Water potential
   (c) Pressure potential (d) Osmotic potential
8. The combination of pressure potential and solute potential is
   (a) Water potential (b) Transpiration potential
   (c) Field potential (d) Stem potential
9. In the absence of transpiration, water moves into and up xylem because of
   (a) Root pressure (b) Turgor pressure
   (c) Evaporation (d) High soil mineral concentration
10. High root pressure can cause water to be lost by leaves through the process of
    (a) Respiration (b) Regurgitation
    (c) Transpiration (d) Guttation
11. Water potential is the
    (a) Combination of turgor pressure and pressure potential
    (b) Difference between pressure potential and osmotic potential
    (c) Combination of pressure potential and solute potential
    (d) Product of pressure potential and osmotic potential
12. Guttation is
    (a) Movement of soluble organic materials through plants
    (b) Movement of water through the apoplast
    (c) Evidence of root pressure
    (d) Negative pressure created by transpiration
13. The term apoplast and symplast were coined by
    (a) Clark (b) Dixon
    (c) Fisher (d) Munch
14. The movement of solutes in the phloem is mainly
    (a) Acropetal (b) Lateral
    (c) Basipetal (d) Lateral-acropetal
15. Translocation of carbohydrates or sugars in the flowering plants occur in the form of
    (a) Maltose (b) Sucrose
    (c) Glucose (d) Starch
16. Supply ends in transport of solutes are
    (a) Green leaves (b) Roots and stem
    (c) Xylem and phloem (d) Hormones and enzymes

17. Scientist who has been associated with work on translocation is
(a) Munch (b) Calvin
(c) Arnon (d) Hill

18. Leaf photosynthates are transported to roots through the cells of
(a) Xylem (b) Pith
(c) Phloem (d) Endodermis

19. Vascular tissue that transports organic nutrients from the leaves to the roots is
(a) Xylem (b) Phloem
(c) Meristem (d) Cambium

20. The photosynthetic cells in the interior of a leaf are what kind of cells?
(a) Parenchyma (b) Phloem
(c) Collenchyma (d) Endodermis

21. Osmosis is the movement of
(a) Solute particles from higher concentration to lower concentration
(b) Solvent particles from higher water potential to lower water potential
(c) Solute particles from higher concentration through a semi-permeable membrane
(d) Solvent particles from lower water potential to higher water potential

22. Sieve tubes are found in
(a) Xylem (b) Phloem
(c) Pith (d) Cortex

23. A component of xylem is
(a) Sieve tube (b) Medullary ray
(c) Selereid (d) Tracheid

24. Sucrose enters a phloem sieve-tube cell because of
(a) Osmosis (b) Water potential
(c) Active transport (d) A process regulated by auxin

25. Blowing water up through a drinking straw is most like
(a) Guttation (b) Diffusion
(c) Mass flow in xylem (d) Mass flow in phloem

26. The term phloem was coined by
(a) Hofmeister (b) Hartig
(c) Strasburger (d) Nageli

27. Guttation is mainly due to
(a) Root pressure (b) Osmosis
(c) Transpiration (d) Imbibition

28. The loss of water by plants in the form of vapor is called
(a) Evaporation (b) Transpiration
(c) Evapotranspiration (d) Vaporization

29. Water molecules enter plant epidermal cells by
(a) Osmosis (b) Active transport
(c) Translocation (d) Transpiration

30. Which portion of the flowering plant anchors the plant in the soil?
(a) Root system (b) Shoot system
(c) Leaves (d) Stem

31. Which portion of a flowering plant absorbs water and minerals?
(a) Flower (b) Blade
(c) Stem (d) Root

32. Which part of a plant is primarily responsible for photosynthesis?
(a) Flower (b) Leaf blade
(c) Stem (d) Root

33. Which part of a flowering plant will store carbohydrate as starch?
(a) Flower (b) Blade
(c) Stem (d) Root

34. Which part of a flowering plant connects the blade of a leaf to the stem?
(a) Petiole (b) Leaf veins
(c) Nodes (d) Internode

35. Which of the following structures provides nutrient molecules for growing plant embryos before the true leaves begin photosynthesizing?
(a) Petiole (b) Leaf veins
(c) Nodes (d) Cotyledon

36. Which tissue forms the outer, protective covering of a plant?
(a) Dermal tissue (b) Ground tissue
(c) Vascular tissue (d) Plant tissue

37. Which tissue system conducts water and nutrients in a plant?
(a) Dermal tissue (b) Plant tissue
(c) Vascular tissue (d) Ground tissue

38. Which of the following cells protect the inner body parts and prevent the plant from drying out?
(a) Epidermal cell (b) Parenchymal cell
(c) Sclerenchymal cell (d) Sieve-tube cell

39. Which of the following cells are relatively unspecialized and correspond best to the generalized plant cell?
(a) Epidermal cell (b) Sclerenchyma cell
(c) Sieve-tube cell (d) Tracheid cell

40. Which of the following cells are hollow and nonliving at maturity?
(a) Epidermal cell (b) Sieve-tube cell
(c) Tracheid cell (d) Sclerenchyma cell

41. Which of the following plant cells transports water and minerals from the roots to the leaves?
    (a) Epidermal cell (b) Parenchyma cell
    (c) Sclerenchyma cell (d) Vessel element

42. Which of the following transports organic nutrients, usually from the leaves to the roots?
    (a) Parenchyma cell (b) Sclerenchyma cell
    (c) Xylem (d) Phloem

43. Which of the following cells will transport water from the roots to the leaves more freely?
    (a) Sieve-tube cell (b) Companion cell
    (c) Tracheids (d) Vessel elements

44. Which of the following cells will always have at least one companion cell associated with it?
    (a) Sclerenchyma cell (b) Tracheid
    (c) Vessel element (d) Sieve-tube cell

45. Which of the following cells have perforated end walls and cytoplasm, but no nuclei?
    (a) Sclerenchyma cell (b) Sieve-tube cell
    (c) Tracheid (d) Vessel element

46. Which of the following is not a function of a plant root?
    (a) Absorb water from the soil (b) Absorb minerals from the soil
    (c) Anchors a plant in the soil (d) Site of photosynthesis

47. Which structure in a leaf transports both water and nutrients from the leaves?
    (a) Palisade mesophyll (b) Spongy mesophyll
    (c) Leaf veins (d) Guard cells

48. Which part of a leaf prevents a leaf from drying out as well as preventing gas exchange?
    (a) Guard cells (b) Spongy mesophyll
    (c) Palisade mesophyll (d) Cuticle

49. Where are the stomatas usually found on a leaf?
    (a) Spongy mesophyll (b) Upper epidermis
    (c) Lower epidermis (d) Vein

50. Which part of a leaf has irregular cells bounded by air spaces?
    (a) Palisade mesophyll (b) Spongy mesophyll
    (c) Upper epidermis (d) Lower epidermis

51. Which part(s) of a leaf carries on most of the photosynthesis?
    (a) Palisade and spongy mesophyll (b) Upper epidermis
    (c) Lower epidermis (d) Cuticle

52. Which part of a leaf allows gases to move into and out of the leaf?
    (a) Palisade and spongy mesophyll (b) Upper epidermis
    (c) Cuticle (d) Stomata

53. Which tissue in a plant transports water from the roots upto the leaves?
(a) Xylem (b) Phloem
(c) Pith (d) Cortex

54. Which of the following cells listed below form(s) a completely hollow pipeline from the roots to the leaves?
(a) Sieve-tube cells (b) Tracheids
(c) Vessel elements (d) Cortex

55. Which of the following is not a factor that allows water to rise in plants?
(a) Atmospheric pressure (b) Cohesion of water molecules
(c) Transpiration
(d) Active transport of sugar into/out of sieve-tube cells

56. What causes water to fill the xylem pipeline completely, from the roots to the leaves, and to resist any separation?
(a) Atmospheric pressure
(b) Cohesion of water molecules
(c) Transpiration
(d) Passive transport of sugar into/out of phloem cells

57. Which of the following creates a pull on water, thereby drawing water up the vessel elements from the roots to the leaves?
(a) Atmospheric pressure
(b) Cohesion of water molecules
(c) Transpiration
(d) Passive transport of sugar into/out of phloem cells

58. Which of the following factors does not play a critical role in the opening and closing of the stomata in a leaf?
(a) ATP (b) Glucose
(c) Potassium ions (d) Water

59. Which process is used to transport potassium ions into the guard cells?
(a) Filtration (b) Diffusion
(c) Osmosis (d) Active transport

60. Which process is used by the guard cells to take up water?
(a) Filtration (b) Diffusion
(c) Osmosis (d) Active transport

61. What ion moves into the guard cell to cause the stomata to open?
(a) Magnesium (b) Chlorine
(c) Sodium (d) Potassium

62. Which vascular tissue in a plant transports sugar?
(a) Phloem (b) Xylem
(c) Pith (d) Cortex

63. Which type of phloem cell is used in moving the products of photosynthesis in a plant?
    (a) Sieve-tube cell (b) Companion cell
    (c) Vessel element cell (d) Guard cell

64. Which phloem cell has a nucleus that helps the other phloem cell to perform its translocation function?
    (a) Sieve-tube cell (b) Companion cell
    (c) Vessel element cell (d) Tracheid

65. Which insect can be used to collect phloem sap for analysis?
    (a) Ants (b) Aphids
    (c) Moths (d) Beetles

66. Which part of a plant is the major source of sugar during the growing season?
    (a) Leaves (b) Petiole
    (c) Stem (d) Bark

67. Which of the following organic molecules is actively transported into the sieve-tube cells?
    (a) Lipids (b) Proteins
    (c) Nucleic acids (d) Sugars

68. Which molecule will build up in the sieve-tube cells to create pressure, which starts a flow of phloem sap?
    (a) Salt (b) Water
    (c) Sugar (d) Protein

♦ Answer Keys ♦

Q.No.	Ans.	Q.No.	Ans.	Q.No.	Ans.	Q.No.	Ans.	Q.No.	Ans.	Q.No.	Ans.
1	a	2	c	3	d	4	b	5	c	6	d
7	c	8	a	9	a	10	d	11	c	12	c
13	d	14	c	15	b	16	a	17	a	18	c
19	b	20	a	21	b	22	b	23	d	24	c
25	a	26	d	27	a	28	c	29	a	30	a
31	d	32	b	33	d	34	a	35	d	36	a
37	c	38	a	39	d	40	b	41	d	42	d
43	a	44	d	45	b	46	d	47	c	48	d
49	c	50	b	51	a	52	d	53	b	54	c
55	d	56	b	57	c	58	b	59	b	60	c
61	d	62	a	63	a	64	b	65	b	66	a
67	d	68	b								

1.3.2. State True or False

1. Carbohydrates move through the sieve tube as a result of the movement of water due to osmosis.
2. The apoplastic system is the system of inter connected protoplasts.

3. Transpiration maintains the shape and structure of the plants by keeping cells turgid.
4. Because of the casparian strip, water and minerals must pass through endodermal cells in order to reach the vascular cylinder.
5. Each stomata has three guard cells, which regulate its opening and closing on the upper epidermis of a leaf.
6. The cohesion-tension theory helps to explain how the phloem moves nutrients.
7. Transpiration creates a pushing effect, which forces a column of water up from the roots to the leaves.
8. Xylem only transports water from the roots to the leaves.
9. When guard cells take up water, the stomata opens.
10. Stomatas are more likely to be opened during the night and closed during the day.

♦ Answer Keys ♦

Q.No.	Ans.	Q.No.	Ans.	Q.No.	Ans.	Q.No.	Ans.	Q.No.	Ans.
1	True	2	False	3	True	4	True	5	False
6	False	7	False	8	False	9	True	10	True

1.3.3. Fill in the blanks

1. The rooted plant, transport in xylem is essentially .... from roots to the stems.
2. Xylem and phloem belong to the .... tissue system.
3. .... are the organs of photosynthesis in flowering plants.
4. The loss of water from a leaf by evaporation is called ....
5. The movement of water from the roots to the leaves is best explained by the ....
6. The ability of water molecules to cling together is called ....
7. When guard cells take up water, they ....
8. A stomata closes when ....
9. Carbon dioxide enters a leaf through the ....
10. The movement of organic substances in phloem is termed the ....

♦ Answer Keys ♦

Q.No.	Ans.	Q.No.	Ans.
1	Unidirectional	2	Vascular
3	Leaves	4	Transpiration
5	Cohesion-tension theory	6	Cohesion
7	Buckle out from their region of attachment and the stomata opens	8	Potassium ions and water move out of the guard cells
9	Stomatas in the lower epidermis	10	Pressure-flow model

1.4. STARCH BIOSYNTHESIS DURING SEED DEVELOPMENT AND MATURATION

1.4.1. Choose the correct answer

1. The reactions of the cell that are carried out for capturing energy are called
   (a) Catabolism (b) Metabolism
   (c) Anabolism (d) Activation energy

2. The reaction, where small precursor molecules are assembled into larger organic molecules is referred as
   (a) Anabolism (b) Catabolism
   (c) Metabolism (d) Activation of energy

3. In the process of photosynthesis, chlorophyll serves as
   (a) An end-product (b) A raw material
   (c) An energy converter (d) A hydrogen acceptor

   Explanation: Chlorophyll is a pigment that captures light energy and transfers it into usable energy in plants.

4. Which color light is least important to a green plant during photosynthetic activities?
   (a) Green (b) Yellow
   (c) Blue (d) Orange

   Explanation: Plants reflect green color because that wavelength of light is least usable. Blue and red wavelengths of light are the best for photosynthesis, and are therefore, seen the least in plants.

5. What does the process of photosynthesis produce?
   (a) Starch, which is metabolized into less complex molecules by dehydration synthesis
   (b) Protein, which is metabolized into less complex molecules by dehydration synthesis
   (c) Glycerol, which is metabolized into more complex carbohydrates by dehydration synthesis
   (d) Glucose, which is metabolized into more complex carbohydrates by dehydration synthesis

   Explanation: Photosynthesis makes carbohydrates, of which starch and glucose are examples. However, glucose is a simple sugar made specifically by photosynthesis and is then, combined with other glucose's to make starch by the process of dehydration synthesis. Starch doesn't get metabolized or broken down into less complex molecules by dehydration synthesis, but by hydrolysis.

6. The light reaction of photosynthesis does not include
   (a) Chemiosmosis (b) Oxygen liberation
   (c) Charge separation (d) Electron transport

   Explanation: The light reaction takes place in four stages: capture of a photon by a pigment, charge separation, electron transport, and chemiosmosis

7. The final product of the Calvin cycle is
   (a) RuBP (b) PGA
   (c) ATP (d) G3P
8. Photosynthesis takes place in the membranes of small sacs called
   (a) Thylakoids (b) Grana
   (c) Photosystems (d) Photons
9. The dark reaction in photosynthesis is limited by
   (a) CO_2, temperature, and light (b) CO_2, light, and water
   (c) Water, temperature, and CO_2 (d) Oxygen, water, and temperature
10. Colors of light most useful in photosynthesis are
    (a) Green, yellow, and orange (b) Red, violet, and blue
    (c) Infrared, red, and yellow (d) Red, white, and blue
11. During what stage of photosynthesis O_2 (is) produced?
    (a) Cyclic photophosphorylation
    (b) The light-dependent reactions involving photosystems I and II
    (c) Carbon fixation (d) The Kreb's cycle
12. The pigment molecules responsible for photosynthesis are located in the
    (a) Mitochondria (b) Cytoplasm of the cell
    (c) Stroma of the chloroplast
    (d) Thylakoid membrane of the chloroplast
13. Both carotenoids and chlorophylls
    (a) Are pigments
    (b) Absorb photons of all energy ranges
    (c) Contain porphyrin rings
    (d) Are nonpigments
14. During what stage of photosynthesis, are ATP and NADPH converted to ADP^{+} Pi and $NADP^{+}$
    (a) The light-dependent reactions (b) The light-independent reactions
    (c) The dark reactions (d) The phytochrome reactions
15. What energy-rich organic compound is produced as a result of the Calvin cycle?
    (a) NADPH (b) CO_2
    (c) ATP (d) Glucose
16. High-energy photons
    (a) Have long wavelengths (b) Have short wavelengths
    (c) Are more likely to produce red light than blue light
    (d) Cannot be absorbed
17. During photosynthesis, photons raise electrons to higher energy levels. These excited electrons belong to what compound?
    (a) ATP (b) RuBP
    (c) Glucose (d) Chlorophyll

18. The oxygen that is released as O_2 during photosynthesis came from .... molecules.
    (a) Carbon dioxide (b) Water
    (c) Glucose (d) Chlorophyll
19. The earliest photosynthesizers probably used what as a source of hydrogen?
    (a) H_2S (b) H_2O
    (c) CO_2 (d) $C_6H_{12}O_6$
20. Production of one molecule of 3-phosphoglyceraldehyde requires how many turns of the Calvin cycle?
    (a) 1 (b) 2
    (c) 3 (d) 6
21. To reduce six molecules of carbon dioxide to glucose via photosynthesis, how many molecules of NADPH and ATP are required?
    (a) 6 NADPH and 6 ATP (b) 12 NADPH and 12 ATP
    (c) 12 NADPH and 18 ATP (d) 18 NADPH and 12 ATP
22. The primary form of sugar transported from the site of photosynthesis to the rest of the plant is
    (a) Glyceraldehyde 3-phosphate (b) Glucose
    (c) Ribulose 1,5-bisphosphate (d) Sucrose
23. Light is required for the light dependent reactions because
    (a) It is the source for electrons
    (b) It splits the water molecule
    (c) It energizes electrons in the reaction center
    (d) It splits ATP molecules which generates the energy necessary to power the light independent reaction
24. The organic molecule produced directly by photosynthesis is
    (a) Lipids (b) Sugar
    (c) Amino acids (d) DNA
25. The photosynthetic process removes .... from the environment.
    (a) Sugar (b) Oxygen
    (c) Chlorophyll (d) Carbon dioxide
26. What are the products of photosynthesis?
    (a) Water and carbon dioxide (b) Water and oxygen
    (c) Oxygen and carbohydrate (d) Carbohydrate and water
27. What organisms are capable of photosynthesis?
    (a) Plants only (b) Plants and algae only
    (c) Plants and some bacteria only (d) Plants, algae, and some bacteria
28. Which of these is not a major photosynthetic pigment in plants?
    (a) Chlorophyll a (b) Chlorophyll b
    (c) Chlorophyll c (d) Carotenoid pigments

29. The term "chloroplast" is derived in part from the Greek root that means
    (a) Plant (b) Light
    (c) Green (d) Photosynthetic
30. To what does the term stroma refer?
    (a) The double membrane of the chloroplast
    (b) A flattened disk or sac in the chloroplast
    (c) A stack of thylakoid membrane structures
    (d) The central fluid-filled space of the chloroplast
31. To what does the term grana refer?
    (a) The double membrane of the chloroplast
    (b) A flattened disk or sac in the chloroplast
    (c) A stack of thylakoid membrane structures
    (d) The central fluid-filled space of the chloroplast
32. Which is most closely associated with the Calvin cycle?
    (a) ATP production (b) Oxygen production
    (c) Carbon dioxide fixation (d) Carbon dioxide production
33. The major enzyme that catalyzes the reduction of carbon dioxide is called
    (a) Ribulose bisphosphate carboxylase (rubisco).
    (b) Glyceraldehyde-3-phospate (G3P).
    (c) PEP carboxylase (PEPcase).
    (d) ATPase
34. The Calvin cycle requires all of the following except .... to occur.
    (a) ATP, NADPH (b) Carbon dioxide
    (c) Light (d) Ribulose bi-phosphate
35. Carbon dioxide enters the leaf through ....
    (a) Chloroplasts (b) Stomata
    (c) Cuticle (d) Mesophyll cells

♦ Answer Keys ♦

Q.No.	Ans.	Q.No.	Ans.	Q.No.	Ans.	Q.No	Ans.	Q.No	Ans.	Q.No	Ans.
1	a	2	a	3	c	4	a	5	d	6	b
7	d	8	a	9	a	10	b	11	b	12	d
13	a	14	b	15	d	16	b	17	d	18	b
19	a	20	c	21	c	22	d	23	c	24	b
25	d	26	c	27	d	28	c	29	c	30	d
31	c	32	c	33	a	34	c	35	b		

1.4.2. State True or False

1. Light-driven electron transport in the chloroplast pumps H+ into the intermembrane space between the outer and inner membranes.
2. CO_2 fixation occurs within the stroma.
3. If you know the absorption spectrum of a pigment, you can predict the pigment's color.
4. For every CO_2 molecule fixed by photosynthesis, one molecule of O_2 is produced.
5. Metabolism is the set of chemical reactions that happen in living organism to maintain life.
6. Metabolism is divided into catabolism and anabolism.
7. Enzymes are not important in metabolism.
8. Starch is the most important carbohydrate stored in seed.
9. Amylose comprises between 11 and 37% of the starch found in plants (depending upon the species and the site of storage); the rest is amylopectin.
10. The hemicelluloses contain mannans.
11. The amorphous regions of the semi-crystalline layers are composed of amylose and non-ordered amylopectin branches.
12. Starch is stored in the form of granules in embryo tissue.
13. Starch is hydrolytically degraded to complex carbohydrates by enzymatic action.
14. α-amylase is not present in air dry seed.
15. Maltase catalyzes the breakdown of maltose to glucose.
16. Starch is synthesized from sucrose by pathways involving both the cytoplasm and the plastid.
17. During seed development, cell division is not followed by elongation, differentiation and storage.
18. DGMI means digalactosyl myo-initiation.
19. MIPS means myo-inositol phosphate synthate.
20. STS means stachyose synthase
21. TGPA means trigalactosyl pinitol A.
22. UDP-gal means uridine diphosphate galactose.
23. VBS means verbascose synthate.
24. LRS means low raffinose and stachyose seeds.

♦ Answer Keys ♦

Q.No.	Ans.	Q.No.	Ans.	Q.No.	Ans.	Q.No.	Ans.	Q.No.	Ans.	Q.No.	Ans.
1	False	2	True	3	True	4	True	5	True	6	True
7	False	8	True	9	True	10	True	11	True	12	False
13	False	14	True	15	True	16	True	17	False	18	False
19	False	20	True	21	True	22	True	23	False	24	False

1.5. PROTEIN BIOSYNTHESIS AND ACCUMULATION DURING SEED DEVELOPMENT AND MATURATION

1.5.1. Choose the correct answer

1. During translation elongation, amino acids are brought to the ribosome with the help of
   (a) EF-T (b) EF-Tu
   (c) EF-Ts (d) EF-Tm
2. Which of the following occurs as a result of translocation?
   (a) The tRNA that was in the P site moves into the A site
   (b) The tRNA that was in the E site moves into the P site
   (c) The tRNA that was in the E site moves into the A site
   (d) The tRNA that was in the P site moves into the E site
3. On the ribosome, mRNA binds
   (a) Between the subunit (b) To the large subunit
   (c) To the small subunit (d) Both small and large subunits
4. Which of the following is not a necessary component of translation?
   (a) Anticodon (b) mRNA
   (c) Ligase (d) Amino acid
5. A polysome could be best described as
   (a) An active site of DNA synthesis
   (b) An active site of protein synthesis
   (c) An active site of lipid synthesis
   (d) An active site of fatty acid synthesis
6. Eukaryotic mRNA binding to the ribosomes is facilitated by ....
   (a) The Shine Dalgarno sequence (b) The 7-methyl guanosine cap
   (c) tRNA (d) Poly A tail
7. Amino acids are joined together into a protein chain by which of the following?
   (a) Transfer RNA (b) DNA polymerase
   (c) Hydrogen bonds (d) Messenger RNA
8. Phosphorus is taken up by the cell during the process of
   (a) Carbohydrate synthesis (b) Protein synthesis
   (c) Lipid synthesis (d) ATP synthesis
9. The site of protein synthesis is
   (a) Ribosome (b) Nucleus
   (c) Endoplasmic reticulum (d) Chromosome
10. The structure in a bacterium that indicates an active site for protein synthesis is
   (a) Chromosome (b) A cell membrane
   (c) Flagellum (d) Polysome

11. Which of the following is not necessary for protein synthesis to occur, once transcription is completed?
    (a) tRNA (b) Ribosomes
    (c) mRNA (d) DNA
12. Ribosomes select the correct tRNAs
    (a) Then bind to the appropriate mRNA
    (b) Solely on the basis of their anticodons
    (c) Depending on their abundance in the cytosol
    (d) With the least abundant anticodons
13. Signal peptide protein removal that is translocated across a membrane is accomplished by
    (a) fMet aminopeptidase (b) Trypsin
    (c) Signal peptidase (d) Chymotrypsin
14. The pathway of a tRNA during polypeptide elongation on the ribosome is
    (a) A site → P site → E site (b) P site → entry site → exit site
    (c) A site → P site → entry site (d) P site → A site → E site
15. Which of the following has unusual bases?
    (a) mRNA (b) tRNA
    (c) rRNA (d) hnRNA
16. The anticodon of tRNA
    (a) Binds to rRNA
    (b) Binds to an amino acid
    (c) Binds to the Shine Dalgarno sequence
    (d) Binds to mRNA codon
17. Processive synthesis is a characteristic feature of
    (a) All DNA polymerases
    (b) DNA Pol III at a replication fork
    (c) Removal of RNA primers on Okasaki fragments
    (d) DNA mismatch repair
18. Which site of the tRNA molecule binds to the mRNA molecule?
    (a) Anticodon (b) Codon
    (c) Amino acid (d) 5 prime end
19. The first step in translation is
    (a) Binding of the two ribosomal subunits to each other
    (b) Binding of mRNA to the two ribosomal subunits
    (c) Binding of tRNA to the small ribosomal subunit
    (d) Binding of mRNA to the small ribosomal subunit
20. Which of the following amino acid starts all proteins synthesis?
    (a) Glycine (b) Proline
    (c) Thymine (d) Methionine

21. The growing polypeptide chain is released from the ribosomes when
    (a) A chain terminating codon is reached
    (b) A chain terminating tRNA binds to the ribosome
    (c) The 7-methyl guanosine cap is reached
    (d) The poly A tail is reached
22. Which of the following RNAs bind to an amino acid?
    (a) mRNA (b) tRNA
    (c) rRNA (d) hnRNA
23. How many bases of nucleic acid determine a single amino acid?
    (a) Four (b) One
    (c) Two (d) Three
24. The ribosomes are composed of
    (a) Proteins (b) RNA
    (c) Both (a) and (b) (d) Lipids
25. In DNA, thymine always pairs with
    (a) Adenine (b) Cytosine
    (c) Guanine (d) Thymine
26. What determines the sequence of amino acids in a protein?
    (a) The sequences of amino acids in another protein
    (b) The sequence of nucleotides in the tRNA
    (c) The sequence of amino acids in the DNA
    (d) The sequence of nucleotides in the mRNA
27. What does the t in tRNA represent?
    (a) Transfer (b) Transpolymerase
    (c) Translation (d) Transcription
28. Which of the following best describes transcription?
    (a) DNA to protein (b) DNA to DNA
    (c) DNA to mRNA (d) mRNA to tRNA
29. Which of the following identify and move amino acids to the site of protein production?
    (a) rRNA (b) iRNA
    (c) tRNA (d) mRNA
30. Which of the following is mismatched?
    (a) mRNA - codon (b) DNA - code
    (c) rRNA - recode (d) tRNA - anticodon
31. Which of the following is true regarding translation?
    (a) tRNA remains attached to the amino acid after the peptide chain is formed
    (b) DNA is being replicated in the nucleus
    (c) mRNA is exposing its anticodon to the ribosome
    (d) mRNA is exposing its codon to the tRNA

32. **When does protein synthesis stop?**
(a) When there is enough protein
(b) When the stop anticodon is located on the mRNA
(c) When the stop codon enters the ribosome
(d) When the intron has been replaced

33. **DNA controls the activities of a cell by coding for the production of**
(a) Proteins (b) Carbohydrates
(c) Amino Acids (d) Bases

34. **The function of tRNA in cell metabolism is to**
(a) Transport amino acids to be used in synthesis
(b) Carry codons to the ribosomes
(c) Synthesise proteins (d) Transcribe the DNA code

35. **The process of tRNA "charging" does not involve ....**
(a) rRNA (b) tRNA
(c) ATP (d) Amino acid

36. **After being formed by the ribosomes located on the endoplasmic reticulum, what is the next organelle to which a protein could be transported?**
(a) Mitochondria (b) Smooth endoplasmic reticulum
(c) Golgi apparatus (d) Chloroplast

37. **Which is a true statement about ribosomes?**
(a) Ribosomes contain DNA and protein.
(b) Ribosomes are active in carbohydrate synthesis.
(c) Ribosomal subunits leave the nucleus after being formed by the nucleolus.
(d) Polyribosomes are the subunits of ribosomes.

38. **Ribosomes are found**
(a) Only in the nucleus (b) In the cytoplasm
(c) Attached to the smooth endoplasmic reticulum
(d) Only in eukaryotic cells

39. **Which nucleic acid molecules are involved in transcription, but not translation in the synthesis of a particular polypeptide?**
(a) tRNA (b) mRNA
(c) rRNA (d) DNA

40. **Which of the following cell organelles serves as the site of protein synthesis?**
(a) Ribosomes (b) Lysosome
(c) Gogi apparatus (d) Peroxisome

41. **Which of the following cell organelles sorts, processes, and packages proteins as well as the membranes of transport vesicles?**
(a) Lysosome (b) Mitochondrion
(c) Golgi apparatus (d) Peroxisome

42. Where is ribosomal RNA produced?
(a) Chloroplasts (b) Lysosomes
(c) Golgi apparatus (d) Nucleoli

43. Which organelle synthesizes proteins for export or repackaging elsewhere?
(a) Smooth endoplasmic reticulum (b) Rough endoplasmic reticulum
(c) Free ribosomes (d) Golgi apparatus

44. What is the correct organelle sequence for the export of material out of a cell?
(a) Rough ER --> smooth ER --> Golgi apparatus
(b) Smooth ER --> rough ER --> Golgi apparatus
(c) Rough ER --> Golgi apparatus --> smooth ER
(d) Smooth ER --> Golgi apparatus --> rough ER

45. Which plant organelles function as storage areas for water, sugars, salts, pigments, and toxic substances?
(a) Lysosomes (b) Vesicles
(c) Vacuoles (d) Golgi apparatus

46. Which of the following organelles and functions are not correctly related?
(a) Endoplasmic reticulum-photosynthesis
(b) Golgi apparatus-secretions
(c) Vacuoles-storage
(d) Lysosomes-enzymes

47. The process of copying the genetic information from DNA to mRNA in the nucleus is called:
(a) Transduction (b) Transformation
(c) Transcription (d) Translation

48. Storage protein soluble in water and moderately soluble in concentrated salt solutions
(a) Albumins (b) Globulins
(c) Prolamins (d) Glutelins

49. Prolamins soluble in
(a) Alcohol (b) Diluted acids
(c) Bases (d) Water

50. Protein responsible for baking properties in bread wheat is
(a) Albumins (b) Globulins
(c) Prolamins (d) Glutelins

51. Hordein, gliadin, zein and secalins are
(a) Albumins (b) Globulins
(c) Prolamins (d) Glutelins

52. Proline content is present highly in
    (a) Albumins (b) Globulins
    (c) Prolamins (d) Glutelins
53. .... prolamin highly present in rice
    (a) Zein (b) Hordein
    (c) Oryzin (d) Kafirin
54. .... prolamin highly present in sorghum
    (a) Zein (b) Hordein
    (c) Kafirin (d) Oryzin
55. .... prolamin highly present in oats
    (a) Zein (b) Hordein
    (c) Avenin (d) Oryzin
56. .... prolamin highly present in wheat
    (a) Zein (b) Hordein
    (c) Gliadin (d) Glutelin
57. .... glutelin highly present in wheat
    (a) Zein (b) Hordein
    (c) Gliadin (d) Glutelin
58. .... glutelin highly present in rice
    (a) Zein (b) Hordein
    (c) Oryzin (d) Oryzenin
59. Amino acids are building blocks of
    (a) Lipids (b) Carbohydrates
    (c) Starch (d) Proteins
60. At which phase of seed development protein accumulation will be maximum
    (a) Early phase (b) Mid phase
    (c) Late phase (d) Immediately after anthesis

♦ Answer Keys ♦

Q.No.	Ans.	Q.No.	Ans.	Q.No.	Ans.	Q.No.	Ans.	Q.No.	Ans.	Q.No.	Ans.
1	b	2	d	3	c	4	c	5	b	6	b
7	a	8	d	9	a	10	d	11	d	12	b
13	c	14	a	15	b	16	d	17	b	18	a
19	d	20	d	21	a	22	b	23	d	24	c
25	a	26	d	27	a	28	c	29	c	30	c
31	d	32	c	33	a	34	a	35	a	36	c
37	c	38	b	39	d	40	a	41	c	42	d
43	b	44	a	45	c	46	a	47	c	48	a
49	a	50	d	51	c	52	c	53	c	54	c
55	c	56	c	57	d	58	d	59	d	60	b

1.5.2. State True or False

1. Translocation is promoted by the elongation factor EF-G.
2. An uncharged tRNA is one that is not carrying an amino acid.
3. A DNA molecule has the same amount of adenine and thymine.
4. Ribosomes are small, dense granules composed of three subunits.
5. Polyribosomes are groups of ribosomes occurring free within the cytoplasm.
6. Vacuoles are more prominent in plant cells than in animal cells.
7. Albumin is soluble in alcohol.
8. Globulin is mainly present in embryo and aleurone layer.
9. Glutelin is soluble in dilute acids or bases, detergents.
10. Oats and rice have high proline content.
11. During early phase of seed development, storage materials are accumulated into storage tissues.
12. Mostly proteins formed during seed development are membrane bound proteins.
13. In cereals, insoluble prolamin storage protiens are synthesised in ER membrane and pass in to lumen of the ER.
14. In dicots, storage proteins are synthesised in ER and transported in vacuole with in golgi vesicles.
15. During endosperm development, non-enzymatic proteins are stored.
16. During embryo development, enzymatic proteins are accumulated.
17. In the dry embryo, the ribosome and messenger RNA are conserved separately within the cell.
18. Seed storage proteins are not synthesized in large quantities during the dry weight accumulation phase of seed development.
19. Seed storage proteins are generally deposited with in organelles called glyoxysomes
20. Amino acids are building blocks of proteins.
21. Polypeptides are short peptides.
22. Oligopeptides are long peptides.
23. In DNA, double helix two strands are bonded by peptide bonds.
24. RNA components necessary for protein synthesis during subsequent germination are stored in the dry embryo.
25. In the endosperm, RNA levels increase during the early stages when the non-storage proteins are mainly synthesized.
26. The decrease in the endospermal RNA levels during maturation is probably due to an increase in endospermal ribonucleases.
27. DNA polymerase elongates RNA strand during transcription.
28. Transcription takes place in cytoplasm.

♦ Answer Keys ♦

Q.No.	Ans.	Q.No.	Ans.	Q.No.	Ans.	Q.No	Ans.	Q.No	Ans.	Q.No	Ans.
1	True	2	True	3	True	4	False	5	True	6	True
7	False	8	True	9	True	10	False	11	False	12	True
13	True	14	True	15	True	16	True	17	True	18	False
19	False	20	True	21	False	22	False	23	False	24	True
25	True	26	True	27	False	28	False				

1.5.3. Fill in the blanks

1. The synthesis of new DNA from existing DNA occurs in the ....
2. The triplet code of CAT in DNA is represented as .... in mRNA and .... in tRNA.
3. The quaternary structure of proteins is represented by the ....
4. Ribosomes are made up of .... subunits.
5. The mRNA start codon, AUG, matches up with a tRNA anticodon ....
6. .... is/are the site of protein synthesis in the cytoplasm.
7. The .... function(s) in the packaging, storage, and distribution of molecules produced by the ER.
8. During endosperm development, ... are accumulated.
9. During embryo development, ... are accumulated.
10. During germination, when level messenger fraction decreases, polysome level ....
11. Polysomes are also called as ....
12. Cytoplasmic mRNA are .... during maturation.
13. Nucleic mRNA are .... during maturation.
14. Ribosomes are present in ....
15. Wheat seeds contain higher amount of .... protein.

♦ Answer Keys ♦

Q.No.	Ans.	Q.No.	Ans.	Q.No.	Ans.
1	Nucleus	2	GUA, CAU	3	Arrangement of subunits
4	2	5	UAC	6	Ribosomes
7	Golgi apparatus	8	Non-enzymatic proteins	9	Enzymatic proteins
10	Increases	11	Polyribosomes	12	Degraded
13	Conserved	14	Rough endoplasmic reticulum	15	Prolamins

1.6. LIPID BIOSYNTHESIS DURING SEED DEVELOPMENT AND MATURATION

1.6.1. Choose the correct answer

1. Which of the following statements concerning fatty acids is correct?
   (a) One of the precursor of prostaglandins
   (b) Phosphatidic acid is a common one
   (c) They all contain one or more double bonds
   (d) They are a constituent of sterols
2. The synthesis of both glycerophospholipids and triacylglycerols involves
   (a) CDP-diacylglycerol (b) Phosphatidate phosphatase
   (c) Phosphatidic acid (d) Phosphoethanolamine
3. The most widely used oils that are in monounsaturated fats are
   (a) Soybean (b) Peanut oil
   (c) Palm oil (d) Corn oil
4. Spherosomes are
   (a) Protein bodies (b) Fat bodies
   (c) Enzymes (d) Hormones
5. Triglycerides are hydrolysed to glycerol and fatty acids by
   (a) Lipase (b) Ligase
   (c) Hexase (d) Kinase
6. Palmitic acid having carbon atoms of
   (a) 12 (b) 14
   (c) 16 (d) 18
7. Olive oils are the richest in
   (a) Diene (b) Triene
   (c) Monoene (d) Protein
8. Diene fatty acid is rich in
   (a) Peanut oil (b) Rape seed oil
   (c) Olive oil (d) Walnut oil
9. Malonyl CoA is produced by
   (a) Acetyl CoA (b) NADPH
   (c) ACP (d) NADP
10. Fatty acid is released from
    (a) ATP (b) ACP
    (c) NADPH (d) ADP
11. Erucic acid is present in
    (a) Castor bean (b) Rape
    (c) Sorghum (d) Cotton
12. Ricinoleic acid is present in
    (a) Rape (b) Tomato
    (c) Sorghum (d) Castor bean

13. In the final stage of seed development, accumulation of triacylglycerols and oleosins in
    (a) Embryo (b) Endosperm
    (c) Seed coat (d) Microphyll
14. Saturated fat is high in
    (a) Pumpkin (b) Coconut
    (c) Almond (d) Sesame
15. Soybean oil contains around 60% of polyunsaturated fatty acids, which are responsible for the
    (a) Low oxidative stability (b) High oxidative stability
    (c) Low reduction stability (d) High reduction stability
16. Myristic acid is present in
    (a) Milk (b) Coconut
    (c) Cocoa butter (d) Safflower

♦ Answer Keys ♦

Q.No.	Ans.	Q.No.	Ans.	Q.No.	Ans.	Q.No	Ans.	Q.No	Ans.	Q.No	Ans.
1	a	2	c	3	b	4	b	5	a	6	c
7	c	8	d	9	a	10	b	11	b	12	d
13	a	14	b	15	a	16	a				

1.6.2. State True or False

1. Fats and oils are organic compounds.
2. In saturated fat, the bonds hold all the hydrogen atoms.
3. If the bonds can hold one additional hydrogen atom, the fat is monounsaturated.
4. Palm oil is a highly saturated oil.
5. Saturated fats are usually liquid at room temperature.
6. Chemically, fats are triesters of glycerol and fatty acids.
7. Seed lipid reserves are stored in the form of spherosomes.
8. Lipids is used to refer to both liquid and solid fats.
9. Unsaturated fatty acids have single bonds.
10. High erucic acid is available in rape seed oil.
11. Arachidonic acid which does not plays a central role as precursor for prostaglandin synthesis.
12. Glucose 6 phosphate dehydrogenase, is the enzyme responsible for the production of the glucanote from glucose-6-phosphate.
13. ACP, which is part of a multi enzyme complex.
14. During the initial or lag phase of seed development, more storage triglyceride is present.
15. Peanut contains high monounsaturated fatty acid.

♦ Answer Keys ♦

Q.No.	Ans.	Q.No.	Ans.	Q.No.	Ans.	Q.No.	Ans.	Q.No.	Ans.	Q.No.	Ans.
1	True	2	True	3	True	4	True	5	False	6	True
7	True	8	True	9	False	10	True	11	False	12	True
13	True	14	False	15	True						

1.7. Hormone Biosynthesis and Accumulation during Seed Development

1.7.1. Choose the correct answer

1. Which of the following ions are affected by gibberellins and are also involved in signal transduction?
   (a) H+ (b) Ca2+
   (c) Na+ (d) Mg2+
2. Pinching the top off a plant to create fuller growth involves which of the following hormones?
   (a) Auxins (b) Ethylene
   (c) Gibberellins (d) Cytokinins
3. Plant hormones are generally produced in .... tissues.
   (a) Cortical (cortex) (b) Epidermal
   (c) Meristem (d) Vascular
4. Which statement about plant hormones is not true?
   (a) Plant hormones include at least one form that is a gas.
   (b) Each naturally occurring hormone has a specific chemical structure.
   (c) Hormones are among the principal internal regulators of plant growth.
   (d) All known growth regulators of plant activities are chemicals produced naturally in plants.
5. Which of these plant hormones is a growth inhibitor?
   (a) Abscisic acid (b) Auxin
   (c) Cytokinin (d) Gibberellin
6. Brassinosteroids effect all of the things listed below except
   (a) Bending stems (b) Hastening senescence
   (c) Membrane polarization (d) Reproductive development
7. Which of these plant hormones was studied in the 1880s by Charles and Francis Darwin in experiments on phototropism?
   (a) Abscisic acid (b) Auxin
   (c) Cytokinin (d) Ethylene
8. The pathway leading to flowering for plants that requires vernalization is the
   (a) Seasonal pathway
   (b) Light-dependent pathway
   (c) Temperature-dependent pathway
   (d) Autonomous pathway

9. **Which plant hormone was discovered in infected rice seedlings that grew extremely tall and slender?**
   (a) Abscisic acid (b) Auxin
   (c) Cytokinin (d) Gibberellin

10. **Cytokinin promotes the growth of**
   (a) Vascular cambium (b) Shoots
   (c) Lateral buds (d) Roots

11. **Which plant hormone, developing in an apical bud, inhibits the growth of lower axillary buds?**
   (a) Abscisic acid (b) Auxin
   (c) Cytokinin (d) Ethylene

12. **Auxin is synthesized in plants from**
   (a) Proline (b) Aspartame
   (c) Phenylalanine (d) Tryptophan

13. **Indoleacetic acid (IAA) is the most commonly occurring form of which plant hormone?**
   (a) Abscisic acid (b) Auxin
   (c) Cytokinin (d) Gibberellin

14. **Environmental signals influence the distribution of an auxin in a plant by**
   (a) Decreasing the cell's sensitivity to the auxin
   (b) Causing auxin to migrate to the lighted portion
   (c) Destroying the auxin
   (d) Causing auxin to migrate into the shaded portion

15. **Dormancy in seeds and buds can be broken by application of which plant hormone?**
   (a) Abscisic acid (b) Auxin
   (c) Cytokinin (d) Gibberellin

16. **A hormone that causes cell division and is found in coconut milk or in dividing root tissue is**
   (a) Abscisic acid (b) Auxin
   (c) Cytokinin (d) Ethylene

17. **How does auxin actually work?**
   (a) In the presence of blue light, it triggers additional photosynthesis and growth.
   (b) In the presence of light, it increases the turgor pressure of the cell and accelerates growth.
   (c) In the presence of light, it moves toward the light source and activates a protein that binds to DNA and starts enzyme production.
   (d) In the presence of unidirectional light, it moves to the shady side and activates an ATP-driven proton pump that results in weakened cell walls and eventual elongation.

18. How does gibberellin actually work?
    (a) In the presence of blue light, it triggers additional photosynthesis and growth.
    (b) In the presence of light, it increases the turgor pressure of the cell and accelerates growth.
    (c) In the presence of light, it increases the formation of additional DNA that codes for additional plant structures.
    (d) It attaches to a plasma membrane receptor and results in gene activation to produce the enzyme amylase that releases sugars for additional cell growth, division, and elongation.

19. A chemical messenger from the embryo that apparently stimulates a seed to digest the endosperm is
    (a) Abscisic acid. (b) Auxin.
    (c) Cytokinin. (d) Gibberellin.

20. Application of which hormone prevents plant tissues from senescing, or aging?
    (a) Abscisic acid (b) Auxin
    (c) Cytokinin (d) Ethylene

21. Which plant hormone is a gas?
    (a) Abscisic acid (b) Auxin
    (c) Cytokinin (d) Ethylene

22. "Foolish seedling" disease in rice is caused by
    (a) Auxins (b) Gibberellins
    (c) Cytokinin (d) Ethylene

23. In vascular plants, most cytokinins are produced in the
    (a) Roots (b) Shoots
    (c) Flowers (d) Leaves

24. Ripening of fruits, such as bananas, is hastened by
    (a) Abiscisic acid (b) Cytokinin
    (c) Indoleacetic acid (d) Ethylene

25. Which of the following plant hormones is incorrectly paired with its function?
    (a) Auxins -- responsible for apical dominance
    (b) Abscisic acid -- regulates the rate of transpiration
    (c) Cytokinins -- delays senescence (aging and decay)
    (d) Gibberellins -- promotes bud and seed dormancy

26. Lateral stem development is controlled by the relative levels of
    (a) Cytokinins and auxins (b) Abscisic acid and auxins
    (c) Auxins and gibberellins (d) Auxins and ethylene

27. The hormone responsible for phototropic responses in the growing tips of plants is:
    (a) Auxin (b) Cytokinin
    (c) Gibberellins (d) Ethylene

28. Tomatoes can be artificially ripened through the use of
    (a) Auxin (b) Cytokinins
    (c) Gibberellins (d) Ethylene
29. Apical dominance in plants is due to which of the following plant hormones?
    (a) Ethylene (b) Gibberellin
    (c) Auxins (d) Cytokinins
30. Gibberellins was named after a fungus called
    (a) Gibberella fujikuroi (b) Gibberella gaditijirrii
    (c) Gibberella acuminate (d) Gibberella africana
31. Gibberellins are chemically
    (a) Phenolic derivatives (b) Terpenoid derivatives
    (c) Adenine derivatives (d) Alkaloid derivatives
32. Gibberella fujikuroi the fungus causes
    (a) Foolish seedling disease of rice or bakanae disease
    (b) Damping off seedling disease of rice or bakanae disease
    (c) Fungal blight disease or rice or bakanae disease
    (d) Rust disease of rice or bakanae disease
33. The most studied gibberellins is
    (a) GA1 (b) GA9
    (c) GA3 (d) GA4
34. All are anti gibberellins except
    (a) Acetosyringone (b) Amo 1618
    (c) Maleic hydrazide (d) Phosphon D
35. All are functions of gibberellins expect
    (a) Promote cell elongation
    (b) Promote dormancy
    (c) Promote elongation of internodes
    (d) Promotes parthenocarpic fruit formation
36. Which of the following bioassay is used to test the presence of gibberellins
    (a) Chlorophyll preservation test
    (b) Dwarf pea elongation technique
    (c) Rice seedling growth inhibition test
    (d) Excised radish cotyledon enlargement test
37. The gibberellins present in the aleurone layer of barley increase transcription
    (a) Lipase and protease (b) Protease and cellulose
    (c) Protease and amylase (d) Lipase and cellulose
38. The hormone which is involved in regulation of florigen synthesis
    (a) Ethylene (b) Zeatin
    (c) Cytokinin (d) Gibberellins

39. Which is the gibberellins involved in elongation responses
    (a) GA1 (b) GA9
    (c) GA3 (d) GA4
40. Which is the gibberellins involved in flowering
    (a) GA1 (b) GA9
    (c) GA3 (d) GA4
41. Early stages of gibberellins synthesis take place in
    (a) Mitochondria (b) Peroxisome
    (c) Plastids (d) Cytoplasm
42. Genetic dwarfism can be nullified by spraying with
    (a) Auxin (b) Zeatin
    (c) Cytokinin (d) Gibberellins
43. .... is an organic molecule synthesized by a plant that has physiological and/or developmental effects.
    (a) Enzyme (b) Plant hormone
    (c) Nutrient (d) Cotyledon
44. Which of the following plant hormones is not a growth promoter?
    (a) Auxins (b) Gibberellins
    (c) Cytokinins (d) Abscisic acid
45. Which of the following plant hormones acts to cause a leaf and/or fruit to drop?
    (a) Gibberellins (b) Cytokinin
    (c) Abscisic acid (d) Ethylene
46. Which of the following plant hormones affects stem elongation?
    (a) Phytochrome (b) Gibberellins
    (c) Cytokinin (d) Abscisic acid
47. Which of the following plant hormones can cause the base of a shoot to form new roots and prevents fruit from dropping too soon?
    (a) Auxin (b) Cytokinin
    (c) Gibberellins (d) Ethylene
48. Which of the following is/are a bound auxins
    (a) Indole-3-acetic acid (b) IAA-glucan
    (c) GA8 glucoside (d) Zeatin
49. Conjugated gibberellin present in Phaseolus vulgaris
    (a) GA8 glucoside (b) GA3
    (c) GA29 (d) GA8
50. The site of gibberellins biosynthesis in Echinocystis is
    (a) Endosperm (b) Embryo
    (c) Epicotyls (d) Hypocotyl

51. Most abundant and widely distributed natural cytokinin in higher plants is
    (a) Zeatin (b) 6-aminopurine
    (c) 6-benzyladenine (d) GA3
52. ABA was first isolated from
    (a) Young cotton fruits (b) Immature maize kernels
    (c) Mature seeds of pea (d) Immature sorghum seeds

♦ Answer Keys ♦

Q.No.	Ans.	Q.No.	Ans.	Q.No.	Ans.	Q.No.	Ans.	Q.No.	Ans.	Q.No.	Ans.
1	b	2	a	3	c	4	d	5	a	6	b
7	b	8	c	9	d	10	c	11	b	12	d
13	b	14	d	15	d	16	c	17	d	18	d
19	d	20	c	21	d	22	b	23	a	24	d
25	d	26	a	27	a	28	d	29	c	30	a
31	b	32	a	33	c	34	a	35	b	36	b
37	c	38	d	39	a	40	b	41	c	42	d
43	b	44	d	45	d	46	b	47	a	48	b
49	a	50	a	51	a	52	a				

1.7.2. State True or False

1. A hormone originating in the terminal bud of a plant suppresses the growth of lateral buds.
2. The response of a plant to touch is called thigmotropism.
3. Plant hormones can stimulate certain physiological processes while inhibiting others.
4. Phytochrome is the photoreceptor for phototropism.
5. In direct sunlight, most phytochrome is in the Pfr form.
6. Plant hormones may promote or inhibit growth.
7. Auxins were the first major plant hormones discovered.
8. Molecules of IAA are chemically liable in aqueous solution.
9. PAA is a naturally occurring gibberellins.
10. Picloram is a synthetic auxin.
11. Auxin do not induce the formation of parthenocarpic fruit.
12. The free auxin rapidly diminishes as the seed reaches full maturity.
13. Conversion to other forms such as ester and bound auxin is known to occur in embryo of maturing cereal kernels and in seeds of certain dicot species.
14. Gibberellins are cyclic diterpenes.
15. The GA biosynthetic pathway can be divided into four stages.
16. Seeds may also contain gibberellins conjugated to a sugar.
17. Gibberellins are synthesized by the seeds themselves.

18. Zeatin is the most abundant and widely distributed natural gibberellin in higher plants and in some bacteria.
19. The initial steps of ABA biosynthesis take place in chloroplasts or other plastids.
20. The initial steps of ABA biosynthesis occur in cytosol.
21. When seeds suffer increasing water stress and when they begin to rehydrate the ABA levels rise as they do in wilting leaves

♦ Answer Keys ♦

Q.No.	Ans.	Q.No.	Ans.	Q.No.	Ans.	Q.No.	Ans.	Q.No.	Ans.	Q.No.	Ans.
1	True	2	True	3	False	4	False	5	True	6	True
7	True	8	True	9	False	10	True	11	False	12	True
13	False	14	True	15	False	16	True	17	True	18	False
19	True	20	True	21	False						

1.7.3. Fill in the blanks

1. .... is produced in large quantities in the climacteric phase of fruit ripening.
2. Stems will form roots more readily if .... are attached than if they were removed.
3. .... stimulates the production of hydrolytic enzymes.
4. ...., in combination with auxin, stimulates cell division in plants and determines the course of differentiation.
5. Auxin increases the .... of cell walls.
6. .... regulates stem elongation in young grass seedlings and most herbs.
7. One of the most important uses of auxins is the .... of abscission.
8. .... is a pigment that is part of a biological clock system that helps control flowering.
9. The naturally occurring auxin is ....
10. The synthetic auxin is ....
11. Bound auxin(s) present in mature kernels of Zea mays are ....
12. The presence of gibberellic acid (GA3) has been reported in immature seeds and seedlings of ....
13. For biosynthesis of gibberellic acid cell compartments required are ....
14. Immature kernels of Zea mays contains ....

♦ Answer Keys ♦

Q.No.	Ans.	Q.No.	Ans.
1	Ethylene	2	Leaves
3	Gibberellins	4	Cytokinin
5	Plasticity	6	Auxin

Q.No.	Ans.	Q.No.	Ans.
7	Prevention	8	Phytochrome
9	IAA	10	NAA,2,4-D Dicamba
11	IAA-glucan, IAA-myo-inositols	12	*Pisum sativum, Ricinus communis, Zea mays*
13	Plastid, Endoplasmic reticulum, Cytosol	14	Zeatin, Zeatin riboside, Zeatin ribotide

1.8. CHANGES IN SECONDARY METABOLITES AND TOXIC COMPOUND DURING SEED DEVELOPMENT

1.8.1 Choose the correct answer

1. The organic compounds that appear to have no direct function of growth and development
   (a) Secondary metabolites (b) Primary metabolites
   (c) Organic acids (d) Enzymes
2. Secondary metabolites are
   (a) Natural products (b) Synthetic products
   (c) Organic products (d) Medicinal products
3. Secondary metabolites influence
   (a) Ecological interaction (b) Biological interaction
   (c) Antogeuistic interaction (d) Mutualistic interaction
4. Absence of secondary metabolites causes
   (a) Immediate effect (b) Long-term impairment
   (c) Pseudo-effects (d) Positive effects
5. Primary metabolite ....
   (a) Lipids (b) Allele chemicals
   (c) Pigments (d) Lignin
6. Secondary metabolites used in preparation of
   (a) Antibiotics (b) Foods
   (c) Blenders (d) Composts
7. Secondary metabolite without nitrogen
   (a) Amines (b) Alkaloids
   (c) Phenylpropanol (d) Glucosinolates
8. Which alkaloid is extremely poisonous to humans
   (a) Nicotine (b) Hyoscamine
   (c) Purine alkaloids (d) Atropine
9. Alkaloids can be purified from extracts by
   (a) Solvent extraction (b) Acid base extraction
   (c) Distillation (d) Expression
10. Terpenes are ....
    (a) 30 carbon units (b) Seven carbon units
    (c) Six carbon units (d) Five carbon units

11. ABA is ....
    (a) Sesquiterpene (b) Diterpenes
    (c) Monoterpenes (d) Triterpenes
12. Glucosinolates occurs in ....
    (a) Tillaceae (b) Convolvulaceae
    (c) Cucurbitaceae (d) Brassicaceae
13. Which inhibitor is a naturally occurring sesquiterpene that regulates plant growth.
    (a) HCN (b) ABA
    (c) Protease inhibitors (d) Trypsin inhibitors
14. Example for secondary metabolites with nitrogen
    (a) Monoterpenes (b) Flavonoids
    (c) Non protein aminoacids (d) Diterpenes
15. The alkaloid which is a powerful insecticide is
    (a) Atropine (b) Morphine
    (c) Cocaine (d) Nicotine
16. 15 carbon terpenes with three C5 units are called as
    (a) Monoterpens (b) Sesquiterpenes
    (c) Diterpenes (d) Triterpenes
17. Which is the toxic compound during seed development?
    (a) Alkaloids (b) Phenols
    (c) HCN (d) Amines
18. Protease inhibitors have been strongly implicated in
    (a) Defense mechanism (b) Growth mechanism
    (c) Inhibiting mechanism (d) Resistance mechanism

♦ Answer Keys ♦

Q.No.	Ans.	Q.No.	Ans.	Q.No.	Ans.	Q.No.	Ans.	Q.No.	Ans.	Q.No.	Ans.
1	a	2	a	3	a	4	b	5	a	6	a
7	c	8	d	9	b	10	d	11	a	12	b
13	b	14	c	15	d	16	b	17	c	18	c

1.8.2 Answer whether the Statement is True or False

1. Terpenes are insoluble in water.
2. Terpenes can decompose at high temperatures to give isopentane.
3. Pyrethroids are diterpenes.
4. Fragrant in roses is due to terpenes.
5. Saponins are named because of their soap like properties.
6. When the plants or seed exposed to stress, the ABA level decreases.
7. Terpenes are biosynthesized from primary metabolites through mevalonic acid pathway.
8. Cocaine is a alkaloid.

9. Alkaloids are nitrogen containing compounds.
10. Solanaceous crops contains alkaloids like cocain.
11. Allelopathic agents are allelochemicals that influence competition among plant species.
12. Terpenes are formed by fusion of four carbon isoprene units.
13. ABA in seed induce the germination of seed and increases the growth of the seed.
14. Larger terpenes are triterpenes, tetraterpenes and polyterpenes.
15. Xanthone is a class of phenolic compound which contains 15 number of C-atoms.
16. HCN is liberated enzymatically from cyanogenic glucosides and inhibits the growth.
17. Protease inhibitors occur widely in roots and leaves of plant kingdom.
18. Tannins are found in testa of cocoa and bean seed.
19. Coumarin is a alkaloid present in pericarp of Beta vulgaris.

♦ Answer Keys ♦

Q.No.	Ans.	Q.No.	Ans.	Q.No.	Ans.	Q.No.	Ans.	Q.No.	Ans.
1	True	2	False	3	False	4	True	5	True
6	False	7	True	8	False	9	Truc	10	Truc
11	True	12	False	13	False	14	True	15	False
16	True	17	False	18	True	19	False		

1.8.3 Fill in the blanks

1. Terpenes orginate from ....
2. The basic structural elements of terpenes are .... units.
3. Secondary metabolites play a significant role in ....
4. Secondary metabolite provide .... and .... protein.
5. .... secondary metabolite in coloured flower attracts pollinators and seed dispersing animals.
6. Gibberellins contains .... C5 units.
7. Terpenes act as a .... of plants.
8. Terpenoids are used extensively for their .... qualities.
9. A well-known terpenoid is ....
10. Terpene which is present in cannabis ....
11. The presence of both .... and .... elements in one molecule gives saponins detergent properties.
12. The simplest phenol is ....
13. .... are used as germicides.
14. .... occurs in (Cocoa and bean) seed particularly in seed coat structures.

15. The most widespread group of pigmented flavonoids ....
16. ABA is not only prevents .... during embrogenesis.
17. Hydrocyanic acid is present in crops like .... and ....
18. .... compound is a major glucosinate of black mustard seeds and horse radish roots.
19. .... is the building block of terpenes.
20. Coumarins are simple phenyl propanoids such as ....
21. The function(s) of flavonoids in the plant is (are) ....
22. HCN is present in following crop ....
23. ABA is a naturally occurring ....

♦ Answer Keys ♦

Q.No.	Ans.	Q.No.	Ans.	Q.No.	Ans.
1	Turpine	2	Isoprene	3	Protection
4	Surface, structural	5	Anthocyanin	6	Four
7	Phytochromes	8	Aromatic	9	Camphor
10	Cannabinoids	11	Lipd soluble, water soluble	12	C_6H_5OH
13	Phenolic compounds	14	Tannin	15	Anthocyanins
16	Premature germination	17	Sunflower, sorghum	18	Sinigrin
19	Isopentenyl diophosphate	20	Trans cinnamic acid, p-coumaric acid, caffeic acid	21	Pigmentation, Defense
22	Oats, Spinach	23	Sesquiterpene		

1.9. POSSIBLE ALTERATION IN METABOLIC PATHWAYS DURING SEED DEVELOPMENT AND MATURATION

1.9.1. Choose the correct answer

1. The plant which became famous due to theory of mutation
   (a) *Triticum vulgare* (b) *Oenothera lamarckiana*
   (c) *Pisum sativum* (d) *Primula sinensis*

2. Mutation is
   (a) A change that affects the parents only, but never inherited
   (b) A change that is inherited
   (c) A factor responsible for plant growth
   (d) A change that affects the off-spring of F2 generation only.

3. Continuous variation is attributed to
   (a) Mutation (b) Crossing over
   (c) Chromosomal aberration (d) Polyploidy

4. Which of the following is physical mutagen
   (a) Gamma ray (b) X ray
   (c) UV ray (d) α ray
5. Mutations are responsible for causing
   (a) Cell division in plants (b) Cell elongation in plants
   (c) Variations in organisms (d) Cell differentiation in plants
6. Which of the following is a strong chemical mutagen?
   (a) Methane (b) Guanine
   (c) Carbon tetrasulphide (d) Caffeine
7. Which of the following is a mutagen?
   (a) NO_2 (b) CO_2
   (c) SO_2 (d) CO
8. A sudden heritable change in a characteristic of an organism is
   (a) Mitosis (b) Hybridization
   (c) Meiosis (d) Mutation
9. Mutation by a change in the DNA molecule of an individual gene
   (a) Forward mutation (b) Point mutation
   (c) Reverse mutation (d) Somatic mutation
10. Agents used for induction of mutations are known as
    (a) Germicides (b) Mutagens
    (c) Steroids (d) Genomes
11. The only one non-ionizing radiation capable of inducing mutation is
    (a) UV rays (b) X rays
    (c) Alpha rays (d) Mega rays
12. Treating a biological material with a mutagen to induce mutation is
    (a) Gametogenesis (b) Mutagenesis
    (c) Sporogenesis (d) Gametocides
13. The process of mRNA degradation that is induced by double-stranded RNA is
    (a) DNA interference (b) RNA interference
    (c) Genome interference (d) Gene interference
14. Expand RISC
    (a) RNA Induced Silencing Complex
    (b) RNA Induced Stimulating Complex
    (c) RNA Induced Sequence Complex
    (d) RNA Induced Segment Complex
15. Alpha rays are
    (a) Chemical mutagen (b) Biological mutagen
    (c) Physical mutagen (d) Physiological mutagen
16. Lysine catabolism is controlled by a bifunctional enzyme
    (a) Dehydrogenase (b) Amylase
    (c) Maltase (d) Lysine ketoglutarate reductase

♦ Answer Keys ♦

Q.No.	Ans.	Q.No.	Ans.	Q.No.	Ans.	Q.No.	Ans.	Q.No.	Ans.	Q.No.	Ans.
1	b	2	b	3	a	4	a	5	c	6	a
7	a	8	d	9	b	10	b	11	c	12	b
13	b	14	a	15	c	16	d				

1.9.2. State True or False

1. Morgan studied mutation in Oenothera lamarkiana.
2. The mutation may increase or decrease the resistance of an organism for the disease.
3. The mutations are always harmful.
4. The non-ionizing radiation like cosmic rays do not cause mutations.
5. Methyl methane sulphonate is a chemical mutagen.
6. Nitrogen and sulphur mustards are not capable of inducing any mutation.
7. The intragenic mutations are also called point mutations.
8. The allopolyploids arise from a heterozygote.
9. Somatic mutations are heritable mutations.
10. X-rays produced mutants in Drosophilla.
11. Ethyl Methane Sulphonate is a chemical mutagen.
12. Cereals are generally high in lysine, but low in methionine.
13. The silencing effect is non heritable.
14. The term mutation was introduced by Morgan.
15. Somatic mutations are mutations occurring in sex chromosomes.
16. Spontaneous mutations occur naturally.
17. Gene mutation is otherwise called as point mutation.
18. DNA contains uracil.

♦ Answer Keys ♦

Q.No.	Ans.	Q.No.	Ans.	Q.No.	Ans.	Q.No.	Ans.	Q.No.	Ans.	Q.No.	Ans.
1	False	2	True	3	False	4	False	5	True	6	False
7	True	8	True	9	False	10	True	11	True	12	False
13	False	14	True	15	False	16	True	17	True	18	False

1.9.3. Fill in the blanks

1. The letters ATP stand for ....
2. Deamination and alkylation induce ....
3. The chromosomal mutation is also called as ....
4. Muller was awarded Nobel prize for inducing mutations by ....
5. Aridine is a chemical ....

6. Metabolic pathways are regulated by a(n) .... enzyme, which is also often the first enzyme in a series.
7. A genetic mutation refers to ....
8. The production of mRNA from a DNA template is termed ....
9. RNAi silencing is triggered by .... RNA leading to the suppression of gene expression.
10. The locus of pea which denotes wrinkled seeds is ....
11. .... is a post-transcriptional gene-silencing phenomenon induced by double-stranded RNA.
12. .... is the most abundant antinutritional phenolic compound in seeds of cruciferous species
13. Mutation caused by change in temperature is ....
14. Mutation caused due to changes in the structure of chromosome is ....
15. The strands in siRNA are ....
16. Lysine is converted to .... through condensation with a-ketoglutarate
17. .... to RNase III family enzyme which cleaves dsRNA to generate siRNAs

♦ Answer Keys ♦

Q.No.	Ans.	Q.No.	Ans.	Q.No.	Ans.
1	Adenosine triphosphate	2	Transition	3	Abberration
4	X-rays	5	Mutagen	6	Rate-limiting
7	A missing or altered nucleotide base or a missing gene	8	Transcription	9	Double stranded
10	r and r_b, rug 3, 4, 5	11	RNA interference	12	Sinapine
13	Exogenous mutation	14	Chromosomal aberration	15	Guide and passenger strand
16	Saccharopine	17	Dicer		

Unit 2

SEED DORMANCY AND GERMINATION

2.1. SEED DORMANCY : TYPES AND CLASSIFICATION

2.1.1. Choose the correct answer

1. Dormancy in seeds and buds can be broken by application of which plant hormone?
   (a) Abscisic acid (b) Auxin
   (c) Ethylene (d) Gibberellin
2. Which plant hormone is called the stress hormone, because it produces dormancy in seeds and buds?
   (a) Abscisic acid (b) Auxin
   (c) Cytokinin (d) Ethylene
3. Which of the following plant hormone is incorrectly paired with its function?
   (a) Auxins -- responsible for apical dominance
   (b) Abscisic acid -- regulates the rate of transpiration
   (c) Cytokinins -- delays senescence (aging and decay)
   (d) Gibberellins -- promotes bud and seed dormancy
4. Bud dormancy is induced by
   (a) ABA (b) IAA
   (c) Ethylene (d) Gibberellic acid
5. Which of the following does not exhibit seed dormancy?
   (a) Phaseolus (b) Xanthium
   (c) Rhizophora (d) Cassia

♦ Answer Keys ♦

Q.No.	Ans.	Q.No.	Ans.	Q.No.	Ans.	Q.No.	Ans.	Q.No.	Ans.
1	d	2	a	3	d	4	a	5	c

2.1.2. Answer whether the Statement is True or False

1. Dormancy is clearly revealed by TZ test
2. Quiescence is also described as imposed dormancy

3. The organ or plant gradually emerges from dormant stage in pre dormancy period
4. The seeds of a few Solanaceae members are impermeable to gases
5. The seed germination is also inhibited by soil fungi and bacteria
6. The seeds of *Ranunculus* contain immature embryo and hence, they remain dormant
7. The seeds of *Iris* and *Populus* require an after ripening period before germination
8. Chemical like cyanides inhibit seed germination
9. Secondary dormancy in seeds is induced on transference under favourable conditions
10. For breaking dormancy, the seeds are never given acid treatment
11. The grains of *Zea* get killed in continued darkness
12. Chemical like ethylene and KNO_3 promote seed germination
13. By giving chilling treatment, the dormant buds are killed
14. Chemicals like ethyl chlorohydrin induce bud dormancy
15. A seed is a dormant embryo waiting for water to stimulate germination

♦ Answer Keys ♦

Q.No.	Ans.	Q.No.	Ans.	Q.No.	Ans.	Q.No.	Ans.	Q.No.	Ans.
1	False	2	True	3	False	4	False	5	True
6	True	7	True	8	True	9	True	10	False
11	True	12	True	13	False	14	False	15	True

2.1.3. Fill in the Blanks

1. .... form of the phytochrome is physiologically active in overcoming seed dormancy.
2. The arrest in growth or development due to environmental factors is called ....
3. The dormancy phase of life history is divided into .... sub phases.
4. The dormancy in the seeds of some members of Chenopodiaceae is due to their impermeability to ....
5. The permeability of seeds to water is also decreased due to soil .... and ....
6. Germination inhibitors occur in the seeds coat, ... and ....
7. The seeds of *Prunus* contain .... embryos.
8. For breaking dormancy, seed coat may be broken and this phenomenon is called as ....
9. Seeds whose germination is affected by light are called as ....

10. In many plants, the bud dormancy phase starts late .... and terminates in ....

11. In any woody plants, bud dormancy is induced by .... days.

♦ Answer Keys ♦

Q.No.	Ans.	Q.No.	Ans.	Q.No.	Ans.
1	Pfr	2	Quiescence	3	Three
4	Water	5	Fungi, Bacteria	6	Embryo, Endosperm
7	Dormant	8	Scarification	9	Photoblastic
10	Summer, spring	11	Short		

2.2. ROLE OF HORMONES ON SEED DORMANCY

2.2.1. Choose the correct answer

1. Which statement about plant hormones is not true?
   (a) Plant hormones include at least one form that is a gas.
   (b) Each naturally occurring hormone has a specific chemical structure.
   (c) Hormones are among the principal internal regulators of plant growth.
   (d) All known growth regulators of plant activities are chemicals produced naturally in plants.

2. Which of these plant hormone is a growth inhibitor?
   (a) Abscisic acid (b) Auxin
   (c) Cytokinin (d) Gibberellin

3. Indole acetic acid (IAA) is the most commonly occurring form of which plant hormone?
   (a) Abscisic acid (b) Auxin
   (c) Cytokinin (d) Ethylene

4. Dormancy in seeds and buds can be broken by application of which plant hormone?
   (a) Abscisic acid (b) Auxin
   (c) Cytokinin (d) Gibberellin

5. Which of the following is/are an effect(s) of auxin?
   (a) Auxin production by seeds promotes the growth of fruit.
   (b) Trees sprayed with auxin will keep mature fruit from falling to the ground.
   (c) Under the effect of gravity, auxin moves to the lower surface of roots and stems, causing roots to curve downward and stems to curve upward.
   (d) All of the choices are effects of auxin.

7. Which plant hormone is a gas?
   (a) Abscisic acid (b) Auxin
   (c) Cytokinin (d) Ethylene

8. Which plant hormone stimulates the ripening of fruit and inhibits plant growth?
   (a) Abscisic acid (b) Auxin
   (c) Cytokinin (d) Ethylene

6. How does auxin actually work?
   (a) In the presence of blue light, it triggers additional photosynthesis and growth.
   (b) In the presence of light, it increases the turgor pressure of the cell and accelerates growth.
   (c) In the presence of light, it increases the formation of hydrolytic enzymes that release energy needed for growth purposes.
   (d) In the presence of unidirectional light, it moves to the shady side and activates an ATP-driven proton pump that results in weakened cell walls and eventual elongation.

♦ Answer Keys ♦

Q.No.	Ans.	Q.No.	Ans.	Q.No.	Ans.	Q.No.	Ans.
1	d	2	a	3	b	4	d
5	d	6	a	7	d	8	d

2.2.2. Match the Following :

1	Auxin	a	Ripens fruit, controls abscission
2	Gibberellin	b	Induces amylase production in seeds
3	Abscisic acid	c	Influences photoperiodism
4	Ethylene	d	Affects stomatal response to water stress
5	Cytokinin	e	Promotes cell division

♦ Answer Keys ♦

Q.No.	Ans.	Q.No.	Ans.	Q.No.	Ans.	Q.No.	Ans.	Q.No.	Ans.
1	c	2	b	3	d	4	a	5	e

2.3. REQUIREMENTS OF GERMINATION

2.3.1. Choose the correct answer

1. Which of the following factors was the most important for seed germination?
   (a) Lots of light (b) Warm temperatures
   (c) Adequate water (d) Paper towels
2. Germination of seed within fruit is called as
   (a) Ovipary (b) Vivipary
   (c) Hypogeal (d) Epigeal
3. Seed planted too deep in the soil usually do not germinate because
   (a) They are under pressure
   (b) They do not get enough light
   (c) They do not get enough oxygen
   (d) The seedlings cannont come up to the soil surface.

4. A seed with fairly unequal cotyledons
   (a) Oak (b) Jack fruit
   (c) Castor (d) Tamarind
5. Vivipary is the characteristic feature of
   (a) Hydrophytes (b) Xerophytes
   (c) Halophytes (d) Mesophytes
6. What are two sources of energy for the physical and chemical changes plants make during germination?
   (a) Water, soil
   (b) The sun, stored chemicals in the seed
   (c) Soil, warmth from the interior of the earth
   (d) Light, organisms in the soil
7. Ability of seed to give rise to a seedling is called as
   (a) Germination (b) Viability
   (c) Dormancy (d) Vigour
8. The presence of cotyledons above the soil surface is refered to
   (a) Epigeal (b) Hypogeal
   (c) Vivipary (d) *In situ*
9. If you expose seeds to a series of red light versus far-red light treatments, which of the following exposure treatments would result in seed germination?
   (a) Red; far-red
   (b) Far-red; red
   (c) Red; far-red; red; far-red; red; far-red; red; far-red
   (d) White light
10. A kind of process by which the dormant embryo wakes up, grows out of the seed coat and establishes itself as seedling
   (a) Germination of seed (b) Fertilization
   (c) Vernalization (d) Plasmolysis

♦ Answer Keys ♦

Q.No.	Ans.	Q.No.	Ans.	Q.No.	Ans.	Q.No.	Ans.	Q.No.	Ans.
1	c	2	b	3	c	4	b	5	c
6	a	7	a	8	a	9	a	10	a

2.3.2. State True or False

1. A seed that germinates after a cold period, will usually grow in the spring
2. Dicotyledons have two seed leaves and always exhibit epigeal emergence
3. When exposed to the right conditions, a seed will always germinate
4. Seed coat dormancy results from chilling period being required to break down the seed coat before germination will occur.
5. Seed viability period is not depend upon storage conditions.

6. Temperature activates the biochemical processes during germination
7. Large seeded species require more water than the small seeded species
8. Water is essential for enzyme activation
9. Germinating seed are metabolically inactive.
10. The process of aerobic respiration yields less energy when compared to the anaerobic respiration process.
11. Third group of plant require cool temperature.

♦ Answer Keys ♦

Q.No.	Ans.	Q.No.	Ans.	Q.No.	Ans.	Q.No.	Ans.	Q.No.	Ans.	Q.No.	Ans.
1	True	2	False	3	True	4	True	5	False	6	False
7	True	8	True	9	False	10	False	11	False		

2.3.3. Fill in the blanks

1. Ferulic acid .... germination of seeds.
2. Parascorbic acid .... germination of seeds.
3. .... is defined as the emergence of the radicle through the seed coat.
4. .... are growth promoting substance
5. .... is needed to activate the biochemical processes during germination.
6. The pH of the water should be in the range of ....
7. .... group of plants requires cool temperatures for seed germination.
8. .... group of plants requires warm temperatures for seed germination
9. Temperature can affect the .... and ....
10. Epiphytes are examples of plants that require .... in order for seed to germinate,
11. .... also affect germination,
12. ...., ... and .... are essential for seed germination.
13. .... and .... influence for seed germination.

♦ Answer Keys ♦

Q.No.	Ans.	Q.No.	Ans.
1	Inhibits	2	Inhibits
3	Germination	4	Auxin
5	Water	6	65 to 75
7	Second	8	Third
9	Percentage and rate of germination	10	Light
11	Photoperiod	12	Moisture, oxygen and favourable temperature
13	Light intensity and light quality		

2.4. IMBIBITION AND ENZYME ACTIVATION

2.4.1. Choose the correct answer

1. Which of the following is not a feature of monocots?
   (a) Flower parts in threes (b) Seed with one cotyledon
   (c) Vascular cambium present (d) Scattered vascular bundles
2. In which of the following structures do pollen grains develop?
   (a) Ovule (b) Stigma
   (c) Anther (d) Carpal
3. The first process by which water enters into the seed coat when a seed is placed in suitable environment for germination is
   (a) Osmosis (b) Imbibition
   (c) Absorption (d) Adsorption
4. When the concentration of the soil solutes is low, the absorption of water is
   (a) Remain normal (b) Decreased
   (c) Increased (d) Stopped
5. When pea seeds and wheat grains are soaked in water, pea seeds showed more swelling than the wheat. The reason is
   (a) Cell membrane of pea seeds is more permeable
   (b) Presence of less hydrophilic colloids in the wheat grains
   (c) Imbibition capacity of proteins is more than that of starch
   (d) Cell walls of wheat grains are less permeable
6. Which one of the following denotes the water potential of the mesophyll cell in wilted condition?
   (a) Greater than the value of its osmotic potential
   (b) Equal to the value of pressure potential
   (c) Equal to the value of osmotic potential
   (d) Equal to zero
7. Which of the following force is responsible for raising water upto 100 m of tall plants?
   (a) Root pressure (b) Capillary action
   (c) Air pressure (d) Transpiration pull
8. Cohesion theory of water movement in plants was putforth by
   (a) Henry Dixon (b) F. F. Blackman
   (c) Melvin cycle (d) Hans. A. Krebs
9. When the conditions are dry, a grass leaf curls inward to minimize water loss due to presence of
   (a) Thick cuticle (b) Large xylem cavities
   (c) Parallel venation (d) Bulliform cells
10. Humidity in atmosphere decreases rate of
    (a) Glycolysis (b) Photosynthesis
    (c) Transpiration (d) Growth

11. Guttation is the result of
    (a) Diffusion (b) Root pressure
    (c) Osmosis (d) Transpiration
12. Hydathodes are also called
    (a) Guard cells (b) Sunken stomata
    (c) Water stomata (d) Subsidiary cells
13. Graham's Law is correlated with
    (a) Osmosis (b) Osmoregulation
    (c) Diffusion (d) Adsorption
14. If water enters in a cell, the pressure exerted by its swollen protoplast is
    (a) Osmotic pressure (b) Diffusion pressure deficit
    (c) Turgor pressure (d) Imbibition
15. What is the name given to the food-storage tissue that surrounds the embryo in a dicot seed?
    (a) Coleorhiza (b) Plumule
    (c) Radicle (d) Cotyledon
16. What is the role of light in the germination of lettuce seeds?
    (a) It stimulates photosynthesis
    (b) It stimulates imbibition
    (c) It inactivates germination inhibitors
    (d) It increases germination temperature.
17. The pressure that develops against the cell wall as a result of water entering the vacuole of the cell is called
    (a) Osmotic potential (b) Plasmolysis
    (c) Pressure potential (d) Diffusion
18. Enzymes are sensitive to
    (a) Cold (b) Cell wall
    (c) Heat (d) Pressure
19. Specificity of protein in enzyme action depends upon
    (a) Active sites (b) Km constant
    (c) Turn over number (d) Linear sequence of amino acids
20. Enzymes exist in the cells as
    (a) Solution (b) Crystals
    (c) Solids (d) Colloids
21. Enzymes functional inside the cells are called
    (a) Endoenzymes (b) Exoenzymes
    (c) Apoenzymes (d) Isoenzymes
22. Enzymes having slightly different molecules structure, but performing identical activity are
    (a) Apoenzymes (b) Isoenzymes
    (c) Holoenzymes (d) Coenzymes

23. Enzymes, vitamins and hormones are common in
    (a) Enhancing oxidative metabolism
    (b) Being synthesised in the body of organisms
    (c) Being proteinaceous
    (d) Regulating metabolism
24. Which of the following influence feedback inhibition of enzyme?
    (a) End product (b) External factors
    (c) Enzyme (d) Substrate
25. Hydrolysis of starch occurs with the help of
    (a) Peptidase (b) Amylase
    (c) Sucrose (d) Lipase
26. Method that renders the seed coat permeable to water so that embryo expansion is not physically retarded is known as
    (a) Scarification (b) Denudation
    (c) Vernalization (d) Stratification
27. Amount of water needed for seed germination of rice seeds is around
    (a) 22-25% (b) 25-27%
    (c) 32-35% (d) 42-45%
28. First process that occurs during germination is
    (a) Imbibition (b) Rupturing of seed coat
    (c) Gibberellin synthesis (d) Amylase synthesis
29. Principle component of seed responsible for imbibition of water is
    (a) Carbohydrate (b) Lipids
    (c) Protein (d) Hormone
30. Extreme impermeability of water through seed coat leads to
    (a) Innate dormancy (b) Seed coat dormancy
    (c) Induced dormancy (d) Embryo dormancy
31. Different permeability of seed coat is due to the ionization of acidic and basic groups of
    (a) Protein membrane nature (b) Carbohydrate membrane nature
    (c) Lipids membrane nature (d) Phospholipid nature
32. In monocots, gibberellins are released from the
    (a) Scutellum (b) Aleurone layer
    (c) Endosperm (d) Embryo
33. Synthesis of hydrolytic enzymes like α-amylase, ribonuclease, endo- β-glucanase and phosphatases are due to
    (a) Auxin (b) Cytokinin
    (c) Gibberellin (d) ABA
34. Hydrolytic enzymes are formed during
    (a) Log phase (b) Lag phase
    (c) Steady state phase (d) Already exists in seed itself

♦ Answer Keys ♦

Q.No.	Ans.	Q.No.	Ans.	Q.No.	Ans.	Q.No.	Ans.	Q.No.	Ans.	Q.No.	Ans.
1	c	2	c	3	b	4	c	5	c	6	c
7	d	8	a	9	d	10	c	11	b	12	c
13	c	14	c	15	d	16	b	17	c	18	c
19	c	20	c	21	a	22	b	23	d	24	a
25	b	26	a	27	c	28	a	29	c	30	b
31	c	32	a	33	c	34	b				

2.4.2 State True or False

1. Seeds that are dispersed by water will have glumes and wooly hairs.
2. Grain seeds have two cotyledons, a hilum, and a micropyle to protect the embryo.
3. Seeds that need scarification can be placed in cold damp sand to break dormancy.
4. Molecules moving from a region of high concentration to a region of lower concentration are said to be moving against a diffusion gradient.
5. During imbibition, large molecules such as cellulose and starch will develop electrical charges and attract water molecules.
6. Essential elements are those nutrients that are needed by the plant in very small amounts.
7. Chemicals like pthalids and cyanides inhibit seed germination.
8. Chemicals like ethylene and KNO_3 promote seed germination.
9. Water stimulates enzyme activity for any metabolic activities of seed germination.
10. Germination is an event of rupture of seed coat alone.
11. Imbibition occurs only in live seed.
12. Imbibition is physical process dependent on metabolic energy.
13. Starch containing seed imbibe more than protein containing seed.
14. Water permeability is greater at micropylar region.
15. Greater the contact of seed with soil leads to greater the amount of water imbibed.
16. Seeds possessing textured seed coats are more likely to have greater seed–soil contact than smooth coated seeds.
17. Smooth coated seeds imbibe water more rapidly than rough coated seeds.
18. Enzyme activation phase is the most essential and dynamic phases in preparation of the seed for embryonic axis elongation.

♦ Answer Keys ♦

Q.No.	Ans.	Q.No.	Ans.	Q.No.	Ans.	Q.No.	Ans.	Q.No.	Ans.	Q.No.	Ans.
1	False	2	False	3	False	4	False	5	True	6	False
7	True	8	True	9	True	10	False	11	False	12	False
13	False	14	True	15	True	16	True	17	False	18	True

2.4.3. Fill in the blanks

1. The permeability of seeds to water is also decreased due to soil .... and ....
2. In conifers, the chilling treatment is given between .... to .... °C.
3. Seeds whose germination is affected by light are called ....
4. Ability to imbibe water is dependent on cell water potential which is a result of ....
5. Composition of seed decides the uptake of ....

♦ Answer Keys ♦

Q.No.	Ans.	Q.No.	Ans.	Q.No.	Ans.
1	Fungi, Bacteria	2	5, 10	3	Photoblastic
4	Cell turgor pressure	5	Water		

2.5. RESPIRATION, MITOCHONDRIAL ACTIVITY AND ATP SYNTHESIS DURING SEED GERMINATION

2.5.1. Choose the correct answer

1. The ultimate electron acceptor of respiration in an aerobic organisms is:
   (a) Cytochrome (b) Oxygen
   (c) Hydrogen (d) Glucose
2. Phosphorylation of glucose during glycolysis is catalysed by
   (a) Phosphoglucomutase (b) Phosphoglucoisomerase
   (c) Hexokinase (d) Phosphorylase
3. Pyruvic acid, the key product of glycolysis can have many metabolic fates. Under aerobic condition it forms
   (a) Lactic acid (b) $CO_2 + H_2O$
   (c) Acetyl CoA + CO_2 (d) Ethanol + CO_2
4. Electron Transport System (ETS) is located in mitochondrial
   (a) Outer membrane (b) Inter membrane space
   (c) Inner membrane (d) Matrix
5. Which of the following exhibits the highest rate of respiration?
   (a) Growing shoot apex (b) Germinating seed
   (c) Root tip (d) Leaf bud

6. Choose the correct statement
   (a) Pyruvate is formed in the mitochondrial matrix.
   (b) During the conversion of succinyl Co-A to succinic acid, a molecule of ATP is synthesized.
   (c) Oxygen is vital in respiration for removal of hydrogen.
   (d) There is complete breakdown of glucose in fermentation.
7. Mitochondria are called power houses of the cell. Which of the following observations support this statement?
   (a) Mitochondria synthesise ATP
   (b) Mitochondria have a double membrane
   (c) The enzymes of the Krebs cycle and the cytochromes are found in mitochondria.
   (d) Mitochondria are found in almost all plants and animal cells.
8. The end product of oxidative phosphorylation is
   (a) NADH (b) Oxygen
   (c) ADP (d) ATP+H_2O
9. When energy-depleted elements associated with a proton are accepted by an organic molecule, the process is called
   (a) Fermentation (b) Anaerobic
   (c) Aerobic (d) Catabolism
10. The end product of glycolysis is
   (a) NADH (b) Acetyl-CoA
   (c) Lactate (d) Pyruvate
11. The final output of the Krebs cycle includes all of the following except
   (a) NADP (b) FADH
   (c) ATP (d) CO_2
12. The usefulness of fermentation as a means of deriving energy is limited because
   (a) It cannot generate enough ATP
   (b) It produces too much NH2
   (c) The end products are toxic to the producer
   (d) It uses more energy than it produces
13. What substance is produced by the oxidation of pyruvate and feeds into the citric acid cycle?
   (a) Pyruvate (b) Glucose
   (c) Acetyl-CoA (d) O_2
14. Glycolysis has an efficiency level of approximately
   (a) 2% (b) 3.5%
   (c) 21.5% (d) 50%
15. Hans Krebs discovered
   (a) Glycolysis (b) Fermentation
   (c) The oxidation of pyruvate (d) The citric acid cycle

16. In aerobic cellular respiration, which generates more ATP, substrate-level phosphorylation or chemiosmosis?
(a) Substrate-level phosphorylation (b) Chemiosmosis
(c) Both generate the same amount of ATP
(d) Neither generates any ATP

17. What role does O_2 play in aerobic respiration?
(a) It combines with acetyl-CoA at the start of the Krebs cycle
(b) It is given off as a by-product during the oxidation of pyruvate
(c) It combines with H_2O to help drive the formation of ATP
(d) It is the final electron acceptor at the end of the electron transport chain

18. During aerobic respiration, FADH2 is produced in
(a) Glycolysis (b) The oxidation of pyruvate
(c) The Krebs cycle (d) The electron transport chain

19. Organisms that do not have the ability to produce or synthesize their own food are called
(a) Anaerobic (b) Exergonic
(c) Catabolic (d) Heterotrophs

20. The proper sequence of stages in glycolysis is
(a) Glucose priming, cleavage and rearrangement, oxidation, ATP generation
(b) Cleavage and rearrangement, glucose priming, ATP generation, oxidation
(c) Glucose priming, oxidation, cleavage and rearrangement, ATP generation
(d) ATP generation, oxidation, glucose priming, cleavage and rearrangement

21. During what stage of cellular respiration is the most ATP synthesized?
(a) Oxidation of pyruvate (b) Krebs cycle
(c) Fermentation (d) Chemiosmosis

22. Catabolic processes
(a) Make complex molecules from simpler ones
(b) Break complex molecules into simpler ones
(c) Occur only in autotrophs (d) Occur only in heterotrophs

23. Which of the following is the reduced form of a temporary electron carrier molecule?
(a) $FADH_2$ (b) ATP
(c) $NADP^+$ (d) CO_2

24. Which one of the following is a product of both cyclic and noncyclic photophosphorylation?
(a) NADPH (b) O_2
(c) ATP (d) Carbohydrate

25. Where does the O_2 come from that is essential for the proper functioning of oxidative phosphorylation?
(a) Fermentation (b) Light reactions of photosynthesis
(c) Dark reactions of photosynthesis (d) Carbon fixation

26. In the Calvin cycle, carbon dioxide is fixed in a reaction with the
    (a) Ribulose diphosphate (b) Ribulose phosphate
    (c) Ribose tri phosphate (d) 3-phosphoglyceric acid
27. The end products of noncyclic photophosphorylation are
    (a) O_2, ATP and NADPH
    (b) CO_2, PGAL (phosphoglyceraldehyde), and H_2
    (c) Water, ADP and $NADP^+$ (d) CO_2, ATP and water
28. The process of respiration in green plants occurs
    (a) Only when stomata are open (b) Only when photosynthesis ceases
    (c) Only when photosynthesis in progress
    (d) At all the times
29. Respiratory enzymes are located in
    (a) Mitochondrial matrix (b) Cristae
    (c) Perimitochondrial space (d) Outer membrane
30. The site of glycolysis in a cell is
    (a) Chloroplast (b) Nucleus
    (c) Cytoplasm (d) Mitochondria
31. Respiration is
    (a) Anabolic process (b) Exothermic process
    (c) Endothermic process (d) Endergonic process
32. The annual plant exchange of gases takes place mainly through
    (a) Leaf scars (b) Lenticels
    (c) Stomata (d) Stem
33. Kreb's cycle takes place in
    (a) Vesicles of ER (b) Mitochondrial matrix
    (c) Lysosomes (d) Dictyosomes
34. Common immediate source of energy in cellular activity
    (a) NAD (b) ATP
    (c) DNA (d) RNA
35. Photorespiration involves
    (a) Glycolate cycle (b) Kreb's cycle
    (c) Calvin cycle (d) CAM cycle
36. In succulent xerophytes, the R.Q. is
    (a) Zero (b) Unity
    (c) Less than one (d) More than one
37. R.Q. of fatty substances is generally
    (a) Zero (b) Unity
    (c) Less than one (d) More than one
38. Conversion of pyruvic acid into ethyl alcohol is facilitated by the enzymes
    (a) Carboxylase (b) Phosphatase
    (c) Dehydrogenase (d) Carboxylase and dehydrogenase

39. During respiration yeast converts glucose to
(a) Ethanol and water (b) Ethanol and oxygen
(c) Ethanol and CO_2 (d) Lactic acid and CO_2

40. The net gain of ATP produced during the oxidation of one glucose molecule in a plant cell
(a) 38 ATP molecules (b) 30 ATP molecules
(c) 36 ATP molecules (d) 24 ATP molecules

41. The final acceptor of electrons in the electron transport chain is
(a) Water (b) Oxygen
(c) Hydrogen (d) Cytochrome b

42. Substrate for photorespiration involves
(a) Serine (b) Glycolate
(c) Malic acid (d) Phosphoglycer

43. Which of the following is produced as a by-product during photorespiration?
(a) CO_2 (b) O_2
(c) H_2O_2 (d) H_2O

44. Transmission reaction occurs during
(a) Respiration (b) Photosynthesis
(c) Photorespiration (d) Glyoxylate cycle

45. NAD is reduced to NADH2 in
(a) Photorespiration (b) Photosynthesis
(c) Respiration (d) Glyoxylate cycle

46. In C3 plants, the photosynthetic yield is reduced due to
(a) Photorespiration (b) Carboxylation
(c) Respiration (d) Neoglucogenesis

47. Which of the following processes protect oxidation of chlorophyll (solarisation)?
(a) Photorespiration (b) Photosynthesis
(c) Respiration (d) Dark respiration

48. Cut surfaces of fruits and vegetables often become dark because
(a) Dust of air makes it dark
(b) Dirty knife makes it dark
(c) Oxidation of tannic acid due to presence of iron in traces from knife
(d) Tannins are formed

49. Which one of the following is a polluting organelle?
(a) Mitochondria (b) Lysosome
(c) Chloroplast (d) Chromatophore

50. The enzyme carboxylase requires a mineral activator. It is
(a) Mg (b) Mo
(c) Fe (d) Mn

51. How many ATP molecules are formed on complete oxidation of a glucose molecule through hexose monophosphate shunt cycle?
(a) 2 (b) 8
(c) 36 (d) 38

52. The correct sequence of Krebs cycle is
(a) Isocitric acid – Oxalosuccinic acid – Citric acid
(b) Isocitric acid – Oxalosuccinic acid – alpha ketoglutari acid
(c) Isocitric acid – Oxalosuccinic acid – Latic acid
(d) Isocitric acid – a ketoglutari acid – Oxalosuccinic acid

53. How many ATP molecules are formed from complete oxidation of acetyl CoA
(a) 12 ATP (b) 15 ATP
(c) 19 ATP (d) 38 ATP

54. The increased rate of respiration during fruit ripening is called
(a) Street metabolism (b) Fermentation
(c) Climacteric (d) Pateur effect

55. Which of the following causes loss of two protons and two electrons?
(a) Deamination (b) Dehydrogenation
(c) Carboxylation (d) CO_2 reduction

56. Where does formation of acetyl CoA from pyruvic acid take place?
(a) Chloroplast (b) Cytoplasm
(c) Mitochondria (d) Glyoxysome

57. An organelle involved with the break down of fats before final oxidation is
(a) Chromatophore (b) Cytosol
(c) Glyoxysome (d) Mitochondria

58. Potato tubers are larger in plants grown at Darjeeling than grown in AP. It is because there is the rate of
(a) Respiration is high (b) Respiration is low
(c) Photosynthesis is high (d) Protein synthesis is high

59. An indispensable role in energy metabolism is played by
(a) Lithium (b) Phosphorus
(c) Calcium (d) Potassium

60. In glycolysis, there is a net gain 2 ATP molecules and two molecules of
(a) $NADPH_2$ (b) $FMNH_2$
(c) $NADH_2$ (d) $FADH_2$

61. In TCA cycle, the hydrogen atoms removed at succinate level are accepted by
(a) CoA (b) FAD
(c) GTP (d) Oxygen

62. R.Q. is higher when respiratory substrate is
    (a) Glucose (b) Protein
    (c) Maleic acid (d) Fat
63. In hexose monophosate shunt, the hydrogen acceptor is
    (a) NAD (b) NADP
    (c) Fd (d) O_2
64. R.Q. of $C_{39}H_{74}O_6$ is
    (a) 0.715 (b) 1.0
    (c) 1.43 (d) 2.14
65. The ratio of energy release between anaerobic and aerobic respiration is
    (a) 1:2 (b) 2:1
    (c) 1:18 (d) 19:1
66. The number of CO_2 molecule released between anaerobic and aerobic respiration is
    (a) 1 (b) 2
    (c) 4 (d) 0
67. How many NADH2 are produced during the aerobic oxidation of alpha ketoglutarate?
    (a) One (b) Two
    (c) Four (d) Eight
68. The connecting link between beta oxidation and Kreb's cycle is
    (a) Pyruvic acid (b) Acetyl CoA
    (c) Acetaldhehyde (d) Citric acid
69. In photosynthesis and respiration, the energy is derived from
    (a) Cytochrome and enzymes (b) Phytochromes
    (c) Electron flow across cytochromes (d) Oxygen

♦ Answer Keys ♦

Q.No.	Ans.	Q.No.	Ans.	Q.No.	Ans.	Q.No.	Ans.	Q.No.	Ans.	Q.No.	Ans.
1	a	2	c	3	c	4	b	5	b	6	c
7	a	8	d	9	a	10	d	11	a	12	c
13	c	14	b	15	d	16	b	17	d	18	c
19	d	20	a	21	d	22	b	23	a	24	c
25	b	26	a	27	a	28	d	29	b	30	c
31	d	32	c	33	b	34	b	35	a	36	a
37	d	38	d	39	c	40	c	41	b	42	c
43	c	44	c	45	a	46	c	47	a	48	c
49	a	50	a	51	d	52	b	53	a	54	c
55	b	56	a	57	c	58	b	59	b	60	c
61	b	62	c	63	b	64	a	65	c	66	d
67	b	68	b	69	c						

2.5.2. Fill in the blanks

1. The most common substrate of respiration is ....
2. During respiration, the substrate is ....
3. The percentage of nitrogen in inhaled air is .... % and exhaled air is .... %.
4. The percentage of carbon dioxide in the inhaled air is .... % and exhaled air is .... %.
5. The process common to aerobic and anaerobic respiration is ....
6. In microbes, the gaseous exchange takes place through ....
7. The energy rich compound generated as final product during respiration is ....
8. Glycolysis takes place in ....
9. Kreb's cycle operates in ....
10. Anaerobic respiration takes place ....
11. Instant source of energy is ....
12. The number of ATP molecules produced during aerobic and anaerobic respiration are .... and .... respectively.
13. The product of glycolysis is ....
14. The energy produced during respiration is stored in ....
15. The breathing roots are the ....
16. The process complementary to respiration is ....
17. More ATP is manufactured during .... than at any other time in all of cellular metabolism.
18. A cyclic electron transport process is the characteristic of ....
19. Oxidative phosphorylation is to respiration as .... is to photosynthesis.
20. In noncyclic photophosphorylation, the ultimate acceptor of electrons that have been produced from the splitting of water is ....
21. The breakdown of glucose occurs by the process known as ....

♦ Answer Keys ♦

Q.No.	Ans.	Q.No.	Ans.	Q.No.	Ans.
1	Glucose	2	Oxidised	3	78 and 78
4	0.04 and 4	5	Glycolysis	6	General surface
7	ATP	8	Cytoplasm	9	Mitochondrial matrix
10	In the cytoplasm	11	Glucose	12	38 and 2
13	Pyruvic acid	14	Adenosine triphosphate	15	Pneumatophores
16	Photosynthesis	17	Oxidative phosphorylation	18	Photosynthesis
19	Electron transport chain	20	Carbon dioxide	21	Glycolysis

2.5.3. Match the following

Column A	Column B
A. Molecular oxygen	i. α – Ketoglutaric acid
B. Electron acceptor	ii. Hydrogen acceptor
C. Pyruvate dehydrogenase	iii. Cytochrome C
D. Decarboxylation	iv. Acetyl Co A

♦ Answer Keys ♦

Q.No.	Ans.	Q.No.	Ans.	Q.No.	Ans.	Q.No.	Ans.
A	ii	B	iii	C	iv	D	i

2.6. PROTEIN AND NUCLEIC ACID SYNTHESIS DURING SEED GERMINATION

2.6.1. Choose the correct answer

1. During seed germination, seed coat ruptures due to
   (a) Differentiation of cotyledons
   (b) Massive glycolysis in endosperm and cotyledons
   (c) Sudden increase in cell division
   (d) Massive imbibition of water
2. An albuminous seed showing hypogeal germination is
   (a) Castor (b) Bean
   (c) Gram (d) Maize
3. Vivipary is
   (a) Seed germination with subterranean cotyledons
   (b) Seed germination with epiterranean cotyledons
   (c) Fruit development without pollination
   (d) Seed germination inside the fruit while attached to the plant.
4. A gas required for germination of pea seed is
   (a) Nitrogen (b) Oxygen
   (c) Hydrogen (d) Water vapours
5. Protective covering over radicle during seed germination is
   (a) Suspensor (b) Coleorhiza
   (c) Epithelium (d) Coleoptile
6. Proteins are made from amino acids by the process of
   (a) Hydrolysis (b) Pinocytosis
   (c) Active transport (d) Dehydration synthesis
7. To which of the following adenine always binds?
   (a) Guanine (b) Cytosine
   (c) Thymine (d) Adenine

8. Guanine always binds to
   (a) Cytosine (b) Guanine
   (c) Thymine (d) Adenine
9. The function of ribosomes within the cell is
   (a) To produce ATP (b) To ferment carbohydrates
   (c) To synthesize proteins (d) To synthesize fat
10. According to base pairing rules the bases of one strand match-up with the bases of the second strand of DNA and the two strands are said to be
   (a) Complementary (b) Oppositely charged
   (c) Identical (d) Supplementary
11. An amino acid that does not derive its carbon skeleton, at least in part, from a-ketoglutarate is:
   (a) Arginine (b) Glutamate
   (c) Glutamine (d) Threonine
12. One amino acid directly involved in the purine biosynthetic pathway is:
   (a) Alanine (b) Aspartate
   (c) Glutamate (d) Leucine
13. .... is a nucleic acid which carries genetic
   (a) mRNA (b) tRNA
   (c) rRNA (d) cRNA
14. .... is a nucleic acid which transports amino acids to the site of protein synthesis
   (a) mRNA (b) tRNA
   (c) rRNA (d) cRNA
15. Adenine pairs with .... in DNA
   (a) Guanine (b) Cytosine
   (c) Thymine (d) Uracil
16. Cytosine pairs with .... in DNA and RNA
   (a) Uracil (b) Guanine
   (c) Thymine (d) Adenine
17. The .... bonds occur between adenine and thymine
   (a) Double (b) Triple
   (c) Tetra (d) Mono
18. There is .... bonding between cytosine and Guanine
   (a) Triple (b) Double
   (c) Tetra (d) Single
19. Adenine pairs with .... in RNA
   (a) Guanine (b) Uracil
   (c) Thymine (d) Cytosine

20. In the synthesis of proteins, what is the function of messenger-RNA molecules?
    (a) They act as a template for the synthesis of DNA.
    (b) They carry information that determines the sequence of amino acids
    (c) They remove amino acids from the nucleus
    (d) They carry specific enzymes for dehydration synthesis
21. Two sugars found in nucleic acids are
    (a) Sucrose and ribose (b) Glucose and fructose
    (c) Deoxyribose and ribose (d) Deoxyribose and glucose
22. The base found in RNA nucleotides but not in DNA nucleotides is
    (a) Uracil (b) Adenine
    (c) Guanine (d) Cytosine
23. The molecule that is responsible for carrying amino acids to ribosomes is
    (a) DNA (b) tRNA
    (c) rRNA (d) mRNA
24. During protein synthesis, peptide bonds are formed at the
    (a) Nucleus (b) Nucleolus
    (c) Lysosomes (d) Ribosomes

♦ Answer Keys ♦

Q.No.	Ans.	Q.No.	Ans.	Q.No.	Ans.	Q.No.	Ans.	Q.No.	Ans.	Q.No.	Ans.
1	d	2	d	3	d	4	b	5	b	6	d
7	c	8	a	9	c	10	a	11	d	12	b
13	a	14	c	15	c	16	b	17	a	18	a
19	b	20	b	21	c	22	a	23	b	24	d

2.6.2. State True or False

1. Following germination, the emerging plant gains nutrients for metabolism from the surrounding soil.
2. From the promoter of the gene, you would expect the coding strand to run 5' to 3'.
3. Regulatory sequences are DNA sequences which encode transcription factors.
4. During transcription, the RNA polymerase would be expected to spend the majority of its time in the open complex rather than the closed complex.
5. The product of a structural gene is mRNA and then a polypeptide.
6. Both types of transcriptional termination rely on stem-loop structures in the newly transcribed RNA.

♦ Answer Keys ♦

Q.No.	Ans.	Q.No.	Ans.	Q.No.	Ans.	Q.No.	Ans.	Q.No.	Ans.	Q.No.	Ans.
1	False	2	True	3	False	4	True	5	True	6	True

2.6.3. Fill in the blanks

1. .... are composed of amino acids.
2. RNA that is complementary to a segment of another RNA molecule to which it will bind is known as ....
3. The two types of nucleic acids are .... and ....
4. The nucleic acid was first isolated in 1868 by ....
5. The term nucleic acid was coined by ....
6. DNA is present in all cells except ....
7. In eukaryotic nucleus, the DNA is in the form of a ....
8. In bacteria, mitochondria and plastids, the DNA molecules are .... in shape
9. The number of DNA molecules in eukaryotic cells corresponds to the number of .... per cell.
10. The sugar present in DNA is ....
11. The molecular formula for pentose sugar in DNA is ....
12. A nucleotide is form of a .... and ....
13. Adenine and guanine belong to a group of compounds called ....
14. .... first established the inter nucleotide bond of nucleic acid.
15. .... forwarded the tetra nucleotide theory.
16. In the cytoplasm, RNA is mainly found in ....
17. In the nucleus, RNA is mainly found in ....
18. .... is the nitrogenous base absent in DNA.
19. .... is the nitrogenous base not present in RNA.
20. Cytosine pairs with .... in DNA and RNA.

♦ Answer Keys ♦

Q.No.	Ans.	Q.No.	Ans.	Q.No.	Ans.
1	Proteins	2	Antisense RNA	3	Deoxyribonucleic acid: Ribo nucleic acid
4	Meischer	5	Altmann	6	Plant virus
7	Double helix	8	Circular	9	Chromosomes
10	Deoxy ribose	11	$C_5H_{10}O_5$	12	Nucleoside and Phosphoric acid
13	Purines	14	Todd	15	Levene
16	Ribosomes	17	Nucleolus	18	Uracil
19	Thymine	20	Guanine		

2.7. DNA SYNTHESIS AND GERMINATION

2.7.1. Choose the correct answer

1. The primary growth phase of a cell is the
   (a) G_0 (b) G_1
   (c) G_2 (d) M

2. Which of the following statement is not correct in regard to cell division?
   (a) Chromatin uncoils to form chromosomes
   (b) Chromatin is found within the nucleus
   (c) Chromatin is made up of the DNA double helix and associated proteins
   (d) Chromosomes can be seen just as cell division is about to occur

3. When does DNA replication occur in the cell cycle?
   (a) Interphase (b) Prophase
   (c) Metaphase (d) Anaphase

4. During which stage of the cell cycle, cell growth and replication of organelles are most significant?
   (a) M phase (b) G_1 phase
   (c) G_2 phase (d) S phase

5. .... phase is the pre DNA replication phase
   (a) S phase (b) G_2 phase
   (c) G_1 phase (d) Prophase

6. .... phase is the post DNA replication phase
   (a) S phase (b) G_1 phase
   (c) G_2 phase (d) Prophase

7. .... is the nitrogenous base absent in DNA
   (a) Guanine (b) Uracil
   (c) Adenine (d) Thymine

8. Dehydrated DNA occurs in
   (a) A form (b) B form
   (c) C form (d) Z form

9. 2-Amino 6-oxypurine is
   (a) Guanine (b) Adenine
   (c) Thymine (d) Uracil

10. .... catalyses the formation of DNA molecule from RNA templates
    (a) RNA dependent DNA polymerase (b) DNA dependent RNA polymerase
    (c) RNA replicase (d) RNA polymerase

11. Double helix describes the structure of a molecule of
    (a) Protein (b) Disaccharide
    (c) Monosaccharide (d) DNA

12. In a DNA molecule, what holds together nitrogenous bases from the two polymer chains?
    (a) Ionic bonds (b) Covalent bonds
    (c) Peptide bonds (d) Hydrogen bonds
13. Which of the following nitrogenous bases is found in DNA, but is not found in RNA?
    (a) Guanine (b) Cytosine
    (c) Thymine (d) Uracil
14. The two strands of a DNA double helix are held together by
    (a) Hydrogen bonds (b) Nonpolar covalent bonds.
    (c) Polar covalent bonds. (d) Hydrophobic exclusions
15. Both deoxyribose and ribose belong to a class of sugars called
    (a) Trioses (b) Hexoses
    (c) Pentoses (d) Polysaccharides
16. In a DNA strand, the nucleotides are linked together by:
    (a) Glycosidic bonds (b) Phosphodiester bonds
    (c) Peptide bonds (d) Hydrogen bonds
17. The fact that a purine base always paired through hydrogen bonds with a pyrimidine base leads to .... in the DNA double helix
    (a) The antiparallel nature (b) The semiconservative nature
    (c) Uniform width throughout DNA (d) Uniform length in all DNA
18. Cytoplasmic division of a cell is called
    (a) Cell plate formation (b) Cytokinesis
    (c) Mitosis (d) Synapsis
19. During which phase of the cell cycle the leading strands are synthesized?
    (a) G_1 (b) S
    (c) G_2 (d) M
20. Combination of a nitrogenous base, a pentose sugar and a phosphate group
    (a) Nucleotide (b) Nucleoside
    (c) Base pair (d) Nucleic acid
21. Chemical interaction of these nucleotides forms
    (a) Phosphodiester linkages (b) Glycosidic bond
    (c) Hydrogen bond (d) Nucleoside bond
22. The gap between DNA synthesis and mitosis is ....
    (a) Gap 0 (b) Gap 1
    (c) Gap 2 (d) S Phase
23. DNA replication occurs during which phase
    (a) G_1 (b) S
    (c) G_2 (d) G_0

24. A family of enzymes that carry out all forms of DNA replication is
    (a) DNA polymerase (b) RNA polymerase
    (c) Hexase (d) Ligase
25. Okazaki fragments is called
    (a) Leading strand (b) Lagging strand
    (c) Lacking strand (d) Logging strad
26. Which enzyme is specifically used in joining the Okazaki fragments
    (a) Topoisomerase (b) SSB proteins
    (c) DNA polymerase (d) DNA ligase
27. DNA synthesis associated with
    (a) Post-germinative stage (b) Development
    (c) Initial phase of germination (d) Pre-germination stage
28. Newly synthesized DNA in radicle tips is associated with
    (a) Nuclear membrane (b) Endosperm
    (c) Cytoplasm (d) Protoplasm
29. DNA synthesis occur in
    (a) Quiescent seed (b) Dry seed
    (c) Imbibed seed (d) Treated seed
30. Which may inhibit germination by inhibiting DNA polymerase
    (a) GA_3 (b) ABA
    (c) Kinetin (d) Auxin

♦ Answer Keys ♦

Q.No.	Ans.	Q.No.	Ans.	Q.No.	Ans.	Q.No.	Ans.	Q.No.	Ans.	Q.No.	Ans.
1	b	2	a	3	a	4	a	5	c	6	c
7	b	8	a	9	a	10	a	11	d	12	d
13	c	14	a	15	b	16	c	17	c	18	b
19	b	20	a	21	a	22	c	23	b	24	a
25	b	26	d	27	a	28	a	29	c	30	b

2.7.2. State True or False

1. Chromosome replication takes place during interphase.
2. Cells can permanently stop midway in the cell cycle and not complete the cycle.
3. DNA must replicate after cell division occurs.
4. During the cell cycle, interphase lasts longer than mitosis.
5. During interphase, the cell is resting.
6. Embryo growth during and following germination in seeds was found to the result of cell expansion and elongation.
7. Deoxyribonucleic acid found in the nuclei of cell.
8. Pentose sugar is deoxyribose.

9. Uracil is present in DNA.
10. Cell growth stops at mitosis stage.
11. DNA synthesis results in a doubling of the chromosomes.
12. To begin synthesis of a new strand, primer is not needed.
13. Dispersive mechanism is not a replication method.
14. SSB proteins stabilize the single stranded structure of DNA.
15. An important feature of a prehydration treatment is the shortening of the time to DNA replication.
16. The embryo of quiescent mature seeds predominantly contains nuclear replication activity.
17. The accumulation of proteins involved with the cell cycle activity is ß-tubulin.
18. When seeds are imbibed in water, high amounts of 4C DNA were detected.
19. The biochemical and electron microscopic techniques which indicate the nuclear DNA.
20. Low temperature pretreatment increase the time required for germination and DNA synthesis.

♦ Answer Keys ♦

Q.No.	Ans.	Q.No.	Ans.	Q.No.	Ans.	Q.No.	Ans.	Q.No.	Ans.
1	True	2	True	3	False	4	True	5	False
6	True	7	True	8	True	9	False	10	True
11	True	12	False	13	False	14	True	15	True
16	False	17	True	18	True	19	True	20	False

2.7.3. Fill in the blanks

1. Auxin stimulates faster elongation of the .... cells.
2. Mitosis is controlled at the .... checkpoint.
3. The spindle forms in the .... phase.
4. Replication of the genome occurs in the .... phase of the cell cycle.
5. DNA replication is controlled at the .... checkpoint.
6. .... consists of mitosis and interphase.
7. RNA differs from DNA in all of the following ways except ....
8. .... contains most of the cell's DNA.
9. The reproduction of new cells from the pre-existing cells is called ....
10. The two types of cell division are .... and ....
11. Mitosis occurs in ....
12. Meiosis occurs in .... or .... cells.

13. Interphase is the phase between .... and ....
14. The process of the division of nucleus is ....
15. The process of the division of cytoplasm is ....
16. .... is the equational division.
17. .... is the reductional division.
18. Recombination occurs due to ....
19. Nucleolus disappears during ....
20. Nucleolus reappears during ....
21. DNA synthesis occurs during .... phase.
22. The number of daughter cells produced from a single cell in meiosis is ....
23. The process of synthesis of mRNA from DNA is called ....
24. The strand of DNA double helix, which is copied in continuous fashion during replication is called ....
25. .... is one of the short segments of RNA –primed DNA that are synthesized during replication of the lagging strand.
26. .... is the place at which DNA replication starts.
27. .... is copied in discontinuous fashion during replication.

♦ Answer Keys ♦

Q.No.	Ans.	Q.No.	Ans.	Q.No.	Ans.
1	Shady side	2	M	3	M
4	S	5	G_2	6	The cell cycle
7	The 5' and 3' orientation of the polynucleotide strand	8	Chromosome	9	Cell division
10	Mitosis, Meiosis	11	Somatic cells	12	Reproductive or germ
13	Telophase and prophase	14	Karyokinesis	15	Cytokinesis
16	Mitosis	17	Meiosis	18	Crossing over
19	Prophase	20	Telophase	21	S
22	4	23	Transcription	24	Leading strand
25	Okazaki fragments	26	Swivel	27	Lagging strand

2.8. CARBOHYDRATE METABOLISM DURING GERMINATION

1.1. Choose the correct answer

1. Which of the following is a simple sugar or monosaccharide?
   (a) Galactose (b) Lactose
   (c) Maltose (d) Sucrose
2. What is the molecular formula for glucose?
   (a) CH_3OH (b) $C_6H_{12}0_6$
   (c) $C_{12}H_{22}O_{11}$ (d) $C_6H_{12}O_5$

3. Maltose is composed of which two sugars?
   (a) Glucose and Glucose (b) Glucose and Galactose
   (c) Glucose and Fructose (d) Fructose and Galactose
4. The process common to aerobic and anaerobic respiration is ....
   (a) Oxidation (b) Glycolysis
   (c) Kreb's cycle (d) Electron transport chain
5. Glycolysis takes place in
   (a) Cyctoplasm (b) Mitochondrial matrix
   (c) Mitochondrial cristae (d) Outside the cell
6. Kreb's cycle operates in
   (a) Cyctoplasm (b) Mitochondrial matrix
   (c) Mitochondrial cristae (d) Outside the cell
7. Instant source of energy is
   (a) Sucrose (b) Glucose
   (c) Fats (d) Amino acids
8. The product of glycolysis is
   (a) Carbon dioxide and water (b) Pyruvic acid
   (c) Acetyl Co A (d) ATP
9. Glycolysis is first discovered by
   (a) Embden-Meyerhof-Parnas (b) T.H. Morgan
   (c) R. Brown (d) Schleiden and Schwann
10. Glucose converted to glucose 6-phosphate by enzyme
    (a) Hexokinase (b) Phosphoglucose isomerase
    (c) Phosphofructokinase (d) Phosphoglycerate kinase
11. Glucose 6 phosphate is converted to fructose 6 phosphate by enzyme
    (a) Hexokinase (b) Phosphoglucose isomerase
    (c) Phosphofructokinase (d) Phosphoglycerate kinase
12. Fructose 6 phosphate is converted to fructose 1,6 phosphate by enzyme
    (a) Hexokinase (b) Phosphoglucose isomerase
    (c) Phosphofructokinase (d) Phosphoglycerate kinase
13. 3- biphosphoglycerate is converted to 3-phosphoglycerate in presence of enzyme
    (a) Hexokinase (b) Phosphoglucose isomerase
    (c) Phosphofructokinase (d) Phosphoglycerate kinase
14. Phosphoenolpyruvate is then, converted to pyruvate by enzyme
    (a) Hexokinase (b) Phosphoglucose isomerase
    (c) Phosphofructokinase (d) Pyruvate kinase
15. Aerobic oxidation of pyruvic acid through a series of reactions was studied by
    (a) Brown Krebs (b) Schleiden krebs
    (c) Hans Krebs (d) Robert Krebs

16. The pyruvic acid produced from glycolytic breakdown of glucose is converted into acetyl Co A by
(a) Decarboxylation reaction (b) Carboxy reaction
(c) Enzyme reaction (d) Tricarboxylation reaction

17. Starch is major food reserve in
(a) Cereal seeds (b) Oil seeds
(c) Pulses seeds (d) Cotton seeds

18. In Brazil nut, carbohydrate is stored in
(a) Radicle/hypocotyl (b) Cotyledon
(c) Embryonic tissues (d) Perisperm

19. In Coffee and yucca seed, carbohydrate is stored in
(a) Radicle/hypocotyl (b) Cotyledon
(c) Embryonic tissues (d) Perisperm

20. Pine seed carbohydrate is stored in
(a) Radicle/hypocotyl (b) Cotyledon
(c) Megagametophyte (d) Perisperm

21. Amylose is converted to glucose and maltose by enzyme
(a) α-Amylase (b) β-Amylase
(c) α-Glucosidase (d) Maltase

22. Amylopectin is converted to glucose and maltose and limit dextrin in the presence of
(a) α-Amylase (b) β-Amylase
(c) α-glucosidase (d) Maltase

23. Maltose is converted to glucose in presence of
(a) α-Amylase (b) β-Amylase
(c) Maltase (d) α-glucosidase

24. Sucrose is converted to fructose and glucose by
(a) α-Amylase (b) β-Amylase
(c) Maltase (d) β-Fructofuranosidase

♦ Answer Keys ♦

Q.No.	Ans.	Q.No.	Ans.	Q.No.	Ans.	Q.No.	Ans.	Q.No.	Ans.	Q.No.	Ans.
1	a	2	b	3	a	4	b	5	a	6	b
7	b	8	b	9	a	10	a	11	b	12	c
13	d	14	d	15	c	16	a	17	a	18	a
19	d	20	c	21	a	22	b	23	c	24	d

2.8.2. State True or False

1. Carbohydrate is natural organic substances that includes sugars, starch and cellulose.
2. The chemical composition of carbohydrates was analyzed in the nineteenth century.
3. Carbohydrate is combination of carbon, hydrogen and water.

4. Glucose and fructose are examples for oligosaccharides.
5. Starch, cellulose and hemicelluloses are examples for polysaccharides.
6. Raffinose and stachyose are examples for oligosaccharides.
7. Sucrose is an example for disaccharide.
8. Glucose and fructose are examples for monosaccharide.
9. Oligosaccharide is insoluble in water.
10. Polysaccharide is insoluble in water.
11. Monosaccharide is soluble in water.
12. Oligosaccharide is soluble in water.
13. Monosaccharide is reducing in nature.
14. Polysaccharide is non reducing in nature.
15. Oligosaccharide is both reducing and non reducing nature.
16. Carbohydrate is rich in groundnut seeds.
17. Raffinose oligosaccharide is nutritionally undesirable or even toxic.
18. Raffinose supports early seedling growth.
19. Starch and sucrose support early seedling growth.
20. In cereal seeds, β – amylase is present in free and bound forms.
21. β – amylase is synthesized *de novo* during early germination of rice.
22. During germination, bound form activated either by disulfide reductase or proteolytic enzyme.
23. α-Amylase is more important than β – amylase during early hours of germination in wheat.
24. α-Amylase enzyme is used for converting amylose in to glucose and maltose.
26. Phosphorolytic action in early stage and shifting to hydrolytic breakdown during later stages of germination in pea.
26. Pyruvate enters the citric acid cycle.
27. Pyruvate enters the electron transport chain.
28. Glucose is metabolized to pyruvic acid via glycolysis.
29. Glucose is metabolized to pyruvic acid via TCA cycle.
30. Sucrose is broken down by this pathway, which is called the sucrose synthase pathway.
31. β – amylase is high in rice.
32. β – amylase is low in pea.

♦ Answer Keys ♦

Q.No.	Ans.	Q.No.	Ans.	Q.No.	Ans.	Q.No.	Ans.	Q.No.	Ans.	Q.No.	Ans.
1	True	2	True	3	True	4	False	5	True	6	True
7	True	8	True	9	False	10	True	11	True	12	True
13	True	14	True	15	True	16	False	17	True	18	False
19	True	20	True	21	True	22	True	23	False	24	True

Q.No.	Ans.	Q.No.	Ans.	Q.No.	Ans.	Q.No.	Ans.	Q.No.	Ans.	Q.No.	Ans.
25	True	26	True	27	False	28	True	29	False	30	True
31	True	32	True								

2.9. BREAKDOWN OF LIPIDS DURING GERMINATION

2.9.1. Choose the correct answer

1. The combination of an amino-alcohol, fatty acid and sialic acid form
   (a) Phospholipids (b) Sulpholipids
   (c) Glycolipids (d) Aminolipids
2. The rate of fatty acid oxidation is increased by
   (a) Phospholipids (b) Glycolipids
   (c) Amino lipids (d) Sulpholipids
3. An example of a glycerophospholipid that is involved in cell signaling is
   (a) Arachidonic acid (b) Ceramide
   (c) Phosphatidylinositol (d) Testosterone
4. What term is based on the Greek root words for "light" and "fat"?
   (a) Glycogen (b) Carbohydrate
   (c) Cellulose (d) Phospholipid
5. Fats, oils, and steroids are
   (a) Proteins (b) Nucleic acids
   (c) Polysaccharides (d) Lipids
6. Which of the following statement is true of lipids?
   (a) Many contain fatty acids in ester or amide linkage
   (b) Most are simply polymers of isoprene
   (c) They are more insoluble in chloroform than in water
   (d) They are more soluble in water than in chloroform
7. β- oxidation of fatty acids is promoted by which of the following?
   (a) ATP (b) NAD +
   (c) FADH2 (d) Acetyl Co A
8. In glycolysis, six-carbon glucose molecules are broken down into two three-carbon molecules of
   (a) Citric acid (b) Pyruvic acid
   (c) Acetyl-CoA (d) Glucose
9. The first step in a sequence of enzymatic reactions in the tricarboxylic acid cycle is the formation of
   (a) Oxaloacetate (b) Citric acid
   (c) Acetyl-CoA (d) Pyruvic acid
10. Beta oxidation breaks up ....
    (a) Glucose molecules (b) RNA
    (c) Proteins (d) Fatty acids

11. **Beta oxidation of fatty acids is inhibited by**
    (a) NADPH (b) Acetyl CoA
    (c) Malonyl CoA (d) Citric acid
12. **The enzyme system for adding double bonds to saturated fatty acids requires all of the following except**
    (a) A mixed-function oxidase (b) ATP
    (c) Cytochrome (d) Molecular oxygen (O_2).
13. **The citric acid cycle is inhibited by which of the following?**
    (a) Fluoroacetate (b) Aerobic conditions
    (c) Malic acid (d) Fluorouracil
14. **Why called beta-oxidation?**
    (a) An intermediate in fatty acid biosynthesis
    (b) Eater-soluble protein albumin
    (c) Its acyl portion transfers to carnitine
    (d) Cleavage takes place at beta-carbon
15. **Where is beta-oxidation done?**
    (a) Mitochondrion (b) Converts Acetyl-CoA to malonyl-CoA
    (c) Acetyl-CoA (d) Acetyl-CoA and malonyl-CoA
16. **What molecule initiates the citric acid cycle by reacting with oxaloacetate?**
    (a) Pyruvate (b) Acetyl CoA
    (c) Oxaloacetate (d) Isocitrate
17. **What is/are the chemical change(s) involved in the conversion of citrate into isocitrate?**
    (a) Hydration followed by dehydration
    (b) Oxidation
    (c) Oxidation followed by reduction
    (d) Dehydration followed by hydration
18. **In which reaction is GTP (or ATP) directly formed in the citric acid cycle?**
    (a) Conversion of succinyl CoA to succinate
    (b) Decarboxylation of α-ketoglutarate
    (c) Conversion of isocitrate to α-ketoglutarate
    (d) Conversion of acetyl CoA to pyruvic acid
19. **The glyoxylate cycle enables plants to survive using only**
    (a) Pyruvate (b) Acetate
    (c) Oxaloacetate (d) Malate
20. **How many high-energy phosphate bonds are expended in gluconeogenesis?**
    (a) Three (b) Six
    (c) Two (d) Four

21. Which of the following compounds cannot serve as the starting material for the synthesis of glucose via gluconeogenesis?
(a) Acetate (b) Glycerol
(c) Lactate (d) Oxaloacetate

22. An enzyme used in both glycolysis and gluconeogenesis is
(a) 3-phosphoglycerate kinase. (b) Glucose 6-phosphatase.
(c) Hexokinase. (d) Phosphofructokinase-1.

23. Beta-oxidation of fatty acids requires all the following coenzymes expect
(a) CoA (b) FAD
(c) NAD (d) NADP

24. During each cycle of β-oxidation
(a) One carbon atom is removed from the carboxyl end of the fatty acid
(b) One carbon atom is removed from the methyl end of the fatty acid
(c) Two carbon atoms are removed from the carboxyl end of the fatty acid
(d) Two carbon are removed from the methyl end of the fatty acid

25. Storage lipids are in the form of
(a) Triglycerides (b) Diglycerides
(c) Monoglycerides (d) Glycerol

26. .... is the first step of lipid reserve mobilization
(a) ß oxidation (b) Lipolysis
(c) Hydration (d) Citric acid cycle

27. Three stage hydrolytic cleavage of the fatty acid ester bonds in triglycerides is catalysed by
(a) Amylase (b) Pectinase
(c) Kitinase (d) Lipase

28. Lipolysis occurs in
(a) Cytosol (b) Oil bodies
(c) Endoplasmic reticulum (d) Mitochondria

29. Glycerol is converted into α-glycerol phosphate
(a) Iso citrate (b) Malate
(c) Glycerol kinase (d) Glyoxolate

30. Active fatty acid is broken down to .... in β oxidation
(a) Acetyl CoA (b) Acetic acid
(c) Acyl Co A (d) Acetaldehyde

31. β-oxidation takes place in
(a) Cell wall (b) Mitochondria
(c) Glyoxysome (d) Cytosol

32. .... cleaves isocitrate to glyoxolate and succinate
(a) Amylase (b) Lipase
(c) Malate synthetase (d) Isocitrate lyase

33. Citric acid cycle takes place in
    (a) Cell wall (b) Glyoxysome
    (c) Mitochondria (d) Cytosol
34. Gluconeogenesis is the reverse process of
    (a) Kreb's cycle (b) Glycolysis
    (c) Citric acid cycle (d) TCA
35. Oxaloacetate is converted in to phosphoenolpyruvate during
    (a) Kreb's cycle (b) Glycolysis
    (c) Citric acid cycle (d) Gluconeogenesis
36. Final product of gluconeogenesis is
    (a) Starch (b) Sucrose
    (c) Glycerol (d) Fat
37. Gluconeogenesis occurs in
    (a) Cytosol (b) Stroma
    (c) Mitochondria (d) Endoplasmic reticulum

♦ Answer Keys ♦

Q.No.	Ans.	Q.No.	Ans.	Q.No.	Ans.	Q.No.	Ans.	Q.No.	Ans.	Q.No.	Ans.
1	c	2	a	3	c	4	d	5	d	6	b
7	b	8	b	9	b	10	d	11	c	12	b
13	a	14	d	15	a	16	b	17	d	18	a
19	b	20	b	21	a	22	a	23	d	24	c
25	a	26	b	27	d	28	b	29	c	30	a
31	c	32	d	33	c	34	b	35	d	36	b
37	a										

2.9.2. State True or False

1. Lipase is the enzyme which hydrolysis fatty acids.
2. Alkaline lipase is capable of hydrolyzing tri, di and monoglycerides.
3. In castor, the endosperm is the main site of lipid accumulation.
4. Potato contains 20% lipid.
5. In plants, stored lipids are generally stored as oil.
6. In jajoba, fats are stored as liquid wax in seeds.
7. Pospholipids are mainly found in seeds.
8. The fatty acid is saturated, if there are single bond between carbon atoms.
9. In β -oxidation, the fatty acid is first activated in a reaction requiring ATP and coenzyme A.
10. Fatty acids with an even number of carbon atoms yield only acetyl CoA.
11. Malate synthetase converts glyoxylate to malate.
12. β-oxidation and glyoxolate cycle takes place in glyoxyzome.
13. Carbohydrates give more energy than fat.

14. 3-ketoacyl-CoA thiolase is the enzyme of β oxidation.
15. Acyl CoA is converted into Acetyl CoA by removal of two carbon atoms.

♦ Answer Keys ♦

Q.No.	Ans.	Q.No.	Ans.	Q.No.	Ans.	Q.No.	Ans.	Q.No.	Ans.
1	True	2	False	3	True	4	False	5	True
6	True	7	False	8	False	9	True	10	True
11	True	12	True	13	False	14	True	15	True

2.9.3. Fill in the Blanks

1. .... is the intermediate between citrate and isocitrate.
2. .... is the location of succinate dehydrogenase.
3. The citric acid cycle take place .... in the cell.
4. .... is the intermediate in both at the beginning and at the end of the citric acid cycle.
5. .... is one of the products of the citric acid cycle.
6. .... is the type of enzyme regulation process that inhibits the pyruvate dehydrogenase complex.
7. Carbons from carbohydrate enter the citric acid cycle in the form of ....
8. In the citric acid cycle, the .... is produced by a substrate-level phosphorylation.
9. During the oxidation of isocitrate, the intermediate that is decarboxylated to form α-ketoglutarate is ....
10. In general, the citric acid cycle is inhibited under .... (high, low) energy conditions.
11. .... is the first citric acid cycle intermediate to be oxidized.
12. The citric acid cycle is also known as the ....
13. Formation of citrate from acetyl CoA and oxaloacetate is a(n) .... reaction.
14. .... is another name for glycolysis.
15. .... substance must be regenerated for glycolysis to proceed.
16. .... molecule is an allosteric inhibitor of phosphofructokinase.
17. Glycolysis produces a net of .... moles of ATP per one mole of glucose.
18. The key enzyme that regulates the pace of glycolysis is ....
19. The first step in gluconeogenesis is the .... of pyruvate to form oxaloacetate.
20. In *Arabidopsis* lipids are stored in ....
21. Fatty acids with odd number of carbon atoms on complete degradation by β-oxidation gives to ....

♦ Answer Keys ♦

Q.No.	Ans.	Q.No.	Ans.	Q.No.	Ans.
1	Cis-aconitate	2	Inner membrane	3	Mitochondria
4	Oxaloacetate	5	Carbon dioxide	6	Phosphorylation
7	Acetyl CoA	8	GTP	9	Oxalosuccinate
10	High	11	Isocitrate	12	Kreb's cycle
13	Condensation	14	Embden Meyerhof pathway	15	NAD^+
16	ATP	17	2	18	Phosphofructokinase
19	Carboxylation	20	Embryo/Endosperm	21	Acetyl CoA/Propionyl CoA

2.10. INITIATION OF EMBRYO GROWTH AND PROTRUSION OF RADICLE

2.10.1. Choose the correct answer

1. The elongation of root and stem forms what is known as the
   (a) Initial plant (b) Protoplant
   (c) Primary plant body (d) Secondary growth

2. Internode elongation in corn results from the growth of the
   (a) Intercalary meristem (b) Protoderm
   (c) Procambium (d) Cork cambium

3. Which of the following is not a kind of epidermal cell?
   (a) Trichome (b) Root hair
   (c) Guard cell (d) Cork

4. Unlike the flow through vessels, water conducted by tracheids must flow through
   (a) Piths (b) Pores
   (c) Sieve plates (d) Tubes only

5. Thee primary walls of the endodermis are impregnated with
   (a) Lignin (b) Wax
   (c) Protein (d) Suberin

6. Which of the following is not a basic plant type
   (a) Pith (b) Ground tissue
   (c) Epidermis (d) Vascular tissue

7. Epiphytic orchid roots are a form of .... root.
   (a) Aerial (b) Adventitious
   (c) Buttress (d) Parasitic

8. Which of the following is not a part of the outer protective layer of mature roots
   (a) Cork cambium (b) Phelloderm
   (c) Periderm (d) Cork

9. The embryonic protoderm is an example of
   (a) Procambium (b) Vascular cambium
   (c) Primary meristem (d) Lateral meristem
10. The form of conduction element found almost exclusively in angiosperm is
   (a) Sieve cell (b) Vessel
   (c) Tracheid (d) Sieve tube
11. The apical meristem gives rise to all of the following except
   (a) Protoderm (b) Vascular cambium
   (c) Procambium (d) Ground meristem
12. The outward cell division of the root apical meristem produces the
   (a) Root hairs (b) Xylem
   (c) Branch roots (d) Root cap
13. Mature sclerenchyma cells are
   (a) Suberized and contain no living protoplasts
   (b) Thin walled and often contain chloroplasts
   (c) Lignified and contain living protoplasts
   (d) Lignified and contain no living protoplast
14. Outer most layer of the stele of a root is the
   (a) Cortex (b) Endodermis
   (c) Pericycle (d) Pith
15. Primary growth in plants is initiated by the
   (a) Apical meristems (b) Vascular cambium
   (c) Lateral meristems (d) Ground tissue
16. The root hairs are found in which of the following areas of the root?
   (a) Root cap (b) Zone of maturation
   (c) Zone of elongation (d) Zone of cell division
17. Which of the statement about phloem tissue is correct?
   (a) Phloem tissue distributes water and minerals absorbed by the roots to the rest of the plant.
   (b) The sieve elements of angiosperms are sieve tube members, while the sieve elements of gymnosperms are sieve cells.
   (c) Phloem may be primary or secondary, but in either case the sieve elements are nonliving cells.
   (d) Primary and secondary phloem is derived from the vascular cambium.
18. Which tissue forms the outer protective covering of a plant?
   (a) Dermal tissue (b) Ground tissue
   (c) Vascular tissue (d) Cortex
19. Xylem and phloem belong to the .... tissue system
   (a) Dermal (b) Ground
   (c) Vascular (d) Cortex

20. Which of the following cells will transport water from the roots to the leaves more freely?
    (a) Sieve tube cell (b) Companion cell
    (c) Tracheids (d) Vessel elements parenchyma cell
21. The cortex found in the dicot stem or a monocot root is made of what type of tissue?
    (a) Ground tissue (b) Epithelial tissue
    (c) Vascular tissue (d) Dermal tissue
22. We are very familiar with two things found in large woody plants, wood and bark. The wood is essentially made of .... and the bark is essentially made of ....
    (a) Xylem; phloem and cork
    (b) Xylem and phloem; epidermal tissue
    (c) Parenchyma; epidermis
    (d) Pith; cork
23. Which of the following is not a modified stem?
    (a) Potato (b) Celery
    (c) Onion (d) Carrot
24. Which function is not shared by both the root system and the shoot system of a plant?
    (a) Absorb water (b) Transport water
    (c) Transport inorganic nutrients (d) Transport organic nutrients
25. Which of the following tissue types in the plant gives rise to the other three types?
    (a) Epidermal tissue (b) Ground tissue
    (c) Meristem (d) Vascular tissue
26. Which of the following tissues makes up the majority of the interior of the plant?
    (a) Epidermal tissue (b) Ground tissue
    (c) Meristem (d) Phloem
27. Which of the following tissues forms the outer protective covering of the plant?
    (a) Epidermal tissue (b) Ground tissue
    (c) Meristem (d) Vascular tissue
28. Which of the following tissues transports water and nutrients within the plant?
    (a) Aaepidermal tissue (b) Aground tissue
    (c) Ameristem (d) Vascular tissue
29. Which of the following tissues is an embryonic tissue type found at the apex of roots and shoots?
    (a) Epidermal tissue (b) Ground tissue
    (c) Meristem (d) Vascular tissue

30. Which statement is not true of the epidermis of a plant?
    (a) The epidermis covers the entire body of an herbaceous plant.
    (b) Guard cells are specialized epidermal cells found on leaf surfaces.
    (c) The epidermis is renewed by meristem on woody plants.
    (d) Root hairs increase the surface area of epidermal cells on roots.
31. Which of these plant tissue types most closely corresponds to the "typical" unspecialized plant cell?
    (a) Xylem (b) Parenchyma
    (c) Collenchyma (d) Sclerenchyma
32. Which of these plant tissue types has cells with thick secondary cell walls, and is made up of fibers and sclereids?
    (a) Xylem (b) Phloem
    (c) Collenchyma (d) Sclerenchyma

♦ Answer Keys ♦

Q.No.	Ans.	Q.No.	Ans.	Q.No.	Ans.	Q.No.	Ans.	Q.No.	Ans.	Q.No.	Ans.
1	a	2	a	3	d	4	a	5	d	6	a
7	a	8	d	9	c	10	b	11	b	12	d
13	d	14	c	15	a	16	b	17	b	18	a
19	c	20	a	21	a	22	a	23	d	24	a
25	c	26	b	27	a	28	d	29	c	30	c
31	b	32	d								

2.10.2. State True or False

1. The conducting elements of the phloem are part of the tracheids
2. An important function of plant parenchyma is to form a thick, hydrophobic layer on the surfaces of stems and leaves which acts as an effective barrier to evaporation
3. In primary roots, phloem is located in the center of the vascular cylinder
4. Tracheids and vessels are found in the xylem of all seed plants
5. Dermal, ground and vascular tissue system in plants arise from apical meristem tissue

♦ Answer Keys ♦

Q.No.	Ans.	Q.No.	Ans.	Q.No.	Ans.	Q.No.	Ans.	Q.No.	Ans.
1	False	2	False	3	False	4	False	5	True

2.11. CELL CYCLE DURING SEED GERMINATION

2.11.1. Choose the correct answer

1. A bivalent consists of
   (a) Two chromatids and one centromere
   (b) Four chromatids and two centromeres
   (c) Two chromatids and two centromeres
   (d) Four chromatids and four centromeres
2. The stage in which daughter chromosomes move toward the poles of the spindle is
   (a) Anaphase (b) Metaphase
   (c) Prophase (d) Telophase
3. Chiasmata are first seen in
   (a) Pachytene (b) Zygotene
   (c) Leptone (d) Diplotene
4. Which of the following cellular structures always disappear during mitosis and meiosis?
   (a) Plasma membrane (b) Nucleolus and nuclear envelope
   (c) Plastids (d) Chloroplasts
5. Chiasma formation occurs in
   (a) Pachytene (b) Zygotene
   (c) Leptone (d) Diplotene
6. Crossing over results in
   (a) Recombination of linked alleles (b) Dominance of alleles
   (c) Segregation of alleles (d) Linkage between genes
7. Cytoplasmic division of a cell is called
   (a) Cell plate formation (b) Cytokinesis
   (c) Mitosis (d) Synapsis
8. Which of the following does not occur by mitosis
   (a) Growth' production of gametes (b) Repair
   (c) Development in the embryo (d) Cleavage
9. Chromosomes exist in an unduplicated form (i.e., unreplicated DNA associated with protein) during what phase(s) of the cell cycle?
   (a) Prophase (b) Telophase
   (c) G1 of interphase (d) G1 of interphase and telophase
10. In which phase of mitosis in plant cells does the nucleolus or nucleoli in the nucleus seem to reappear?
    (a) Prophase (b) Metaphase
    (c) Anaphase (d) Telophase
11. The term that refers to division of the cytoplasm is
    (a) Cytokinesis (b) Anaphase
    (c) Karyokinesis (d) Fission

12. Which of the following is an example of a characteristic of life known as metabolism found in plants?
    (a) An unattended potted plant that gets knocked over in the greenhouse produces a shoot that bends toward the light and away from the pull of gravity.
    (b) Cell division occurring in the apical meristem of the stem and root extends the length of these modules.
    (c) A plant attracts a pollinating insect by producing brightly colored flowers to increase the likelihood of successful mating event leading to seed production.
    (d) Sunlight energy is captured by a plant leaf through the process called photosynthesis.

13. The structure that contains the genetic information in a bacterial cell is called the
    (a) Nucleus (b) Nucleoid
    (c) Nucleosome (d) Nucleoprotein

14. The term used for DNA in a nucleus before it becomes condensed in preparation for mitosis is
    (a) Eukaryote (b) Chromosome
    (c) Chromatin (d) Centromere

15. In prokaryotes, the chromosome is held in a region called the
    (a) Nucleoid (b) Centriole
    (c) Centrosome (d) Kinetochore

16. Which occurs in metaphase?
    (a) Centrioles move to opposite poles
    (b) Chromosomes line up along the equator of the dividing cell
    (c) Chromosomes move to opposite poles
    (d) The nuclear envelope disappears

17. Which occurs in anaphase?
    (a) Centrioles move to opposite poles
    (b) Chromosomes line up along the equator of the dividing cell
    (c) Chromosomes move to opposite poles
    (d) The nuclear envelope disappears

18. Which does not occur in telophase?
    (a) Cytokinesis usually gets under way
    (b) The nuclear envelope is being constructed
    (c) The centromeres split apart
    (d) Chromosomes decondense into chromatin

19. .... phase is the post replication phase
    (a) S phase (b) G1 phase
    (c) G2 phase (d) Prophase

20. .... phase is the pre DNA replication phase
    (a) S phase (b) G2 phase
    (c) G1 phase (d) Prophase
21. Cell grows and becomes larger in which phase of cell cycle activity
    (a) G0 (b) G1
    (c) S (d) M
22. Division of cytoplasm and cell membrane is called as
    (a) Plasmolysis (b) Protolysis
    (c) Cytokinesis (d) Meiosis
23. Cell elongation is promoted by which hormone
    (a) Cytokinin (b) ABA
    (c) Auxin (d) Ethylene
24. Osmotic solution used for demonstration of cell elongation and radicle protrusion
    (a) PEP (b) PEG and mannitol
    (c) Glycin betaine (d) Sorbitol
25. DNA synthesis is associated with
    (a) Accumultion of β-tubulin (b) Degradation of β-tubulin
    (c) Accumulation of α-tubulin (d) Degradation of α-tubulin
26. Protein accumulate during Arabidopsis seed germination
    (a) Gluten (b) Actin -7
    (c) Proline (d) Zeatin
27. Key element involved during cell expansion and radicle protrusion
    (a) β-tubulin (b) Actin -7
    (c) β-2,4 tubulin (d) β-3,5 tubulin
28. Which of the following element involved in cell cycle activity
    (a) α-amylase (b) β-amylase
    (c) Dehydrogenase (d) Cyclin dependant lanase
29. Cell cycle genes present only after germination in Arabidopsis and cabbage seeds
    (a) Before and after germination (b) Seed development
    (c) Seed maturation (d) Seed dormancy
30. Inhibitor present in cabbage seeds
    (a) ABA (b) Coumarin
    (c) Hydroxy urea (d) Methyl urea
31. Mitosis takes place in which phase of cell cycle activity
    (a) G_0 phase (b) G_1 phase
    (c) S phase (d) M phase
32. Cell cycle activity is suppressed in
    (a) Developing seed (b) Dormant seed
    (c) Maturing seed (d) Germinating seed

♦ Answer Keys ♦

Q.No.	Ans.	Q.No.	Ans.	Q.No.	Ans.	Q.No.	Ans.	Q.No.	Ans.	Q.No.	Ans.
1	b	2	a	3	d	4	c	5	d	6	a
7	b	8	b	9	d	10	d	11	a	12	d
13	b	14	c	15	a	16	b	17	c	18	c
19	c	20	c	21	b	22	c	23	c	24	b
25	a	26	b	27	c	28	d	29	a	30	c
31	d	32	b								

2.11.2. State True or False

1. Cell growth and cell division proceeds through a well defined series of steps and stages called the cell cycle
2. A replicating fork is where the synthesis of new DNA strands takes place.
3. RNA synthesis is needed to prime DNA synthesis.
4. Prokaryotic cells divide by binary fission.
5. Cell division is a pre-germination phenomenon.
6. Hydroxyl urea present in cabbage seed will affect radicle protrusion.
7. DNA synthesis takes place in S phase of cell cycle activity.
8. Mitosis takes place in M phase of cell cycle activity.
9. G0 is called as first phase of cell cycle activity.
10. Nucleolus disappear in metaphase of cell division.
11. Cell elongation is promoted by auxin.
12. DNA replication takes place in G2 phase of cell cycle activity
13. β-tubulin accumulate during germination.
14. Mitosis inhibitor in cabbage is hydroxyl urea.
15. Cell cycle activity is suppressed in dormant seeds.

♦ Answer Keys ♦

Q.No.	Ans.	Q.No.	Ans.	Q.No.	Ans.	Q.No.	Ans.	Q.No.	Ans.
1	True	2	True	3	True	4	True	5	False
6	False	7	True	8	True	9	False	10	False
11	True	12	False	13	True	14	True	15	True

2.11.3. Fill in the blanks

1. The reproduction of new cells from the pre-existing cells is called ....
2. The two types of cell division are .... and ....
3. Mitosis occurs in ....
4. Meiosis occurs in .... or .... cells
5. Interphase is the phase between .... and ....

6. The process of the division of nucleus is ....
7. The process of the division of cytoplasm is ....
8. .... is the equational division
9. .... is the reductional division
10. Recombination occurs due to ....
11. During .... phase, synthesis of RNA takes place
12. Nucleolus disappears during ....
13. Nucleolus reappears during ....
14. DNA synthesis occurs during .... phase
15. The number of daughter cells produced from a single cell in meiosis is ....

♦ Answer Keys ♦

Q.No.	Ans.	Q.No.	Ans.	Q.No.	Ans.
1	Cell division	2	Mitosis Meiosis	3	Somatic cells
4	Reproductive or germ	5	Telophase and prophase	6	Karyokinesis
7	Cytokinesis	8	Mitosis	9	Meiosis
10	Crossing over	11	G	12	Prophase
13	Telophase	14	S	15	4

Unit 3

SEED DETERIORATION

3.1. ULTRASTRUCTURAL CHANGES DURING SEED DETERIORATION

3.1.1 Choose the correct answer

1. Seed deterioration is
   (a) Multiple process (b) Reversible process
   (c) Inexorable process (d) Physical changes
2. Seed deterioration is
   (a) Irreversible process (b) Weight loss
   (c) Moisture loss (d) Exorable process
3. Loss of membrane integrity is due to
   (a) Increase in protein (b) Increase in leachate
   (c) Decrease in amino acids (d) Decrease in fatty acids
4. Loss of phospholipids in cell membrane is caused by
   (a) Protein peroxidation (b) Phosphokinase enzyme
   (c) Phosphofructase enzyme (d) Lipid peroxidation
5. Decrease in respiration rate during deterioration is due to
   (a) Mitochondria damage (b) Mitochondria rearrangement
   (c) Poor oxygen supply (d) More CO_2
6. During seed deterioration, protein synthesis is retarded because of
   (a) Deorganization of mitochondria (b) Loss of membrane integrity
   (b) Disassociation of ribosomes (d) Loss of phospholipids
7. In deteriorated seed, the endoplasmic reticulum is
   (a) Short (b) Long
   (c) Small (d) Large
8. Which one of the following is the physiological manifestation of seed deterioration
   (a) Decreased seed weight (b) Decreased respiration
   (c) Decreased abnormal seedling (d) Delayed germination
9. Which one of the following is the biochemical manifestation of seed deterioration
   (a) Decreased enzyme activity (b) Increased respiration
   (c) Reduced seedling growth (d) Decreased tolerance to stress

10. The internal factor that influence the seed deterioration is
    (a) Seed moisture (b) RH
    (c) Temperature (d) Stress

♦ Answer Keys ♦

Q.No.	Ans.	Q.No.	Ans.	Q.No.	Ans.	Q.No.	Ans.	Q.No.	Ans.
1	c	2	a	3	b	4	d	5	a
6	c	7	a	8	d	9	a	10	a

3.1.2 Fill in the blanks

1. Seed deterioration is an .... process
2. Loss of membrane integrity leads to increase in ....
3. Seed leachate increase is attributed to loss of membrane ....
4. Disfunction of mitochondria during seed deterioration results in decrease in ....
5. At later stages of seed deterioration, mitochondria becomes pigmented and ....
6. At later stages of seed deterioration, mitochondria changed to .... and fragmented
7. In non viable seed, ribosome fail to ....
8. .... synthesis is retarded due to dissociation of ribosome
9. Protein synthesis during seed deterioration is terminated because of .... dissociation
10. Dictyosomes (Golgi bodies) became unstacked and finally ....
11. During seed ...., the endoplasmic reticulum is short
12. .... develops a peculiar lobed condition in deteriorated seed
13. Increased abnormal seedling is the .... changes during deterioration
14. .... and .... are biotic factors that influence seed deterioration
15. Reduced enzyme activity is the .... changes during deterioration

♦ Answer Keys ♦

Q.No.	Ans.	Q.No.	Ans.	Q.No.	Ans.
1	Inexorable/irreversible	2	Seed leachates	3	Phospholipids
4	Respiration	5	Fragmented	6	Pigmented
7	Dissociate	8	Protein	9	Ribosome
10	Disappeared	11	Deterioration	12	Nucleus
13	Physiological	14	Insects, pests	15	Biochemical

3.1.3 State True or False

1. Seed deterioration is irreversible process
2. Coalescence of protein bodies influence the cell membrane damage
3. Loss of phospholipids in cell membrane is caused by phospholipase enzyme activity
4. Mitochondria becomes permanently shrinked during deterioration
5. Mitochondrial disfunction and fragmentation leads to decrease in respiration rate
6. At early stages of seed deterioration, mitochondria becomes pigmented and fragmented
7. At later stages of seed deterioration, mitochondria becomes pigmented and fragmented
8. In non viable seed, ribosome dissociation takes place
9. Protein synthesis in deteriorated seed is stopped due to dissociation of ribosomes
10. In deteriorated seed, dictyosomes (Golgi bodies) became stacked and finally disappeared
11. Dictyosoems (Golgi bodies) did not unstacke in deteriorated seed
12. Deteriorated seed has a short endoplasmic reticulum
13. The nucleus of the cell in the deteriorated seed does not show any change
14. Seed moisture relationships is the internal factor that influences the seed deterioration
15. Temperature is the internal factor that influences the seed deterioration

15. Physiological state of seed is the internal factor that influences the seed deterioration

17. Environmental stress is the external factors that influences the seed deterioration

♦ Answer Keys ♦

Q.No.	Ans.	Q.No.	Ans.	Q.No.	Ans.	Q.No.	Ans.	Q.No.	Ans.	Q.No.	Ans.
1	True	2	False	3	True	4	False	5	True	6	False
7	True	8	False	9	True	10	False	11	False	12	True
13	False	14	True	15	False	16	True	17	True		

3.2. FOOD RESERVES CHANGES DURING SEED DETERIORATION

3.2.1 Choose the correct answer

1. Which one of the following seed has greater surface area to volume ratio that uptake more water and more prone to deterioration

   (a) Larger seed (b) Smaller seed

   (c) Thicker seed (d) Thinner seed

2. In accelerated ageing test, the seeds are exposed to
   (a) Low temperature and RH
   (b) High temperature and low RH
   (c) High temperature and high moisture
   (d) High temperature and high RH

3. Which test is engaged to study physiological and biochemical changes during seed deterioration?
   (a) Cold test (b) Brick gravel test
   (c) Accelerated ageing test (d) Chilling test

4. Which is the substrate for respiration?
   (a) Carbohydrate (b) Sucrose
   (c) Glucose (d) Fructose

5. Which enzyme is used for carbohydrate degradation?
   (a) Lipase (b) Proteinase
   (c) Phosphatase (d) a - amylase

6. Which one of the following biochemical process helps to reduce the carbohydrates
   (a) Respiration (b) Photorespiration
   (c) Peroxidation (d) Reduction

7. Which one of the following increases as the protein content decreases during seed deterioration
   (a) Protein content (b) Fatty acids
   (c) Amino acids (d) Lipid content

8. Which enzyme is involved in the catabolic process of lipids
   (a) Proteinase (b) Amylase
   (c) Catalase (d) Lipase

9. Which test is used to know the extent of food reserve disintegration in aged seed
   (a) Electrical conductivity test (b) pH of leachate
   (c) Ferric chloride test (d) Phenol test

10. An decrease in germination had a good correlation with
    (a) Increase in protein (b) Increase in electrolyte leakage
    (c) Increase in fat (d) Increase in carbohydrate

11. Which one of the reducing sugars is accumulated in higher amount due to carbohydrate degradation?
    (a) Glucose (b) Fructose
    (c) Raffinose (d) Sucrose

12. Which one of the following is the monosaccharide that is degraded from carbohydrates during ageing
    (a) Sucrose (b) Glucose
    (c) Raffinose (d) Starch

♦ Answer Keys ♦

Q.No.	Ans.	Q.No.	Ans.	Q.No.	Ans.	Q.No.	Ans.	Q.No.	Ans.	Q.No.	Ans.
1	b	2	d	3	c	4	c	5	d	6	a
7	c	8	d	9	a	10	b	11	d	12	b

3.2.2 Fill in the blanks

1. .... seeds have smaller surface area to volume ratio that are less exposed to water uptake; and they are less prone to deterioration than smaller seed
2. .... test has been developed as self ageing technique
3. .... temperature and 95±5% RH is used for accelerated ageing test
4. 40±1°C temperature and .... RH is used for accelerated ageing test
5. Loss of enzyme activities could be responsible for higher .... accumulation in aged seed
6. The presence of reducing sugar may lead to deterioration of protein components through .... reactions
7. During seed deterioration, long chain fatty acids are broken into .... compounds
8. The accumulation of free fatty acids has deleterious effect on .... damage
9. The accumulation of free fatty acids increased fusion of plant vesicles which lead to an increase in ....
10. An increase in amylase activity indicates the fast degradation of ....
11. An increase in total free fatty acids reveals the degradation of ....
12. In seed deterioration, protein degradation results in more accumulation of ....
13. .... bodies in aged cotyledons are severely diminished
14. In deteriorated seed, among soluble sugars, ... is very predominant
15. ...., fructose and galactose are monosaccharides that are degraded from carbohydrates during deterioration
16. A depletion in raffinose content of deteriorated seed had a close relationship with a decrease in .... per cent
17. Fructose, glucose and .... are monosaccharides that are degraded from carbohydrates during deterioration
18. In deteriorated seeds, the monosaccharides like glucose, ... and galactose contents are in smaller amount when compared to sucrose
19. .... leakage is more in deteriorated seed than sound live seed
20. In the deteriorated seed, the .... perturbation results in an increased electrolyte leakage

♦ Answer Keys ♦

Q.No.	Ans.	Q.No.	Ans.	Q.No.	Ans.	Q.No.	Ans.
1	Larger	2	Accelerated aging	3	40°±1°C	4	95±5%
5	H_2O_2	6	Amadori and Maillard	7	Smaller	8	Membrane
9	Membrane leakage	10	Carbohydrates	11	Lipids	12	Free amino acids
13	Lipid	14	Sucrose	15	Glucose	16	Germination
17	Galactose	18	Fructose	19	Electrolyte	20	Membrane

3.2.3 State True or False

1. Smaller seeds are prone to more deterioration than larger seed
2. High temperature of 40°±1°C and high RH of 95±5% is used for accelerated ageing test
3. High temperature of 40°±5°C and high RH of 95±1% is used for accelerated ageing test
4. The only cause for membrane disruption is an increase in free fatty acids
5. One of the major cause for membrane damage is an increase in free fatty acid content during seed deterioration
6. Carbohydrates are degraded because of an increase in amylase as they are utilized for respiration
7. An increase in peroxidase during seed deterioration shows a decrease in protein
8. An increase in lipases indicates an increase in lipid content
9. An increase in lipases indicates an increase in lipid peroxidation
10. An increase in lipases showed a decrease in lipid content
11. An increase in lipases expressed that there is an increase in free fatty acids
12. An sharp increase in total free amino acid is noticed during seed deterioration because of faster degradation of protein
13. Total soluble sugars content decreased during seed deterioration
14. Raffinose content is predominant among total reducing sugars in the deteriorated seed
15. In the deteriorated seed, among the total reducing sugars, the sucrose content was higher
16. Sub division of lipid bodies into smaller ones takes place during seed deterioration that indicates non-disorganization of lipids
17. A decrease in germination was observed in deteriorated seed which had a good correlation with a decrease in electrolyte leakage because of degradation of complex food reserves into simpler ones

18. An increase in electrolyte leakage leads to a decrease in germination because of membrane disruption
19. The monosaccharides such as raffinose, glucose and fructose are present in small amounts in the deteriorated seed
20. As raffinose content depleted in the aged seed, germination and vigour also retarded

♦ Answer Keys ♦

Q.No.	Ans.	Q.No.	Ans.	Q.No.	Ans.	Q.No.	Ans.	Q.No.	Ans.
1	True	2	True	3	False	4	True	5	True
6	True	7	False	8	False	9	False	10	True
11	True	12	True	13	False	14	False	15	True
16	False	17	False	18	True	19	False	20	True

3.3. ENETIC CHANGES DURING SEED DETERIORATION

3.3.1 Choose the correct answer

1. Which part of the seed is more prone to genetic damage during deterioration?
   (a) Radicle meristem (b) Plumule meristem
   (c) Scutellum meristem (d) Cotyledon meristem
2. Genetic changes take place in which kind of seed?
   (a) Fresh seed (b) Dry seed
   (c) Old seed (d) Imbibed seed
3. What is the effect of genetic changes during ageing on seed quality
   (a) Germinate and grow
   (b) Germinate and produce abnormal seedling
   (c) Imbibe and remain as such
   (d) Imbibe and remain hard
4. Seed longevity is a
   (a) Qualitative trait (b) Quantitative trait
   (c) Dominant (d) Recessive
5. Seed longevity is a quantitative trait as it is controlled by
   (a) Single loci (b) Double loci
   (c) Single gene (d) Multiple gene
6. What type of damage to DNA and protein is likely to be involved during ageing?
   (a) Peroxidative damage (b) Reductive damage
   (c) Oxidative damage (d) Reactive damage
7. What kind of damage occur in chromosome in deteriorating seed
   (a) Rubbing (b) Deteriorating
   (c) Aberration (d) Rupturing

8. As per old consensus of seed deterioration, which production is failed?
   (a) rRNA (b) tRNA
   (c) mRNA (d) mtRNA
9. Which one of the following is available more numbers in the meristematic tissues?
   (a) Golgi bodies (b) Vesicles
   (c) Chloroplast (d) Mitochondria
10. As per new consensus of seed deteriorations which processes is hindered?
   (a) DNA replication (b) DNA degradation
   (c) mRNA transcription (d) mRNA translation
11. According to new consensus of seed deterioration, which is more prone for free radical attack?
   (a) Nuclear DNA (b) Chloroplast DNA
   (c) Mitochondria DNA (d) Leucoplast DNA
12. Which substance is more dangerous for seed deterioration as per new consensus
   (a) Free radicle (b) Free radical
   (c) Anion (d) Cation
13. Which kind of reaction takes place for free radical production during ageing as per new consensus
   (a) Lipid oxidation (b) Lipid denaturation
   (c) Lipid reduction (d) Lipid peroxidation
14. If there is specific damage to bases of DNA, what happens?
   (a) Weak bond (b) Strand intact
   (c) Strand breakage (d) Removal of H band
15. During lipid peroxidation of seed deterioration, besides production of free radicals, another product is also produced that is called as
   (a) Dialdehydes (b) Volatile aldehydes
   (c) Monoaldehydes (d) Polyaldehydes
16. Which one of the following is attacked by free radicals which lead to increased genetic mutation
   (a) Nucleotide bases (b) Sugar residues
   (c) Phosphate residues (d) Hydrogen bonds
17. Which DNA is highly conserved against free radical attack?
   (a) Nuclear DNA (b) Chromosomal DNA
   (c) Chloroplast DNA (d) Mitochondria DNA
18. Which process of protein synthesis is affected by nuclear DNA degradation due to free radical attack?
   (a) Translation (b) Transcription factor
   (c) Transcription factor (d) Transcription

19. Which event is late for radicle protrusion and growth due to DNA degradation by free radicals?
    (a) Cell enlargement (b) Cell division
    (c) Cell differentiation (d) Cell multiplication
20. Which event is delayed for cell division in deteriorated seed
    (a) DNA multiplication (b) DNA bifurcation
    (c) DNA replication (d) DNA duplication
21. Which part is protecting the free radical damage to nuclear DNA?
    (a) Cell membrane (b) Plasma membrane
    (c) Plasmadesmata (d) Nuclear membrane
22. Which one of the following is surrounding the nuclear DNA and that is destroyed by free radicals before damaging mDNA
    (a) Histones (b) Aldehydes
    (c) Lipids (d) Free fatty acids
23. Mutations are first detected as chromosomal aberrations which delays the onset of a process necessary for cell division and that process is known as
    (a) Mitosis (b) Meiosis
    (c) Osmosis (d) Plasmosis
24. Which factor can influence an increase in the rate of accumulation of chromosome aberration?
    (a) An increase in Temperature (b) An increase in Relative humidity
    (c) An increase in moisture (d) An increase in oxygen
25. Which stage of the cell cycle of embryo cells remains for extended period of time due to chromosomal aberrations
    (a) G0 stage (b) G0/G1 stage
    (c) M stage (d) M/S stage

♦ Answer Keys ♦

Q.No.	Ans.	Q.No.	Ans.	Q.No.	Ans.	Q.No.	Ans.	Q.No.	Ans.	Q.No.	Ans.
1	a	2	c	3	b	4	b	5	d	6	c
7	c	8	c	9	d	10	a	11	c	12	b
13	d	14	c	15	b	16	a	17	a	18	d
19	b	20	c	21	d	22	a	23	a	24	c
25	b										

3.3.2 Fill in the blanks

1. .... tissue is more prone to genetic damages
2. In ageing, meristematic tissues are susceptible to .... damages
3. Genetic damages occur in the embryo which is unable to .... and grow
4. Seedlings from aged seeds may show high .... rates due to genetic changes

5. Seed longevity is a quantitative trait, since it is likely to be controlled by .... genes
6. Loss of viability during ageing is a .... trait, as it is probably controlled by multiple genes
7. .... damage occur to DNA and protein during ageing
8. In aging, ... and proteins are subjected to oxidative damage
9. A close correlation between seed ageing and the accumulation of .... aberrations is observed
10. According to old consensus of seed deterioration, DNA is degraded which fail to produce ....
11. There is a .... degradation which results in failure of mRNA production as per old consensus of seed deterioration
12. Because of DNA degradation, the mRNA fails to produce which results in faulty .... synthesis
13. As per new consensus of seed aging, production of .... causes profound damage to cell membrane
14. According to new consensus of seed deterioration, ... peroxidation is the main cause
15. Lipid .... is the major cause for seed deterioration as per new consensus
16. Many mitochondria is present in .... cells as it is active location for seed deterioration
17. Mitochondrial .... replication is hindered due to new consensus of seed deterioration
18. .... DNA replication is completely inhibited during deterioration as per new consensus
19. The important genetic change in the deteriorating seed is a decrease in .... acid synthesis
20. Modification of .... residue which leads to strand breakage of DNA
21. One of the genetic changes is a specific damage to the .... of DNA
22. The products of lipid peroxidation are volatile .... and free radicals
23. During lipid peroxidation, the products formed are volatile aldehydes and ....
24. The volatile .... cross link with macromolecules like sugars, amino acids and polypeptide chains thereby altering their structure
25. The volatile aldehydes cross link with macromolecules like ...., amino acids and polypeptide chains thereby altering their structure
26. .... DNA degradation is caused by free radicals
27. Lipid peroxidation produces free radicals that destroys .... membrane and ultimately causes DNA degradation

28. During lipid peroxidation free radicals are liberated which destroy nuclear membrane and ultimately causes .... degradation
29. .... DNA is highly conserved against free radical attack
30. Nuclear DNA, when compared to mitochondrial DNA, is difficult to be attacked by free ....
31. Nuclear DNA is enclosed by a protective .... membrane and through which only the free radicals reaches histones and then to DNA
32. The nuclear DNA is surrounded by a protein called .... which are embedded in the nuclear membrane
33. The free radicals produced by lipid peroxidation must first penetrate nuclear membrane and destroy .... before damaging nuclear DNA
34. Lipid peroxidation releases free radicals which first penetrate .... membrane and then destroy histones and finally DNA
35. .... synthesis is severely impaired in imbibed deteriorated seed and DNA replication is delayed
36. In imbibed deteriorated seed .... replication is delayed
37. Free radicals attack .... sugar besides the purines and pyrimidine
38. Besides .... and pyrimidines, the deoxyribose sugar is also attacked by free radicals
39. Free radicals attack purines and .... as well as deoxyribose sugar
40. Mutations are first detected as chromosomal aberrations that delay the onset of .... necessary for cell division
41. Because of chromosomal aberrations, the .... occur which delays the mitosis for cell division
42. Detection of mutation due to .... aberrations which affects mitosis needed for cell division
43. Chromosomal aberrations result in mutation that delays the mitosis for .... division
44. An increase in .... of seed, the rate of accumulation of chromosome aberration also increased
45. If there is an increase in moisture content of seed during aging, the accumulation of .... aberrations also increased
46. The embryo cells are non-dividing and are maintained in the .... stage of the cell cycle for extended period of time
47. The mitochondria of embryo cells also contains .... in a small quantity
48. Enzymes encoded by mtDNA are absolutely essential for oxidative ....
49. Maintenance of mtDNA is vital for active ....
50. When cell divides, the .... and mtDNA are separately replicated
51. During cell division, the nuclear and .... DNA are separately replicated
52. The nuclear and mtDNA are separately .... during mitosis

♦ Answer Keys ♦

Q.No.	Ans.	Q.No.	Ans.	Q.No.	Ans.
1	Meristematic	2	Genetic	3	Germinate
4	Mutation	5	Multiple	6	Quantitative
7	Oxidataive	8	DNA	9	Chromosome
10	mRNA	11	DNA	12	Protein
13	Free radical	14	Lipid	15	Peroxidation
16	Meristematic	17	DNA	18	Mitochondrial
19	Nucleic	20	Sugar	21	Bases
22	Aldehydes	23	Free radicals	24	Aldehydes
25	Sugars	26	Nuclear/ chromosome/ mitochondrial	27	Nuclear
28	DNA	29	Nuclear	30	Radicals
31	Nuclear	32	Histones	33	Histones
34	Nuclear	35	DNA	36	DNA
37	Deoxyribose	38	Purines	39	Pyrimidine
40	Mitosis	41	Mutation	42	Chromosomal
43	Cell	44	Moisture	45	Chromosome
46	G 0/G 1	47	mtDNA	48	Phosphorylation
49	Respiration	50	Nuclear	51	Mitochondrial
52	Replicated				

3.3.3 State True or False

1. In aged seeds, meristem appears most vulnerable to deterioration
2. Seedlings from old seeds show high mutation rates
3. When compared to sound live seeds, seedlings grown from deteriorated seeds expressed less rate of mutation
4. Embryos from deteriorated seed unable to germinate and grow
5. Seed longevity is a quantitative trait as it is controlled by multiple genes
6. Seed longevity is a qualitative trait since it is probably governed by a single gene
7. Ageing of seed is governed by a single gene which contributed for quantitative trait
8. Chromosomal aberrations have a close relationship with seed longevity
9. As per old consensus, the DNA degradation occurs in aged seed which results in tRNA production failure
10. In deteriorated seed, as per the old consensus, the mRNA production is curtailed which has an ill effect on protein synthesis
11. The seed deterioration, as per new consensus, is caused by lipid oxidation
12. Lipid peroxidation results in free radical production which hinders the mtDNA replication and that is the major reason for seed deterioration as per new consensus

13. Lipid peroxidation helps in free radical synthesis which hinders nuclear DNA replication that is the main cause for seed deterioration as per new consensus
14. According to new consensus of seed deterioration, mtDNA replication is hindered which results in less ATP synthesis
15. Modification of phosphate residues may lead to strand breakage of DNA
16. An increase in nucleic acid synthesis which lead to DNA damage
17. During genetic changes of seed deterioration, modification of sugar residues may lead to strand breakage of DNA
18. One of the genetic changes during seed deterioration is an decrease in nucleic acid synthesis
19. Free radicals attack easily mtDNA than nDNA
20. Nuclear DNA degradation by free radical is somewhat difficult when compared to chromosomal DNA degradation
21. It is difficult for free radicals to damage mtDNA when compared to nuclear DNA
22. Because the nuclear DNA is protected by nuclear membrane and surrounded by lipids, the free radicals easily damage the nuclear DNA
23. RNA replication is completely delayed in deteriorated seed due to nuclear DNA damage
24. DNA synthesis is severely impaired in imbibed deteriorated seed and DNA replication is delayed
25. Specific damage to the bases may leave the strand breakage, but modification of sugar residue lead to strand intact
26. Mutation due to chromosomal aberration delays the onset of mitosis needed for cell division
27. Chromosomal aberrations lead to mutation that delays the initiation of meiosis needed for cell division
28. An increase in temperature during seed aging may lead to reduction in longevity due to high rate of chromosomal aberrations
29. High temperature and relative humidity during aging may cause chromosomal aberrations
30. Because of chromosomal aberration G 1/M stage of cell cycle in the embryo cells is maintained for extended time period
31. Age related mutation caused by chromosome aberration will result in maintaining extended time period of G0/G 1 stage of cell cycle in embryo cells
32. During seed deterioration, accumulation of chromosome aberration will lead to maintain G 1/M phases for extended time period of cell cycle in the embryonic cells
33. Mitochondrial matrix also contains a small amount of mtDNA
34. RNA is also present in the mitochondrial matrix

35. Maintenance of mtDNA is vital for photorespiration of cells
36. Enzymes encoded by mtDNA are not essential for oxidative phosphorlyation

♦ Answer Keys ♦

Q.No.	Ans.	Q.No.	Ans.	Q.No.	Ans.	Q.No.	Ans.	Q.No.	Ans.	Q.No.	Ans.
1	True	2	True	3	False	4	True	5	True	6	False
7	False	8	True	9	False	10	True	11	False	12	True
13	False	14	True	15	False	16	False	17	True	18	True
19	True	20	True	21	False	22	False	23	False	24	True
25	False	26	True	27	False	28	False	29	True	30	False
31	True	32	False	33	True	34	False	35	False	36	False

3.4. ENZYME ACTIVITY CHANGES DURING SEED DETERIORATION

3.4.1 Choose the correct answer

1. Which biochemical energy is reduced during seed deterioration
   (a) ADP (b) ATP
   (c) NADP (d) NADPH
2. Which one of the enzyme decreases during seed deterioration
   (a) α-amylase (b) β-amylase
   (c) Oxidase (d) Peroxidase
3. Which one of the enzyme found in the cellular cytoplasm and matrix space of mitochondria
   (a) SOD (b) POD
   (c) CAT (d) APX

♦ Answer Keys ♦

Q.No.	Ans.	Q.No.	Ans.	Q.No.	Ans.
1	b	2	d	3	a

3.4.2 Fill in the blanks

1. Biodeterioration is due to activity of .... present in seed
2. The enzyme activity of deteriorating seeds is dependant on moisture and ....
3. In deteriorated seeds, if respiration is reduced, the production of biochemical energy .... is also reduced
4. A decrease in ...., the ATP production also decreases
5. .... is an enzyme catalyzing oxidation process in different substrates using hydrogen peroxide and producing water and free radical

♦ Answer Keys ♦

Q.No.	Ans.	Q.No.	Ans.	Q.No.	Ans.	Q.No.	Ans.
1	Enzymes	2	Temperature	3	ATP	4	Respiration
5	Peroxidase						

3.4.3 State True or False

1. The extent of deterioration depends upon the level of enzyme activity, which in turn depends on moisture and light
2. The seed moisture and surrounding temperature affects the enzyme activity which determines the seed deterioration
3. The reduction in enzyme activity will result in respiration decrease
4. If respiration decreases during deterioration, the ATP production also reduces
5. SOD activity increased along with the decrease in seed viability level
6. Peroxidase in an enzyme calayzing oxidation process

♦ Answer Keys ♦

Q.No.	Ans.	Q.No.	Ans.	Q.No.	Ans.	Q.No.	Ans.	Q.No.	Ans.	Q.No.	Ans.
1	False	2	True	3	True	4	True	5	False	6	True

3.5. LIPID PEROXIDATION

3.5.1 Choose the correct answer

1. Oxygen dependent deterioration of oilseeds is known as
   (a) Alkalinity (b) Acidity
   (c) Decaying (d) Rancidity
2. Rancidity has been noticed since ancient times as a major problem in storage of
   (a) Pulse seed (b) Oilseed
   (c) Cereal seed (d) Millet seed
3. Non-radical intermediate formed from unsaturated fatty acids during deterioration is called as
   (a) Hydroperoxide (b) Aldehydes
   (c) Hexanal (d) Alkanes
4. ROS are group of atoms behave as a unit and now named as
   (a) Free radicle (b) Intact radicle
   (c) Free radical (d) Intact radical
5. Free radicals capable of independent existence that contains one or more
   (a) Unpaired electrons (b) Unpaired neutrons
   (c) Paired electrons (d) Paired neutrons

6. Free radicals are formed by the loss or gain a type of atom by non-radical reaction
   (a) Single electron (b) Double electron
   (c) Single bond (d) Double bond
7. What kind of fatty acids is more often affected by free radicals?
   (a) Monosaturated (b) Polysaturated
   (c) Unsaturated (d) Polyunsaturated
8. How many bonds are available in polyunsaturated fatty acids?
   (a) Single bond (b) Double bonds
   (c) Triplet bonds (d) Multiple double bonds
9. Besides multiple double bonds, the polyunsaturated fatty acids also contain another compound that possess reactive hydrogens, what is that?
   (a) Ethylene - CH_2 - group (b) Ethylene - CH - group
   (c) Methylene - CH_2 - group (d) Methylene - CH - group
10. Which one of the following is the non-enzymatic lipid peroxidation reaction?
    (a) Photo-hydrogenation (b) Autooxidation
    (c) Peroxidation (d) Antioxidation
11. Non-enzymatic peroxidation reaction is named as
    (a) Photo-oxygenation (b) Photo-hydrogenation
    (c) Peroxidation (d) Antioxidation
12. Which of the following free radicle is produced by enzymatic peroxidation reaction?
    (a) Superoxide (b) Hydroxyl
    (c) Peroxnitrate (d) Hydrogen peroxide
13. Which one of the following free radical is produced only at mitochondria
    (a) Hydroxyl (b) Hydrogen peroxide
    (c) Superoxide (d) Peroxinitrate
14. The site of production of hydrogen peroxide besides mitochondria is
    (a) Peroxysome (b) Chloroplast
    (c) Leucoplast (d) Vesicles
15. What is the source of singlet O_2?
    (a) Hydrogen peroxide (b) Lipid peroxide
    (c) Hydroxyl radical (d) Superoxide radical
16. Which one of the process is not involved in autooxidation
    (a) Initiation (b) Propagation
    (c) Peroxidation (d) Termination
17. Which product is produced at initiation step of autooxidation
    (a) Hydroxyl radicle (b) Singlet oxygen
    (c) Aldehydes (d) Fatty acid radical

18. Which one of the ROS is most reactive to abstract a hydrogen atom from a methylene group
    (a) O_2° (b) $^1O_2$
    (c) $^{\circ}OH$ (d) H_2O_2

19. Which one of the following ROS is insufficient reactive to abstract a hydrogen atom from a methylene group
    (a) O_2° (b) $^1O_2$
    (c) $^{\circ}OH$ (d) H_2O_2

20. Which compound is formed when the carbon radical bond, that is released during initiation of autooxidation, tends to be stabilized by a molecular rearrangement
    (a) Corrugated diene (b) Complicated diene
    (c) Congested diene (d) Conjugated diene

21. What radical is produced when the conjugated diene, that is formed during initiation step of autooxidation, is able to combine, with O_2 under aerobic conditions?
    (a) Hydroxyl (b) Peroxyl
    (c) Epoxyl (d) Acyl

22. Which is the second step of autooxidation
    (a) Initiation (b) Propagation
    (c) Termination (d) Photooxidation

23. Which compound is produced during propagation of autooxidation
    (a) $^{\circ}OH$ (b) HO_2°
    (c) Peroxyl (d) Lipid peroxide

24. When autooxidation stops?
    (a) When two non radical reacts to produce a radical
    (b) When two non radicle reacts to produce radicle
    (c) When two radicals reacts to produce non radical
    (d) When two radicles reacts to produce non radicle

25. Which is an antioxidant?
    (a) Vitamin A (b) Vitamin B
    (c) Vitamin D (d) Vitamin E

26. Which one of the following is an antioxidant made within the plant?
    (a) Amylase (b) Protease
    (c) Lipase (d) Catalase

27. Which one of the following is an antioxidant?
    (a) Amylase (b) Protease
    (c) Lipase (d) Peroxidase

28. Which one of the following enzyme is an antioxidant?
    (a) Amylase (b) Protease
    (c) Lipase (d) Superoxide dismutase

29. In photooxidation of lipid peroxidation, which one of the following is the sensitizer that helps the double bond interacts with singlet oxygen?
    (a) Methylene red (b) Raffinose
    (c) Riboflavin (d) Vitamin E
30. Which one of the following is the sensitizer that helps for an interaction between double bond and singlet oxygen during photooxidation of lipid peroxidation?
    (a) Methylene blue (b) Raffinose
    (c) Vitamin A (d) Vitamin D
31. Which factor helps for production of singlet oxygen during photooxidation of lipid peroxidation
    (a) Temperature (b) RH
    (c) Water (d) Light
32. What kind of product is obtained when singlet oxygen reacts with a double bond between C_{12} and C_{13} of unsaturated fatty acid during photooxidation of lipid
    (a) C_{12} and C_{13} hydroperoxide (b) C_{12} and C_{13} hydroxyl radical
    (c) C_{12} and C_{13} conjugated diene (d) C_{12} and C_{13} aldehydes
33. Photooxidation of oleic acid is quicker than autooxidation. How many times?
    (a) 30 times (b) 300 times
    (c) 3000 times (d) 30,000 times
34. Photooxidation of polyenes is quicker than autooxidation. How many times?
    (a) 10 - 15 times (b) 100 - 150 times
    (c) 1000 - 1500 times (d) 10,000 - 15,000 times

♦ Answer Keys ♦

Q.No.	Ans.	Q.No.	Ans.	Q.No.	Ans.	Q.No.	Ans.	Q.No.	Ans.	Q.No.	Ans.
1	d	2	b	3	a	4	c	5	a	6	a
7	d	8	d	9	c	10	b	11	a	12	d
13	c	14	a	15	b	16	c	17	d	18	c
19	c	20	a	21	d	22	b	23	b	24	d
25	c	26	d	27	d	28	d	29	d	30	c
31	a	32	d	33	a	34	d				

3.5.2 Fill in the blank

1. The oxygen dependent deterioration of lipids, known as .... has been noticed since ancient times as a major problem in the storage of oilseeds
2. Rancidity has been observed to be a major problem in .... since ancient time

3. Lipid .... are non radical derivates from unsaturated fatty acids
4. Lipid peroxide is an non radical derivative from .... fatty acids
5. Lipid hydro peroxide is an non radical intermediate derived from ....
6. Lipid .... is an non radical derivative from glycolipids
7. ROS are now named as ....
8. .... is a ROS
9. Free radicals are capable of independent existence that contain .... unpaired electrons
10. .... are capable of independent existence that contains one or more unpaired electrons
11. Free radicals are capable of independent existence that contain one or more .... electrons
12. Free radicals are capable of independent existence that contain on ore more unpaired ....
13. Free radicals are formed either by the .... of a single electron by non radical reaction
14. Non radical reaction helps in the loss or gain of a .... to form free radicals
15. During lipid peroxidation, free radicals .... electrons from the lipids in cell membranes
16. Lipid peroxidation refers to the .... degradation of lipids
17. Free radicals most often affect .... fatty acids
18. During lipid peroxidation, free radicals most often affects polyunsaturated fatty acids because they contain .... double bond
19. Polyunsaturated fatty acids contain multiple double bonds in between which lie .... groups that possess expecially reactive hydrogen
20. Polyunsaturated fatty acid contains multiple double bonds in between which lie methylene -CH_2- groups that possess especially reactive ....
21. The site of production of superoxide radical is ....
22. .... radical is produced in mitochondria
23. Hydrogen peroxide is produced in ....
24. Besides mitochondria and peroxysomes, the hydrogen peroxide is also produced in ....
25. .... radical is produced by non-enzymatic reaction
26. .... radical is produced by enzymatic reaction
27. Besides seperoxide radical, ... radical is also produced by non enzymatic reaction
28. Besides hydroxyl radical, ... radical is also produced by non enzymatic reaction

29. Superoxide after production, reduced and forms .... peroxide
30. .... radical is reduced and formed into hydrogen peroxide
31. In the presence of Fe_2^+, H_2O_2 produces the very active species .... by the Fenton reaction
32. .... in the presence of Fe_2^+ produces *OH by Fenton reaction
33. H_2O_2 in the presence of .... produces *OH by .... reaction
34. In the presence of Fe_2^+, H_2O_2 produces *OH by .... reaction
35. Singlet O_2 may be formed during degradation of .... peroxide
36. Singlet .... may be formed during degradation of lipid peroxide
37. Single $^1O_2$ may be formed during degradation of lipid ....
38. Autooxidation is a radical chain process involving 3 steps namely ...., propagation and termination
39. The three steps involved in autooxidation are initiation, ... and termination
40. During autooxidation, there are three sequences namely initiation, propagation and ....
41. Initiation is the step of autooxidation whereby a .... radical is produced
42. .... is the step of autooxidation whereby a fatty acid radical is produced
43. Initiation is the step of .... whereby a fatty acid radical is produced
44. During initiation step of autooxidation, ROS such as .... which combines with a hydrogen atom to make water and a fatty acid radical
45. During initiation step of autooxidation, ... such as OH or HO_2 which combines with a hydrogen atom to make water and a fatty acid radical
46. During initiation step of autooxidation, ROS such as OH° or HO_2 which combines with a hydrogen atom to make .... and a fatty acid radical
47. During initiation step of autooxidation, ROS such as OH or HO_2 which combines with a hydrogen atom to make water and a fatty acid ....
48. During initiation step of autooxidation, ROS such as OH or HO_2 which combines with a hydrogen atom to make water and a .... radical
49. During initiation step of autooxidation, ROS such as OH or HO_2 which combines with .... atom to make water and a fatty acid radical
50. .... is the most efficient ROS that attack polyunsaturated fatty acid during initiation step of autooxidation, while O_2. is insufficiently reactive
51. °OH is the most efficient ROS that attack polyunsaturated fatty acid during initiation step of autooxidation, while .... is insufficiently reactive

52. °OH is the most efficient ROS that attack .... fatty acid during initiation step of autooxidation, while O_2 is insufficiently reactive
53. ROS, with sufficient reactivity, is able to abstract .... atom from methylene group of polyunsaturated fatty acid during autooxidation
54. ROS, with sufficient reactivity, is able to abstract hydrogen atom from .... group of polyunsaturated fatty acid during autooxidation
55. ROS, with sufficient reactivity, is able to abstract hydrogen atom from methylene group of .... fatty acid during autooxidation
56. During initiation step of autooxidation, the presence of .... bond in the polyunsaturated fatty acid weakens the C-H bonds on the carbon atom adjacent to the double bond so makes H removal easier
57. During initiation step of autooxidation, the presence of double bond in the .... fatty acid weakens the C-H bonds on the carbon atom adjacent to the double bond so makes H removal easier
58. During initiation step of autooxidation, the presence of double bond in the polyunsaturated fatty acid weakness .... bonds on the carbon atom adjacent to the double bond so makes H removal easier
59. During initiation step of autooxidation, the presence of double bond in the polyunsaturated fatty acid weaken C-H bonds on the carbon atom adjacent to the double bond so makes .... removal easier
60. Under aerobic conditions, conjugated dienes, formed during initiation of autooxidation, are able to combine with O_2 to give a .... radical
61. Under aerobic conditions, conjugated dienes, formed during initiations of autooxidation, are able to combine with .... to give a peroxyl radical
62. Under aerobic conditions, conjugated ...., formed during initiation of autooxidation, are able to combine with O_2 to give a peroxyl radical
63. During propagation step of autooxidation, peroxyl-fatty acid radical reacts with another free fatty acid producing a different fatty acid radical and a .... peroxide, if it had reacted with itself
64. The autooxidation stops when two radicals react to produce a .... species
65. .... is an antioxidant
66. .... is an enzyme which act as an antioxidant
67. During photooxidation of lipid, in the presence of sensitizers ...., a double bond interacts with singlet oxygen produced from O_2 by light
68. In photooxidation, one possible reaction of singlet oxygen with a double bond between C_{12} and C_{13} of one fatty acid is to produce 12 and 13 ....

69. Photooxidation of oleic acid is .... times quicker than autooxidation
70. Photooxidation of polyene is .... times quicker than autooxidation

♦ Answer Keys ♦

Q.No.	Ans.	Q.No.	Ans.	Q.No.	Ans.
1	Rancidity	2	Oilseeds	3	Hydroperoxide
4	Unsaturated	5	Unsaturated fatty acids/phospholipids / glycolipids	6	Hydroperoxide
7	Free radical	8	Superoxide/hydroxyl / hydrogen peroxide	9	One or more
10	Free radicals	11	Unpaired	12	Electrons
13	Loss/Gain	14	Single electron	15	Steal
16	Oxidative	17	Polyunsaturated	18	Multiple
19	Methylene - CH_2 -	20	Hydrogens	21	Mitochondria
22	Superoxide/Hydrogen peroxide	23	Mitochondria	24	Microsome
25	Hydroxyl/Superoxide	26	Hydrogen peroxide	27	Hydroxyl
28	Superoxide	29	Hydrogen	30	Superoxide
31	°OH	32	H_2O_2, Fenton	33	Fe_2^+
34	Fenton	35	Lipid	36	$^1O_2$
37	Peroxide	38	Initiation	39	Propagation
40	Termination	41	Fatty acid	42	Initiation
43	Autooxidation	44	OH° or HO_2	45	ROS
46	Water	47	Radical	48	Fatty acid
49	Hydrogen	50	OH	51	O_2°
52	Polyunsaturated	53	Hydrogen	54	Methylene
55	Polyunsaturated	56	Double	57	Polyunsaturated
58	C-H	59	H	60	Peroxyl
61	O_2	62	Dienes	63	Lipid
64	Non-radical	65	Vitamin E	66	Catalase/ Peroxidase/ Superoxide dismutase
67	Chlorophyll/ Porphyrins/ Myoglobin/ riboflavin/ erythrosine/ methylene blue	68	Hydroperoxide	69	30,000
70	1000 - 1500				

3.5.3 State True or False

1. Rancidity is a major problem in the storage of pulse seed
2. Oxygen-dependent deterioration known as acidity, takes place in oilseeds during storage

3. Oxygen dependent deterioration is known as rancidity that is a major problem in oilseeds storage
4. Lipid hydro peroxides are non radical intermediates derived from unsaturated fatty acid
5. Hydroperoxide is an non-radical derivatives from phospholipids
6. During lipid peroxidation, formation of non-radical intermediates called hydroxyl radical take place from glycolipids
7. ROS are now named as free radicals
8. A group of atoms that behave as a unit, known as ROS, are now named as free radicals
9. Free radicals are capable of interdependent existence that contains one or more unpaired electrons
10. Free radicles are formed either by a loss or gain single electron by non radical reaction
11. Loss or gain of single electron by non-radical reaction may lead to production of free radicals
12. Lipid peroxidation refers to oxidative phosphorylation of lipid
13. Lipid peroxidation is the process whereby free radicals steal protons from the lipids in cell membrane that damages cell
14. Free radicals most often affects polyunsaturated free fatty acids, because they contain multiple double bonds in between which lie methylene $-CH_2-$ groups that possess reactive hydrogens
15. Polyunsaturated fatty acids contain multiple double bonds in between which lie methylene $-CH_2-$ groups that possess reactive hydrogen and because of this, the polyunsaturated fatty acids are more liked by free radicals for attack
16. Superoxide radicals first production site is the internal mitochondrial membrane
17. Superoxide radical is reduced and forms hydrogen peroxide
18. Hydrogen peroxide is mainly produced by non-enzymatic reaction
19. In the presence of Fe_2^+, H_2O produces the °OH by the Fenton reaction
20. H_2O_2 in the presence of Fe_2^+, produces a very active species OH by Fenton reaction
21. Singlet $^1O_2$ may be formed during degradation of lipid peroxidation
22. Singlet O_2 may be formed during degradation of lipid peroxide
23. Autooxidation reaction involves three steps and they are initiation, propagation and termination
24. Peroxidation is one among three steps of autooxidation
25. Initiation step of autooxidation is able to produce free fatty acid radical
26. Initiation is the first step of autooxidation where fatty acid radical is produced
27. During initiation of autooxidation, ROS such as O_2 or H_2O combine with a hydrogen atom to make water and fatty acid radical

28. In initiation of autooxidation, ROS such as OH or H_2O combine with a hydrogen atom to make water and fatty acid radical
29. ROS such as OH° or HO_2 produced during initiation of autooxidation combine with a hydrogen atom to make water and fatty acid radical
30. °OH is the most efficient ROS that attack polyunsaturated fatty acid than O_2°
31. °O_2 is the most efficient ROS that attack polyunsaturated fatty acid, whereas *OH is insufficiently reactive
32. ROS formed during autooxidation is able to abstract carbon atom from a methylene group (-CH_2-)
33. ROS formed during autooxidation is able to abstract hydrogen atom from a methylene group (-CH_2-)
34. During initiation step of autooxidation, the presence of a double bond in the fatty acid weakens C-H bonds on the carbon atom adjacent to the double bond and so makes removal easier
35. Under aerobic conditions, conjugated, dienes produced during initiation step of autooxidation are able to combine with OH to give a peroxyl radical
36. Under aerobic conditions, conjugated dienes produced during initiation step of autooxidation, are able to combine with O_2 to give a peroxyl radical
37. In propagation of autooxidation, peroxyl radical reacts with another free fatty acid producing a different fatty acid radical and a hydroxy peroxide
38. The unstable hydroxy radical, during propagation of autooxidation, reacts with another free fatty acid producing an different radical and a lipid peroxide
39. Lipid peroxide is formed when the peroxyl radical reacts with a another fatty acid during propagation of autooxidation
40. Termination is a step where radical reaction stops when two radicals react and produce non-radical species
41. In termination of autooxidation, radical reaction stops when two radicals reacts without producing non-radical species
42. Vitamin A is an antioxidant that breaks the auto oxidation reaction in the cell membrane
43. In cell membranes, the antioxidant, Vitamin E is the chain breaking molecule
44. The natural antioxidants used to break the chain reaction mechanism are superoxide dismutase, catalase and peroxidase
45. Superoxide dismutase, catalase and oxidases are natural antioxidant that breaks chain reaction mechanism
46. Give possible reaction of photooxidation of singlet oxygen with a double bond between C_{12} and C_{13} of one fatty acid is to produce 14 and 15 hydroperoxide

47. In the presence of sensitizer, erythrosine, a double bond interacts with singlet oxygen produced from O_2 by light
48. In the presence of sensitizer, ethylene blue, a double bond interacts with singlet oxygen produced from O_2 by light
49. Photooxidation of oleic acid is 3000 times quicker than autooxidation
50. The photooxidation of polyenes is 1000 - 1500 times quicker

♦ Answer Keys ♦

Q.No.	Ans.	Q.No.	Ans.	Q.No.	Ans.	Q.No.	Ans.	Q.No.	Ans.
1	False	2	False	3	True	4	True	5	True
6	False	7	False	8	True	9	False	10	False
11	True	12	False	13	False	14	False	15	True
16	True	17	True	18	False	19	False	20	True
21	False	22	True	23	True	24	False	25	False
26	True	27	False	28	False	29	True	30	True
31	False	32	False	33	True	34	False	35	False
36	True	37	False	38	False	39	True	40	True
41	False	42	False	43	True	44	True	45	False
46	False	47	True	48	False	49	False	50	True

3.6. DETECTION OF LIPID LOSS DURING SEED DETERIORATION

3.6.1 Choose the correct answer

1. Phospholipid is a .... type of lipid
   (a) Functional lipids (b) Storage lipids
   (c) Solid lipids (d) Liquid lipids
2. During ageing, the lipids are broken into
   (a) Free fatty acid (b) Acid value
   (c) Iodine value (d) Amino acids
3. Fats become rancid when expose to water, and the water found frozen is called as
   (a) Hydrolytic rancidity (b) Oxidative rancidity
   (c) Flavour reversion (d) Free fatty acids
4. Fats become rancid when they expose to oxygen in air and is called as
   (a) Hydrolytic rancidity (b) Oxidative rancidity
   (c) Flavour reversion (d) Free fatty acids
5. Chemical deterioration of fats occurs when the fatty acids attached to glycerol molecule are broken into smaller units that yield off flavor and odour is called as
   (a) Flavour reversion (b) Odour reversion
   (c) Rancidity (d) Free fatty acids

◆ **Answer Keys** ◆

Q.No.	Ans.	Q.No.	Ans.	Q.No.	Ans.	Q.No.	Ans.	Q.No.	Ans.
1	a	2	a	3	a	4	b	5	c

3.6.2 Fill in the blanks

1. The lipids are stored in the form of ....
2. The iodine value is the measure of .... fatty acid
3. TBA test is used for detection of ....
4. .... and .... are the types of rancidity
5. The most commonly used oilseeds for study of flavor reversion is ....
6. The breakdown of an essential fatty acid to leading to an undesirable flavor change prior to the start of rancidity is called as ....
7. In industries, they are incorporating .... fatty acids for reducing the risk of rancidity

◆ **Answer Keys** ◆

Q.No.	Ans.	Q.No.	Ans.	Q.No.	Ans.	Q.No.	Ans.
1	Triglycerides	2	Unsaturated	3	MDA	4	Hydrolytic and oxidative rancidity
5.	Cotton and corn	6	Flavour reversion	7	Unsaturated		

3.6.3 State True or False

1. Rancid products have more shelf life due to poor quality
2. The increase in ageing, increases the free fatty acids content
3. Increasing ageing, increase the acid value
4. In cotton, the peroxidase value increases with increasing ageing
5. The iodine value is a measure of saturated fatty acid

◆ **Answer Keys** ◆

Q.No.	Ans.	Q.No.	Ans.	Q.No.	Ans.	Q.No.	Ans.	Q.No.	Ans.
1	False	2	True	3	True	4	True	5	False

3.7. BIOLOGICAL EFFECT OF LIPID PEROXIDATION

3.7.1 Choose the correct answer

1. Site of free radical formation
   (a) Nucleus (b) Rhizome
   (c) Mitochondria (d) Golgi bodies
2. Biological effect of lipid peroxidation
   (a) Mitochondria disfunction (b) Enzyme activation
   (c) Membrane stability (d) Membrane activation
3. The site of free radical formation in mitochondria
   (a) Inner layer (b) Outer layer
   (c) Matrix (d) Cristae

4. The factor which is responsible for more prone to free radical attack of mitochondrial DNA than nuclear DNA is ....
   (a) Lack of histone proteins　　b. Presence of histone proteins
   (c) Non-formation of ATP　　d. None of the above
5. Degraded DNA leads to poor .... transcription
   (a) mRNA　　(b) t-RNA
   (c) r-RNA　　(d) n-RNA
6. Richest oxygen sink in the cells
   (a) Mitochondria　　(b) Ribosome
   (c) Nucleus　　(d) Cytoplasm

♦ Answer Keys ♦

Q.No.	Ans.	Q.No.	Ans.	Q.No.	Ans.	Q.No.	Ans.	Q.No.	Ans.	Q.No.	Ans.
1	c	2	a	3	d	4	a	5	a	6	a

3.7.2 Fill in the blanks

1. Earliest stages of germination are pre-programmed on ....
2. During storage, electrical conductivity is increased due to ....
3. The richest oxygen sink in the cells is ....
4. The presence of .... can reduce the lipid peroxidation
5. Free radicles attack causes ....
6. Mitochondrial disfunction leads to leaking of electrons to form .... radicals

♦ Answer Keys ♦

Q.No.	Ans.	Q.No.	Ans.	Q.No.	Ans.
1	m-RNA	2	Membrane damage	3	Mitochondria
4	Antioxidants	5	Lipid peroxidation	6	O_2^-

3.7.3 State True or False

1. ATP necessary for rapid seedling growth during germination will be provided by golgi bodies
2. The presence of histone proteins makes mitochondrial DNA more prone to free radical attack
3. Seed deterioration reduced seedling growth as a consequence of increased respiration and increased mitochondrial cells
4. Free radical production and lipid peroxidation culminate in reduced synthesis of many enzymes

♦ Answer Keys ♦

Q.No.	Ans.	Q.No.	Ans.	Q.No.	Ans.	Q.No.	Ans.
1	False	2	False	3	False	4	True

GLOSSARY

A

ABA: Abscisic acid (ABA), also known as abscisin II and dormin, is a plant hormone. ABA functions in many plant developmental processes, including bud dormancy.

Abscission: The shedding of parts of a plant.

Accelerated ageing (AA): A method of exposing seed to an unfavourable storage environment of high temperature and humidity in order to speed up the natural ageing process. The method is used as a vigour test and to predict storage life of seed.

Accumulation: The act or process of collecting together or becoming collected.

Actinomycin D: Actinomycin D is the most significant member of actinomycines, which are a class of polypeptide antibiotics isolated from soil bacteria of the genus *Streptomyces*.

Adenine: Adenine is a nucleobase with a variety of roles in biochemistry including cellular respiration, in the form of both the energy-rich adenosine triphosphate (ATP) and the cofactors nicotinamide adenine dinucleotide (NAD) and flavin adenine dinucleotide (FAD), and protein synthesis, as a chemical component of DNA and RNA.

Adenosine triphosphate (ATP): A nucleotide, $C_{10}H_{16}N_5O_{13}P_3$, that is composed of adenosine and three phosphate groups and releases energy when hydrolyzed to ADP. It is present in all cells, where it is used to store and transport energy needed for biochemical reactions.

ADP: Adenosine diphosphate, abbreviated ADP, is a nucleoside diphosphate. It is an ester of pyrophosphoric acid with the nucleoside adenosine.

ADP-glucose: ADP-glucose (Adenosine diphosphate) is an intermediate in cellulose and starch synthesis in some plants and in glycogen in some bacteria.

Aerobic respiration: The pyruvate produced in glycolysis undergoes further breakdown through a process called aerobic respiration in most organisms.

Aggregate: An aggregate is a collection of items that are gathered together to form a total quantity.

Albumin: Albumin is an umbrella term for a type of protein which is water soluble.

Aldehyde (RCHO): A functional group characterized by a carbonyl group flanked by a hydrogen atom on one side and a carbon on the other.

Aleurone layer: The aleurone layer is the outermost layer of the endosperm, followed by the inner starchy endosperm.

Alkaloid: Any of various organic compounds normally with basic chemical properties and usually containing at least one nitrogen atom in a heterocyclic ring, occurring chiefly in many vascular plants and some fungi.

Alkanes: Alkenes are acyclic branched or unbranched hydrocarbons having one or more double carbon-carbon bonds in their molecules.

Allosteric regulation: Allosteric regulation is the regulation of an enzyme or other protein by binding an effector molecule at the protein's allosteric site.

Alpha Amylase: Alpha Amylase is an enzyme which aids in the breakdown of starch to maltose.

Amino acid: A molecule consisting of the basic amino group (NH2), the acidic carboxylic group (COOH), a hydrogen atom (-H), and an organic side group (R) attached to the carbon atom, thus, having the basic formula of $NH_2CHRCOOH$. The building block of protein in which each is coded for by a codon and linked together through peptide bonds.

Amylase: Amylase is an enzyme that catalyses the breakdown of starch into sugars.

Amylopectin: Amylopectin is a soluble polysaccharide and highly branched polymer of glucose found in plants.

Amyloplasts: Non-pigmented organelles found in some plant cells. They are responsible for the synthesis and storage of starch granules, through the polymerization of glucose.

Amylose: Amylose is a straight-chain polymer of glucose units joined by alpha-1, 4 glycosidic linkages

Anabolism: Anabolism is the set of metabolic pathways that construct molecules from smaller units.

Anaerobic respiration: Anaerobic respiration is a form of respiration using electron acceptors other than oxygen.

Anaphase: Stage of mitosis during which the two sets of chromosomes separate and move away from each other. Composed of anaphase A (chromosomes move toward the two spindle poles) and anaphase B (spindle poles move apart).

ANT: Adenine translocator

Anther: The pollen bearing portion of a stamen.

Anthesis: The period of flower development when the stigma is ready to receive the pollen is known as anthesis.

Anticodon: An anticodon is a unit made up of three nucleotides that correspond to the three bases of the codon on the mRNA.

Antinutritional factors: The antinutritional factors (ANFs) may be defined as those substances generated in natural feed stuffs by the normal metabolism of species and different mechanisms which exert effects contrary to optimum nutrition.

Antioxidants: The most common antioxidants (sometimes called primary antioxidants) are the phenolic or aromatic amine types. These are defined as chain-breaking types, because the mechanism for their reaction involves the antioxidant absorbing a free radical to form a stable radical.

Antitranspirants: A substance sprayed on the plant leaves to reduce the rate of transpiration and conserve moisture.

Apex: The shoot tip or terminal point on the lobe of a leaf; plural is apices.

Apical dominance: The presence of auxins in the lateral areas of the plant (in between the root and shoot tip) prevent lateral growth. If you cut off the shoot tip of a plant, the lack of 'diffusable' auxins means that they cannot inhibit growth in these lateral areas. This is known as apical dominance.

Apical meristem: A region of actively dividing cells at the tip of a root or stem, growth results in an increase in length.

Apoplast: Movement through Apparent Free Space (AFS) present in the cell wall.

***Arabidopsis*:** *Arabidopsis* is a genus in the family Brassicaceae.

Arachidonic acid: An essential fatty acid, $C_{20}H_{32}O_2$, found in the lipids of most tissues, that is a precursor in the synthesis of prostaglandins, prostacyclins, and related effectors.

Arginine: Arginine, an essential amino acid, has a positively charged guanidino group. Arginine is well designed to bind the phosphate anion, and is often found in the active centers of proteins that bind phosphorylated substrates.

Ascomycota fungus: Any fungus of the class *Ascomycetes* (or subdivision Ascomycota) in which the spores are formed inside an *ascus*.

Ascorbate peroxidases (or APX): Enzymes that detoxify peroxides such as hydrogen peroxide using ascorbate as a substrate. The reaction they catalyse is the transfer of electrons from ascorbate to a peroxide, producing dehydroascorbate and water as products.

Ascorbic acid: Ascorbic acid (vitamin C) is a useful water-soluble antioxidant. To increase its fat solubility it is often used in the form of its palmitic ester. E-numbersare 304 (ascorbyl palmitate), 300 (L-ascorbic acid), 301 (sodium ascorbate), and 302 (calcium ascorbate).

Asparagines: Asparagine is the amide of aspartic acid.

Assimilation: The processes of photosynthesis and the absorption of raw materials by which plants derive their nutrition.

Asterad: Both the basal and terminal cells contribute to embryo development.

ATP: Adenosine triphosphate (ATP), coenzyme used as an energy carrier in the cells of all known organisms.

Atropine: Atropine is a naturally occurring tropane alkaloid extracted from deadly nightshade (*Atropa belladonna*), jimson weed (*Datura stramonium*), mandrake (*Mandragora officinarum*) and other plants of the family solanaceae.

AUG: The initiator codon (AUG) codes for the amino acid N-formylmethionin.

Autosomal: Mutation occurring in autosomes.

Autooxidation: The most common form of oxidative deterioration. Reaction occurs by a radical chain process and olefinic hydroperoxides are formed. It is promoted by trace metals (especially copper and iron) which act as pro-oxidants and is hindered by a range of materials which serve as antioxidants.

Auxin: Auxin is a generic term for characterized by their capacity to induce elongation in shoot cells. (or) Auxins are compounds that cause enlargement of plant cells. Auxins are a class of plant hormones (or plant growth substances) with some morphogen-like characteristics.

Axil: The upper (and usually smaller) angle between a stem and the petiole of a leaf.

Axillary meristem: The meristematic region in the axil of a leaf that gives rise to an axillary bud.

B

Bakane disease: It is a disease that infects the rice plant. It is caused by the fungus *Gibberella fujikuroi*, the metabolism of which produces a surplus of gibberellic acid.

Beta- oxidation: Beta-oxidation is the process by which fatty acid molecules are broken down in the mitochondria to generate acetyl-coA, which enters the citric acid cycle, and NADH, $FADH_2$ to be used by the electron transport chain.

Biosynthesis: Biosynthesis is an enzyme-catalyzed process in cells of living organisms by which substrates are converted to more complex products.

Biosynthetic pathway: Biosynthetic pathways are energy requiring, this energy usually being supplied either as phosphate bond energy by ATP or as reducing power by NADPH.

Bivalent: A pair of synapsed homologous chromosomes.

Brassinosteroids (BRs): Brassinosteroides are steroidal hormones that regulate the growth and development of plants. Brassins, after the genus Brassica, comprise a complex mixture of lipids purified from rape (*Brassica napus* L.) pollen. The characteristics biological activity-elongation, curvature, and splitting of the treated internode in the bean second-internode bioassay-is called brassin activity.

C

C- terminus: The C-terminus (also known as the carboxyl-terminus, carboxy-terminus, C-terminal tail, C-terminal end, or COOH-terminus) is the end of an amino acid chain (protein or polypeptide), terminated by a free carboxyl group (-COOH).

(or) A series of related enzymic reactions involved in the synthesis and catabolism of carbohydrates.

Calvin cycle: The Calvin cycle is a metabolic pathway found in the stroma of the chloroplast in which carbon enters in the form of CO_2 and leaves in the form of sugar.

Calyx: All the sepals together form the calyx.

Cambial initials: Cells of the vascular cambium which give rise through periclinal divisions to either phloem or xylem (fusiform initials), or to rays (ray initials).

Cambium: Lateral meristematic cells of either vascular cambium or cork cambium.

Carbohydrate metabolism: The sum of the anabolic and catabolic processes of the body involved in the synthesis and breakdown of carbohydrates, principally glucose, fructose, and galactose. Some of the processes are glycogenesis and glycolysis. Energy-rich phosphate bonds are produced in many metabolic reactions requiring carbohydrates.

Carbohydrate: A sugar molecule. Carbohydrates can be small and simple (for example, glucose) or they can be large and complex (for example, polysaccharides such as starch, chitin or cellulose).

Carotenoids: Carotenoids are organic pigments that are found in the chloroplasts and chromoplasts of plants and some other photosynthetic organisms like algae, some bacteria, and some fungi.

Carpel/Pistil: Female part of the flower. Consisting of the stigma, style and ovary.

Caryophyllad: Basal cell undergoes no further division, and the suspensor, if present, is always derived from the terminal cell.

Catabolism: Catabolism is the set of metabolic pathways that break down molecules into smaller units and release energy.

Catalase: An oxidative catalyst enzyme found in all organisms which require oxygen or can survive in its presence, especially in the peroxisomes. catalase is a haem-based enzyme which catalyses the breakdown of hydrogen peroxide into oxygen and water $(2H_2O_2 - 2H_{20}+O_2)$.

Cd: Cadmium

Cell adhesion: Cell adhesion is the binding of a cell to a surface, extracellular matrix or another cell using cell adhesion molecules such as selectins, integrins, and cadherins.

Cell cycle: A cell cycle is a series of events that takes place in a cell as it grows and divides. The cycle of growth and asexual reproduction of a cell, consisting of interphase followed in actively dividing cells by prophase, metaphase, anaphase, and telophase.

Cell division: The moment when a cell divides in two separate daughter cells. The process of reproduction or formation of new cell from the pre-existing cell in living organisms.

Cell membranes: The cell membrane is a biological membrane that separates the interior of all cells from the outside. The cell membrane is selectively permeable to ions and organic molecules and controls the movement of substances in and out of cells. It consists of the lipid bilayer with embedded proteins.

Cell signalling: Cell signalling is part of a complex system of communication that governs basic cellular activities and coordinates cell actions.

Cell wall: Outer covering of the cell that protects the plant cell and gives it shape

Cell: The basic unit of life.

Cellular cytoplasm: In eukaryotic cells, the cytoplasm is that part of the cell between the cell membrane and the nuclear envelope. It is the jelly-like substance in a cell that contains the cytosol, organelles, and inclusions, but not including the nucleus. In fact, the cytoplasm and the nucleus make up the protoplasm of a eukaryotic cell.In prokaryotic cells that do not have a well-defined nucleus, the cytoplasm is simply everything enclosed by the cell membrane. It therefore contains the cytosol, and all the other cellular components, including the chromosome in the nucleoid region.

Cellular respiration: Cellular respiration is the set of the metabolic reactions and processes that take place in the cells of organisms to convert biochemical energy from nutrients into adenosine triphosphate (ATP), and then release waste products. The most prevalent and efficient catabolic pathway for the production of ATP, in which oxygen is consumed as a reactant along with the organic fuel.

Centriole: A cylinder shaped cell structure found in most eukaryotic cells, though it is absent in higher plants and most fungi.

Cellulose: Cellulose is a molecule comprised of carbon, hydrogen, and oxygen, and is found in the cellular structure of virtually all plant matter. A cylinder shaped cell structure found in most eukaryotic cells, though it is absent in higher plants and most fungi.

Centrioles: Centrioles are cylindrical structures that are composed of groupings of microtubules. Centrioles are found in animal cells and play a role in cell.

Chalaza: Non-micropylar end of the seed. The base of an ovule, bearing an embryo sac surrounded by integuments. The site of an ovule opposite the micropyle, and adjacent to the stalk region.

Chemical dormancy: Chemical dormancy refers to chemical germination inhibitors in seed covering tissues preventing germination.

Chemical mutagen: Chemical mutagens are defined as those compounds that increase the frequency of some types of mutations.

Chemiosmosis: Chemiosmosis is the movement of ions across a selectively permeable membrane, down their electrochemical gradient.

Chemotropism: Enables plant roots to avoid some toxins and grow toward water and nutrients.

Chenopodiad: Basal cell and terminal cell both contribute to embryo development.

Chlorophyll: Chlorophyll is a green pigment found in cyanobacteria and the chloroplasts of algae and plants.

Chloroplasts: The site of photosynthesis in a plant cell. They contain chlorophyll, a green pigment that absorbs energy from sunlight.

Christae: The multiple-folded inner membrane of a cell's mitochondrion that are finger-like projections.

Chromatid: The chromosomes condense into compact structures. Each replicated chromosome can now be seen to consist of two identical chromatids (or sister chromatids) held together by a structure known as the centromere.

Chromoplasts: Contain different colored pigments. The most important type of chromoplast is the chloroplast, which contains the green pigment chlorophyll.

Chromosomal aberration: Any change in the normal structure or number of chromosomes; often results in physical or mental abnormalities.

Chromosomal DNA: Structure found in the nucleus of a cell, which contains the genes.

Chromosomal variation: Change in number of chromosomes per cell.

Chromosome: A chromosome is an organized structure of DNA and protein found in cells.

Chromosomes: Contain the centrioles, which are involved in the process of mitosis.

Citrate: A citrate is a derivative of citric acid; that is, the salts, esters, and the polyatomic anion found in solution.

Closed vascular bundle: A vascular bundle with no cambium tissue.

CoA: Coenzyme A

Coat dormancy (coat component of non-deep PD): The combination of an embryo with low growth potential and mechanical constraint from the seed covering layers can result in dormancy.

Cocaine: Cocaine (benzoylmethylecgonine) (INN) is a crystalline tropane alkaloid that is obtained from the leaves of the cocoa plant.

Codon AAA: The codon AAA specifies the amino acid phenylalanine.

Codon: A codon is defined by the initial nucleotide from which translation starts.

Coleoptiles: Coleoptile is the pointed protective sheath covering the emerging shoot in monocotyledons such as oats and grasses.

Coleorhiza: In grasses, a sheath that encloses the radicle of the embryo.

Collateral vascular bundle: A vascular bundle with phloem on the abaxial side of the xylem.

Collenchyma: Elongated cells with uneven primary cell walls and containing no lignin. Usually found in early development of stems and leaf petioles.

Companion cell: A specialized parenchyma cell, located in the phloem of flowering plants and closely associated with the development and function of a sieve-tube element.

Complementary DNA (cDNA): A DNA sequence which was produced from mRNA by reverse transcription. A cDNA is so-called because its sequence is the complement of the original mRNA sequence. However, when double-stranded cDNA is synthesized, it contains both the original sequence and its complement.

Compounds: Can also be measured directly by HPLC.

Condensed tannin: Condensed tannins are polymers formed by the condensation of flavans. They do not contain sugar residues.

Conjugated diene: A diene is a hydrocarbon chain that has two double bonds that may or may not be adjacent to each other. This section focuses on the delocalization of pi systems by comparing two neighboring double bonds. The arrangements of these double bonds can have varying affects on the compounds reactivity and stability. Conjugated dienes are two double bonds separated by a single bond. Nonconjugated (Isolated) Dienes are two double bonds are separated by more than one single bond. Cumulated Dienes are two double bond connected to a similar atom.

Conservative method: The old DNA strands are conserved while a new copy is synthesized from the old molecule.

Convolvulaceae: Convolvulaceae, known commonly as the bindweed or morning glory family, are a group of about 60 genera and more than 1,650 species of mostly herbaceous vines, but also trees, shrubs and herbs.

Cork cambium: A lateral meristem producing cork centripetally in stems and sometimes roots.

Corolla: All the petals together form the corolla.

Cortex: Large zone, many layered, cells thin-walled parenchymatous with intercellular spaces, stores food and water. Region found between the epidermis and the vascular system in roots and stems. These cells are involved in food storage and the movement of water from the epidermis.

Cotyledon: An embryonic seed leaf; there is characteristically 1 for monocots and 2 for dicots

Cotyledonary trace: A vascular connection between the root and hypocotyl in angiosperms, resembling a leaf trace.

CPS: *ent*-copalyl diphosphate synthase

Cristae: The infoldings or inward projections of the inner membrane of the mitochondrion, which are studded with proteins and increase the surface area for chemical reactions to occur, like cellular respiration.

Crucifer: Basal cell plays only a minor role (or name) in embryo development.

Cuticle: A layer of water-impervious material (cutin) deposited on the outer surfaces of epidermal cell walls, particularly in leaves.

Cyclosis: Cyclosis is the circulation or streaming of the cytoplasm within some living cells.

Cysteine: Cysteine is a sulfur-containing amino acid with the chemical formula $HO_2CCH(NH_2)CH_2SH$.

Cytokinesis: The final cellular division to form two new cells. In plants a cell plate forms along the line of the metaphase plate; in animals there is a constriction of the cytoplasm. The cell then enters interphase - the interval between mitotic divisions.

Cytokinins (kinins): A group of chemical substances that have a decisive influence in the stimulation of cell division.

Cytoplasm: Gel-like substance within the cell membrane containing water, enzymes, salts, organelles, and various organic molecules.

Cytosine: Cytosine is one of the four main bases found in DNA and RNA, along with adenine, guanine, and thymine (uracil in RNA).

Cytoskeleton: A network of fibers throughout the cytoplasm that helps the cell to maintain its shape and gives support to the cell.

Cytosol: Part of the cytoplasm that is not held within organelles is called the cytosol. The cytosol or intracellular fluid (ICF) or cytoplasmic matrix is the liquid found inside cells.

D

DAG: Diacylglycerol

Deamination: Deamination is the removal of an amine group from a molecule. Enzymes which catalyse this reaction are called deaminases.

Decene: Decene is an alkene with the formula $C_{10}H_{20}$. Decene contains a chain of ten carbon atoms with one double bond.

Degradation: The processes by which organic substances are broken down by living organisms.

Dehydrin: Dehydrin (DHN) is a multi-family of proteins present in plants that is produced in response to cold and drought stress.

Deoxyribonucleotide: A deoxyribonucleotide is the monomer, or single unit of DNA, or deoxyribonucleic acid.

Deoxyribose: Deoxyribose, or more precisely 2-deoxyribose, is a monosaccharide with idealized formula H-(C=O)-(CH_2)-($CHOH$)$_3$-H. Its name indicates that it is a deoxy sugar, meaning that it is derived from the sugar ribose by loss of an oxygen atom.

Desaturation: Conversion of a saturated compound into an unsaturated compound by a removal of two hydrogen molecules and the creation of a double bond.

Diacylglycerol: A diglyceride, or a diacylglycerol (DAG), is a glyceride consisting of two fatty acid chains covalently bonded to a glycerol molecule through ester linkages.

Diakinesis: Homologous chromosomes continue to separate, and chiasmata move to the ends of the chromosomes.

Dicer: Dicer is an endoribonuclease in the RNase III family that cleaves double-stranded RNA (dsRNA) and pre-microRNA (miRNA) into short double-stranded RNA fragments called small interfering RNA (siRNA) about 20-25 nucleotides long, usually with a two-base overhang on the 3' end.

Differentiation: The process of cells or tissue becoming structurally specialized, *e.g.* floral differentiation.

Diffusion: The process by which molecules spread from areas of high concentration to areas of low concentration.

Dioxygenase: Dioxygenase is an enzyme which catalyses the incorporation of both atoms of molecular oxygen into substrates using a variety of reaction mechanisms.

Diplotene: Homologous chromosomes start to separate but remain attached by chiasmata.

Disaccharides: Disaccharides are compounds in which two monosaccharides are joined by a glycosidic bond. A glycosidic bond to the anomeric carbon can be either α or β. For example, maltose, the disaccharide obtained by enzyme-catalyzed hydrolysis of starch, consists of two D-glucopyranose units joined by a 1,4'-α-glycoside bond. The "prime" superscript indicates that C-4 is not in the same ring as C-1. Unlike the other disaccharides, sucrose is not a reducing sugar and does not exhibit mutarotation because the glycosidic bond is between the anomeric carbon of glucose and the anomeric carbon of fructose.

Disaccharides: Disaccharides, on hydrolysis, yield two monosaccharide molecules. Eg. Sucrose, maltose and lactose.

Dispersive method: The old molecule disintegrated in the process of replication and the two new molecules are synthesized.

Disruption: A structural defect of an organ, part of an organ, or a larger region of the body resulting from the extrinsic breakdown of, or an interference with, an originally normal developmental process.

Diterpene: Diterpene, a type of terpene, is an organic compound composed of four isoprene units and has the molecular formula $C_{20}H_{32}$.

DMAPP: Dimethylallyl pyrophosphate (or -diphosphate) (DMAPP) is an intermediate product of both mevalonic acid (MVA) pathway and DOXP/MEP pathway.

DNA: Deoxyribonucleic acid (DNA) is a molecule encoding the genetic instructions used in the development and functioning of all known living organisms and many viruses.

DNA polymerase: This is an enzyme that catalyze the synthesis of new DNA strand alongside a template DNA strand. All DNA polymerases synthesize DNA in the 5' to 3' direction and can add a nucleotide only onto preexisting 3'-OH group. For this reason, DNA polymerase needs a primer at which it can add the first nucleotide.

DNA replication: DNA replication is a biological process that occurs in all living organisms and copies their DNA; it is the basis for biological inheritance. The process starts when one double-stranded DNA molecule produces two identical copies of the molecule.

Dormancy, Embryo (= Endogenous): Dormancy as a result of conditions within the embryo itself, *e.g.* inhibiting substances or incompletely developed embryo.

Dormancy, Thermo: Primary or secondary dormancy in which the seed must be subjected to a high, low or fluctuating temperature prior to germination. In temperate regions thermo dormancy is overcome by pre-chilling or stratification; in tropical regions by exposing the seeds to normal diurnal fluctuations in temperature prior to germination.

Dormancy: Lack of growth due to any external or internal cause.

Dormant Seed: Seeds does not have the capacity to germinate in a specified period of time under any combination of normal physical environmental factors that are otherwise favorable for its germination, *i.e.* after the seed becomes non-dormant.

Double bond: Carbon atoms are usually held together by a two-electron bond (a single bond). When bound by four electrons the linkage is described as a double bond. Compounds containing double bonds are said to be unsaturated and can generally exist in stereoisomeric forms . They undergo addition reactions such as hydrogenation and are prone to deterioration through autoxidation.

Double dormancy or Combined dormancy: It is defined that where two or more dormancy types occur in the seed *e.g.* fleshy fruits with chemical inhibitors combines with hard endocarp.

Double fertilization: A characteristic of angiosperms in which a pollen tube carries two sperm cells to the female gametophyte in the ovule.

dsRNA: Double stranded RNA

E

Ecodormancy: Dormancy caused by physical or biological factors externally imposed

Efflux: Process of flowing out.

Eg. Starch and glycogen.

Egg cell: The egg cell is a haploid nucleus, with surrounding cytoplasm enclosed in a thin wall or follicle.

Electron spin resonance (ESR): A spectroscopic technique, similar to nuclear magnetic resonance, used to detect atoms having orbitals containing unpaired electrons, as found in free radicals. The molecules are exposed to high-frequency microwave radiation, causing the electrons in unpaired orbitals to resonate between the two available spin states of that orbital and to release distinct patterns of radiation.

Electron: Electrons are negatively charged particles that surround the atom's nucleus. Electrons were discovered by J.J. Thomson in 1897.

Electrophoregram: An electropherogram is a plot of results from an analysis done by electrophoresis automatic sequencing.

Elongation: Part of transcription of DNA into mRNA

Embryo sac: The mature female gametophyte of angiosperms containing the egg cell.

Embryo: An embryo is a multicellular diploid eukaryote in its earliest stage of development, from the time of first cell division until birth, hatching, or germination.

Embryogenesis: Embryogenesis is the process that produces a plant embryo from a fertilized ovule by asymmetric cell division and the differentiation of undifferentiated cells into tissues and organs.

Embryogeny: After the fertilization of an ovule, the embryo develops to maturity by a complex series of stages known as embryogeny.

Endodermis: The endodermis is a single layer of cells that forms a boundary between the cortex and the inner vascular cylinder. The endodermis is the central, innermost layer of cortex in some land plants.

Endodormancy: Dormancy caused by internal physiological factors

Endogenous dormancy: Any cause of dormancy related to the embryo *e.g.* immature development or chemical inhibitors located in the embryo is collectively referred as endogenous dormancy or embryo dormancy.

Endogenous mutation: Caused by certain internal factors like change in metabolism, nutrition etc.

Endoplasmic reticulum: A network of membranous canals filled with fluid. They carry materials throughout the cell. ER is the "transport system" of the cell. The endoplasmic reticulum (ER) is an organelle of cells in eukaryotic organisms that forms an interconnected network of membrane vesicles.

Endopolyploidy: An increase in the number of chromosome sets caused by replication without cell division.

Endosperm dormancy: Endosperm weakening can be either part of the coat dormancy release or part of the germination program. Since the endosperm is in most cases living tissue it can actively participate in regulating embryo constraint by influencing both the ABA/GA ratio and sensitivity to these hormones.

Endosperm: Endosperm is the tissue produced inside the seeds of most flowering plants around the time of fertilization.

Energy of activation: The energy required to initiate a (bio) chemical reaction.

Enzymes: Enzymes are large biological molecules responsible for the thousands of chemical inter conversions that sustain life.

Epiblema: Single, outermost layer of thin-walled cells.

Epicotyl: The axis above the attachment of the cotyledons is the epicotyl, which also ends in an apical meristem.

Epidermis: This layer of cells surrounding the root is responsible for absorption of water and minerals dissolved in water.

Epistase: Epistase is the development of well-defined nucellar or integumentary tissue in the micropylar region of the seed.

Epoxide: An organic compound whose molecule contains a three-membered ring involving an oxygen atom and two carbon atoms.

Equational division: The second meiotic division is an equational division because it does not reduce chromosome numbers.

ER: Endoplasmic reticulum

Essential fatty acids: These are polyunsaturated fatty acids the human body needs and cannot make for itself or derive from other fatty acids. Linoleic acid (omega-6) and alpha-linolenic acid (omega-3) are usually considered essential because they cannot be made in the body, but can be converted to long-chain forms, which are usually the most critical to the body's needs.

Ethyl methanesulphonate (EMS): Ethyl methanesulphonate (EMS) is a mutagenic, teratogenic, and possibly carcinogenic organic compound with formula $CH_3SO_3C_2H_5$. It produces random mutations in genetic material by nucleotide substitution; particularly by guanine alkylation.

Ethylene: A plant growth regulator produced by ripening tissues, stimulates cell wall to soften. Ethylene is an organic compound, a hydrocarbon with the formula C_2H_4 or $H_2C{=}CH_2$. It is a colorless flammable gas with a faint "sweet and musky" odor when pure.

Etiolation: The condition in which a plant has pale, underdeveloped leaves and extended internodes; ususally caused by insufficient light.

Eukaryote: An organism whose cells are characterized by the presence of membrane bound nuclei.

Exodermis: The exodermis is a layer of cells from the outermost layer of the cortex of many angiosperms.

Exogenous Dormancy: Mechanical resistance, physical impermeability, inhibitors or light sensitivity associated with seed coat is called as seed coat dormancy.

Exogenous mutation: Caused by external factors like change in temperature, climate etc.

5' (Five-prime) end: The 5′-end designates the end of the DNA or RNA strand that has the fifth carbon in the sugar-ring of the deoxyribose or ribose at its terminus.

F

Fad: Fatty acid desaturases

FAD: Flavin adenine dinucleotide

Fat: Fats consist of a wide group of compounds that are generally soluble in organic solvents and generally insoluble in water.

Fatty acid: A chain of carbon atoms with hydrogen atoms attached to them. Each fatty acid has an acidic or carboxyl end and an omega end containing only hydrogens. Three fatty acids attached to glycerol make a triglyceride or fat.

Fermentation: A catabolic process that makes a limited amount of ATP from glucose without an electron transport chain and that produces a characteristic end-product, such as ethyl alcohol or lactic acid.

Fertilization: Union of male and female gametes.

FFA: Free fatty acid

Filament: The stalk of the stamen which bears the anther.

Filial: Denoting the generation or generations after the parental generation.

Flavonoids: Flavonoids are a class of plant secondary metabolites.

Floating respiration: If, carbohydrates are used as respiratory substrate that is called floating respirator.

Flower: The sexual reproductive structure of the angiosperms, typically with a gynoecium, androecium, perianth and an axis.

Forward mutation: Development of new mutant type from a wild type.

Fosmidomycin: Fosmidomycin is an antibiotic that was originally isolated from culture broths of bacteria of the genus *Streptomyces*.

FPP: Farnesyl Pyrophosphate (FPP)

Free (unesterified) fatty acids: Fatty acids in unbound (unesterified) form. Oils and fats which are mainly triacylglycerols contain in their natural (crude) state small amounts of free acids which may be removed by processing.

Free radical: Radicals (often referred to as free radicals) are atoms, molecules, or ions with unpaired electrons on an otherwise open shell configuration. These unpaired electrons are usually highly reactive, so radicals are likely to take part in chemical reactions. They may damage proteins, DNA, lipids, and cell membranes.

Fructans: A fructan is a polymer of fructose molecules.

Fructose: Fructose (fruit sugar) is a ketohexose (a six-carbon ketonic sugar), which occurs in sweet fruits and honey. Glucose and fructose have the same molecular formula, $C_6H_{12}O_6$, but have different structures. Pure, dry fructose is a very sweet, white, odorless, crystalline solid. Fructose is one of the sweetest of all sugars and is combined with glucose to make sucrose, or common table sugar. An older common name for fructose is levulose, after its levorotatory property of rotating plane polarized light to the left (in contrast to glucose which is dextrorotatory). The polysaccharide inulin is a polymer of fructose.

Fruit: The fruit is the structure on a plant that is derived from the flower.

Fumarate: A dicarboxylic acid intermediate in the tricarboxylic acid cycle. Can be derived from aspartate, phenylalanine and tyrosine for input to the krebs Cycle.

Funicle: A filamentous stalk attaching a seed or ovule to the placenta.

G

G0 phase: Arresting phase where the cell has left the cycle and has stopped dividing.

G1 Phase: Gap 1 phase of the eucaryotic cell cycle, between the end of Cytokinesis and the start of DNA synthesis.

G1P: Glucose 1-phosphate (also called cori ester) is a glucose molecule with a phosphate group on the 1'-carbon.

G2 Phase: Gap 2 phase of the eucaryotic cell cycle, between the end of DNA synthesis and the beginning of mitosis.

Galactosylglycerides: A class of glycolipids structurally analogous to phospholipids; they are the major glycolipids of plants.

Gap period: One of two intermediate periods within the cell cycle. Gap 1 occurs between mitosis and the DNA synthesis phase, gap 2 occurs between DNA synthesis and next mitosis.

GDP: Guanosine diphosphate

Gene mutation: A gene mutation is a permanent change in the DNA sequence that makes up a gene. Mutations range in size from a single DNA building block (DNA base) to a large segment of a chromosome.

Genetic code: The genetic code is the set of rules by which information encoded within genetic material (DNA or mRNA sequences) is translated into proteins (amino acid sequences) by living cells.

Geotropism: It is the movement of a plant organ in response to gravity. If a plant is placed horizontally the shoots will bend upward against gravity (negative geotropism), wheareas roots will bend downward along with the force of gravity (positive geotropism).

Germinal mutation: Mutations occurring in reproductive cells and such mutations are heritable and passed on next generations.

Germination capacity (=germination percentage): Proportion of a seed sample that has germinated normally within a specified test period, usually expressed as a percentage; cf. viability.

Germination energy: The proportion of germination which has occurred up to the time of peak germination, or the time of maximum germination rate, or up to some pre-selected point, usually seven test days. The critical time of measurement can be chosen by several means.

Germination: It is the reactivation of metabolic activity in the seed, an awakening from dormancy. It culminates with the emergence of the radicle (embryonic root) and finally, the plummule (first shoot).

or

Seed germination is the resumption of active growth of the embryo that results in the rupture of the seed coat and the emergence of the young plant under favourable conditions.

GGPP: Geranylgeranyl Diphosphate. GGPP is a precursor of many diterpenoid compounds, including the phytol side chain of chlorophyll, and tetraterpenoids (40 carbons), including carotenoids.

Gibberellins: Gibberellins (GAs) are plant hormones that regulate growth and influence various developmental processes, including stem elongation, germination, dormancy, flowering, sex expression, enzyme induction, and leaf and fruit senescence.

Gibberellins-like activity: Substances which have biological activity as gibberellins together with some of their chemical or physical properties, but which have not been chemically characterized.

GL: Glycolipid

Globoid: A globoid body, such as any of those occurring in certain plant granules.

Globulin: The globulins are a family of globular proteins that have higher molecular weights and water solubility values than the albumins.

Gluconeogenesis: Glucose can also be synthesized from noncarbohydrate precur- sors by reactions referred to as gluconeogenesis.

Glucose: Glucose (grape sugar, blood sugar), $C_6H_{12}O_6$, is an aldohexose (a monosaccharide sugar having six carbon atoms and an aldehyde group). An older common name for glucose is dextrose, after its dextrorotatory property of rotating plane polarized light to the right. Glucose in free (in sweet fruits and honey) or combined form (sucrose, starch, cellulose, glycogen) is is probably the most abundant organic compound in nature. During the photosynthesis process, plants use energy from the sun, water from the soil and carbon dioxide gas from the air to make glucose. In cellular respiration, glucose is ultimately broken down to yield carbon dioxide and water, and the energy from this process is stored as ATP molecules (36 molecules of ATP across all processes).

Glucosinolate: The glucosinolates are a class of organic compounds that contain sulfur and nitrogen and are derived from glucose and an amino acid.

Glutamate: The carboxylate anions and salts of glutamic acid are known as glutamates.

Glutamine: Glutamine is one of the 20 amino acids encoded by the standard genetic code.

Glutathione reductase: An enzyme catalyzing the reduction of glutathione disulfide to glutathione in the presence of NADPH. The antioxidant glutathione is regenerated for protecting the cell from free radicals.

Glutathione: A natural sulfur-bearing peptide formed from the linking of three amino acids: glutamic acid, cysteine and glycine. Glutathione acts as an antioxidant and detoxicant and is involved with the selenium-containing enzyme glutathione peroxidase. Glutathione is also involved in amino acid transport across cell membranes and is a powerful free radical scavenger and antioxidant.

Glutathione-ascorbate cycle: It is a metabolic pathway that detoxifies hydrogen peroxide (H_2O_2), which is a reactive oxygen species that is produced as a waste product in metabolism. The cycle involves the antioxidant metabolites: ascorbate, glutathione and NADPH and the enzymes linking these metabolites.

Glutelin: A simple protein found in the seeds of cereals.

Glycerolipids: Glycerolipids are composed of mono-, di-, and tri-substituted glycerols, the best-known being the fatty acid triesters of glycerol, called triglycerides.

Glycine: Glycine is a non-essential, neutral, genetically coded amino acid. It is the only protein-forming amino acid without a center of chirality. Glycine is an organic compound with the formula NH_2CH_2COOH.

Glycolipid: Compound containing one or more monosaccharide residues bound by a glycosidic linkage to a hydrophobic moiety such as an acylglycerol, a sphingoid, a ceramide, or a prenyl phosphate.

Glycolysis: Glycolysis is the metabolic pathway that converts glucose $C_6H_{12}O_6$, into pyruvate, $CH_3COCOO^- + H^+$.

Glycosidase: Glycoside hydrolases (also called glycosidases or glycosyl hydrolases) catalyze the hydrolysis of the glycosidic linkage to release smaller sugars

Glycosome: A membrane-enclosed organelle that contains the glycolytic enzymes.

Glyosome: Glyoxysomes are specialized peroxisomes found in plants.

Glyoxylate cycle: The glyoxylate cycle, a variation of the tricarboxylic acid cycle, is an anabolic pathway occurring in plants.

Golgi bodies: Golgi bodies are stacks of flattened membranous stacks. Responsible for manufacturing, storing and shipping certain cellular products.

Golgi vesicles: Within the golgi body the endoplasmic reticulum is smooth, and it pinches together, at its terminus to produce membrane-bounded vesicles. These vesicles contain whatever enzymes or other substances have been transported through the endoplasmic reticulum to the golgi body.

GPP: Geranyl Pyrophosphate (GPP)

Granum: Stack of thylakoids is called as granum

Gravitropism: The bending of a plant organ in response to gravity; also called geotropism.

Ground meristem: The central primary meristem, produces cells which will become ground tissue.

Growth regulators: Growth regulators are used to modify a crop by changing the rate or pattern, or both, of its response to the internal and external factors that govern development from germination through vegetative growth, reproductive development, maturity, and senescence or aging, as well as postharvest preservation.

GTP: Guanosine triphosphate.

Guanine: Guanine is one of the four main nucleobases found in the nucleic acids DNA and RNA, the others being adenine, cytosine, and thymine (uracil in RNA).

H

Hard seed coat: These are impermeable to water, gases so restrict water uptake and oxygen exchange.

Heptone: Any one of several isometric hydrocarbons, C_7H_{16}, of the paraffin series (nine are possible, four are known); -- so called because the molecule has seven carbon atoms.

Herbivore: A herbivore is an organism anatomically and physiologically adapted to plant material

Heterosis: Heterosis, hybrid vigor, or outbreeding enhancement, is the improved or increased function of any biological quality in a hybrid offspring.

Heterozygote: A nucleus, cell or organism possessing two different alleles for a particular gene.

Hexokinase: A hexokinase is an enzyme that phosphorylates hexoses (six-carbon sugars), forming hexose phosphate.

Hexose: A hexose is a monosaccharide with six carbon atoms, having the chemical formula $C_6H_{12}O_6$.

Hilum: The scar on the ovule made where the funiculus detaches at maturity is known as the hilum.

Histidine: Histidine (abbreviated as His or H) is an α-amino acid with an imidazole functional group.

Histone: A histone is a protein that provides structural support to a chromosome. In order for very long DNA molecules to fit into the cell nucleus, they wrap around complexes of histone proteins, giving the chromosome a more compact shape. Some variants of histones are associated with the regulation of gene expression.

HMBPP: Hydroxy-3-methyl-but-2-enyl pyrophosphate (HMB-PP)

hnRNA: Heterogeneous nuclear RNA

Hormones: Hormones initiates the reaction (initiation of cell expansion and multiplication by auxin and cytokinin.

Hybridation: The process of combining different varieties of organisms to create a hybrid.

Hydrocarbon: A molecule composed only of hydrogen and carbon.

Hydrocyanic acid: A highly toxic liquid that has the odor of bitter almonds and boils at 25.6°C; used to manufacture cyanide salts, acrylonitrile, and dyes, and as a fumigant in agriculture. Also known as formonitrile; hydrogen cyanide; prussic acid.

Hydrogen peroxide: Hydrogen peroxide is a highly reactive chemical containing the elements hydrogen and oxygen.

Hydrogenosome: A membrane-enclosed organelle of some anaerobic ciliates, trichomonads and fungi.

Hydrolzable tannins: A hydrolyzable tannin or pyrogallol-type tannin is a type of tannin that, on heating with hydrochloric or sulfuric acids, yields gallic or ellagic acids.

Hydroperoxyeicosatetraenoic acids (HpETE): Hydroperoxides produced from arachidonic (eicosatetraenoic) acid. Six isomers are known with the hydroperoxy group at C-5, 8, 9, 11, 12 or 15 and appropriate tetraene systems containing one conjugated diene unit, e.g. 5-OOH 6, 8,11,14-eicosatetraenoic acid.

Hydroxy ketone: A hydroxy ketone (often referred to simply as a ketol) is a functional group consisting of a ketone flanked by a hydroxyl group.

Hydroxyl radical: An oxygen and hydrogen atom occurring as a group (OH-).

Hydroxylation: Hydroxylation is a chemical process that introduces a hydroxyl group (-OH) into an organic compound.

Hyoscamine: Hyoscyamine is a tropane alkaloid. It is a secondary metabolite found in certain plants of the solanaceae family, including henbane, mandrake, jimsonweed, tomato and deadly nightshade.

Hypogeal: When the epicotyls, elongates quickly leaving the cotyledons on the ground, the germination is said to be hypogeal.

I

Imbibition: Imbibition is the phenomenon of adsorption of water by the solid particles of a substance without forming a solution.

Immature embryo: Seed embryo, which has not attained a development stage to make it capable of germination.

Impermeable: Barrier restricting passage of motile molecules, *e.g.* a seed or fruit coat obstructing water passage and hence imbibition.

Incomplete dominance: The relationship between two genes, neither of which fully masks the expression of the other.

Indole-3-acetic acid: A substance that acts as a growth hormone or auxin in plants, where it controls cell enlargement and, through interaction with other plant hormones, also influences cytokinesis.

Indole-3-butyric acid: A synthetic plant hormone promoting elongation of stems and roots.

Induced mutations: Mutations can be induced in a variety of ways, such as by exposure to ultraviolet or ionizing radiation or chemical mutagens.

Inexorable process: A process which cannot be prevented from continuing or progressing.

Influx: Process of flowing in

Inhibition: Here, restraint or repression of a seed function, *e.g.* inhibitory substances in a fruit or seed restraining or repressing seed germination.

Inhibitor: Chemical compound which inhibits germination, *e.g.* coumarin and abscisic acid (ABA).

Initiation: The initiation step consists of the formation of a free radical. A free radical (R) is a molecular fragment having one or more unpaired electrons accessible to easily react with other hydrocarbons (RH). Free radicals are usually short-lived and highly reactive.

Innate Dormancy: When it is due to internal conditions of the plant or its organ.

Inositol: Inositol is a carbohydrate, though not a classical sugar. It has a taste which has been assayed at half the sweetness of table sugar.

Integuments: The protective layer around an ovule that becomes the seed coat.

Intercalary meristem: A meristem located between non-dividing tissues, such as near the base of a blade of grass.

Interfascicular cambium: Vascular cambium that develops between the sites of vascular bundles and in the ground tissue.

Interfascicular region: Sites of ground tissue located between the vascular bundles in a stem.

Intermediate (Seed): In relation to storability, seeds that can be dried to a moisture content as that of orthodox seeds, but are sensitive to low temperatures typically employed for orthodox seeds.

Intermembrane space: The intermembrane space (IMS) is the region between the inner membrane and the outer membrane of a mitochondrion or a chloroplast.

Internode: Regions of a stem between nodes.

Interphase: The period between cell division. It is a metabolically very active stage as DNA synthesis occurs during this phase.

Interxylarly: Within, or enclosed by, the xylem tissue.

Invertase: Invertase is an enzyme that catalyzes the hydrolysis of sucrose.

Ionizing radiation: Ionizing (or ionising) radiation is radiation composed of particles that individually carry enough kinetic energy to liberate an electron from an atom or molecule, ionizing it.

IPA (Indole-3-pyruvic acid) pathway: Tryptophan is deaminated to form indole-3-pyruvic acid (IPA) followed by decarboxylating of the latter resulting in the formation of indole-3-acetaldehyde (IA1d).

IPP: Isopentenyl Pyrophosphate (IPP)

Irradiation: Exposure of a biological material to radiation (x-rays, gamma rays etc.,) is known as irradiation.

Irreversible process: If a system is taken from one state to another but cannot be brought back to the same initial state, then the process is called irreversible.

Isocitrate lyase: Isocitrate lyase, or ICL, is an enzyme in the glyoxylate cycle that catalyzes the cleavage of isocitrate to succinate and glyoxylate.

Isocitrate: Isocitrate is formed from citrate with the help of the enzyme aconitase, and is acted upon by isocitrate dehydrogenase. Salts and esters of isocitric acid are known as isocitrates.

Isoleucine: Isoleucine (abbreviated as Ile or I) is an α-amino acid with the chemical formula $HO_2CCH(NH_2)CH(CH_3)CH_2CH_3$.

Isomerisation: The conversion of a compound into an isomer of itself.

Isoprene: Isoprene, or 2-methyl-1, 3-butadiene, is a common organic compound with the formula $CH_2{=}C(CH_3)CH{=}CH_2$. It is a colorless volatile liquid. Isoprene is produced by many plants.

J

Jasmonate: Jasmonate (JA) and its derivatives are lipid-based hormone signals that regulate a wide range of processes in plants, ranging from growth and photosynthesis to reproductive development.

K

KAO: *ent*-kaurenoic acid oxidase

Kinase: A kinase is a type of enzyme that transfers phosphate groups from high-energy donor molecules, such as ATP, to specific substrates, a process referred to as phosphorylation.

KO: *ent*-kaurene oxidase

L

LAM: Low amylase

Lateral meristem: A region of actively-dividing cells located along the length of a root or stem; growth results in an increase in girth.

Legumin: A protein obtained mainly from the seeds of leguminous plants.

Leptotene: A substage of prophase I of meiosis I during which the thread like chromosomes appear in the nucleus.

Leucine: Leucine is a branched-chain α-amino acid with the chemical formula $HO_2CCH(NH_2)CH_2CH(CH_3)_2$. Leucine is classified as a hydrophobic amino acid due to its aliphatic isobutyl side chain.

Leucoplasts: Colorless organelles which store starch or other plant nutrients. (example - starch stored in a potato)

Light reaction: The first stage of photosynthesis, occurring only in the presence of light, during which energy captured from light drives the production of ATP.

Light: Some species required light for seed germination. Both light intensity (lux) and light quality (colour and wavelength) influence seed germination.

Lignin: Lignin or lignen is a complex chemical compound most commonly derived from wood, and an integral part of the secondary cell walls of plants and some algae.

Lipase: Enzyme that catalyzes the cleavage of fatty acids from the glycerol moiety of a triglyceride.

Lipid biosynthesis: Protein involved in the synthesis of lipids, a diverse class of compounds which are insoluble in water, but soluble in organic solvents.

Lipid metabolism: Lipid metabolism refers to the processes that involve the intercourse and degradation of lipids.

Lipid peroxidation: Lipid peroxidation refers to the oxidative degradation of lipids. It is the process whereby free radicals "steal" electrons from the lipids in cell membranes, resulting in cell damage. This process proceeds by a free radical chain reaction mechanism. The oxidation of polyunsaturated fatty acid side chains, initiated by a free radical such as the hydroxyl radical (OH•), to form a relatively stable carbon free radical, which reacts further with molecular oxygen to form a peroxy radical, and further yet with another fatty acid side chain to generate another carbon free radical.

Lipid peroxides: Lipid peroxides are produced in the presence of a free radical by the oxidation of unsaturated fatty acids. This can lead to a damage of the tissue by changes of the cell membranes.

Lipid: An essential structural component of living cells, lipids are a class of oily organic compounds which are insoluble in water but soluble in fats and oils. The lipid class of molecules mainly consists of fats, oils and waxes.

Lipolysis: Lipolysis is the breakdown of lipids and involves hydrolysis of triglycerides into free fatty acids, followed by degradation into acetyl units by beta oxidation.

Lipoxygenase: An enzyme of the oxidoreductase class that catalyzes reactions between linoleate and other fatty acids and oxygen to form hydroperoxy-fatty acid derivatives.

LKR: Lysine ketoglutarate reductase

Lysine: Lysine is an α-amino acid with the chemical formula $HO_2CCH(NH_2)(CH_2)_4NH_2$.

Lysosomes: A small sac-like structures surrounded by a single membrane and containing strong digestive enzymes which when released can break down worn out organelles or food. The lysosome is also known as a suicide sac.

M

M Phase: Period of the eucaryotic cell cycle during which the chromosomes are condensed and the nucleus and cytoplasm divide.

Macromolecule: A macromolecule is a very large molecule commonly created by polymerization of smaller subunits (monomers). They are typically composed of thousands or more atoms. The most common macromolecules in biochemistry are biopolymers (nucleic acids, proteins, carbohydrates and polyphenols) and large non-polymeric molecules (such as lipids and macrocycles).

MAG: Monoacylglycerol

Malate: The salts and esters of malic acid are known as malates. The malate anion is an intermediate in the citric acid cycle.

Malondialdehyde (MDA): Malondialdehyde is a product of polyunsaturated fatty acid peroxidation and used as marker of oxidative stress.

Malonic acid pathway: The malonic acid pathway, although an important source of phenolic secondary products in fungi and bacteria, is of less significance in higher plants.

Maltase: Maltase catalyzes the breakdown of maltose to glucose it is apparently produced during the early stages of seed development under stimulation by GA.

Matrix (of mitochondria): The mitochondrion consists of an outer membrane, an inner membrane and a gel-like material called the matrix. The matrix contains soluble enzymes that catalyze the oxidation of pyruvate and other small organic molecules. The mitochondrial matrix also contains the mitochondria's DNA and ribosomes. The word "matrix" stems from the fact that this space is viscous, compared to the relatively aqueous cytoplasm.

Maturation: Maturation is the process of becoming mature; the emergence of personal and behavioral characteristics through growth processes.

Mechanical Dormancy: Mechanical dormancy is said to be caused by seed covering tissues mechanically restricting embryo expansion and development.

Megagametogenesis: The development of the female gametophyte (embryo sac) from the megaspore mother cell is called as megagametogenesis,

Megagametophyte: The haploid megaspore inside the nucellus gives rise to the female gametophyte, called the megagametophyte.

Megaspore: The formation of megaspores inside the ovules of seed plants. A diploid cell in the ovule, called a megasporophyte or a megaspore mother cell.

Megasporogenesis: The formation of megaspores inside the ovules of seed plants.

Megasporophyte: A diploid cell in the ovule, called a megasporophyte.

Meiosis: Meiosis is the form of eukaryotic cell division that produces haploid sex cells or gametes (which contain a single copy of each chromosome) from diploid cells (which contain two copies of each chromosome).

Membrane lipids: Membrane lipids are a group of compounds (structurally similar to fats and oils) which form the double-layered surface of all cells.

Membrane: A membrane is a thin, film-like structure that separates two fluids.

Meristem: Tissue composed of undifferentiated, embryonnary cells, able to divide into new specialized tissues. These cells will be responsible for growth in length of stems and roots (primary meristem) or for the increase in thickness of the trunk, stems and roots of many plants (secondary meristem).

Mesophyll: The photosynthetic tissue of a leaf, located between the upper and lower epidermis.

Messenger RNA (mRNA): A transcript of a protein-encoding gene.

Metabolism: Metabolism is the set of chemical reactions that happen in living organisms to maintain life. The totality of an organism's chemical processes, consisting of catabolic and anabolic pathways.

Metabolite: Metabolites are the intermediates and products of metabolism.

Metaphase: Stage of mitosis at which chromosomes are firmly attached to the mitotic spindle at its equator but have not yet segregated toward opposite poles.

Methionine: Methionine is an α-amino acid with the chemical formula $HO_2CCH(NH_2)CH_2CH_2SCH_3$. This essential amino acid is classified as nonpolar.

Methyl erythritol Phosphate Pathway (MEP): The methyl erythritol phosphate pathway provides a route to isoprenoid biosynthesis in the plastids of plants, in protozoa, and in prokaryotes.

Methyl methane sulphonate: Methyl methane sulphonate (MMS) is an alkylating agent and a carcinogen. It is also a suspected reproductive toxicant, and may also be a skin/sense organ toxicant.

Methylene group: A methylene group is any part of a molecule that consists of two hydrogen atoms bound to a carbon atom, which is connected to the remainder of the molecule by a double bond. (-CH2-) The group may be written 2, where the '=' denotes the double bond. This structural element is also called methylidene.

Mevalonate pathway (MVA): The mevalonate pathway or HMG-CoA reductase pathway or mevalonate-dependent (MAD) route or isoprenoid pathway, is an important cellular metabolic pathway present in all higher eukaryotes and many bacteria.

Microfilaments: Part of the cytoskeleton and are made of solid rods of globular proteins.

Micropyle: A micropyle is a minute opening in the ovule wall (integument) through which the pollen tube enter during fertilization.

Microtubules: Hollow rods that function primarily to help support and shape the cell.

Mid dormancy: From pre-dormancy, a plant organ may pass into full dormancy. At this stage, it is difficult to induce the resumption of growth by making alteration in the environment.

Miotic phase (M): When the cell is dividing. The miotic phase of the "cell cycle" consists of two stages namely mitosis and cytokinensis.

miRNA: MicroRNAs (miRNAs) are small RNAs that typically are partially complementary to sequences in metazoan messenger RNAs.

Mitochondria (mitochondrion, singular (Syn:chondriosome): Any of various round or long cellular organelles of most eukaryotes that are found outside the nucleus, produce energy (adenosine triphosphate- ATP). for the cell through cellular respiration, and are rich in fats, proteins, and enzymes -- called also chondriosome. Mitochondria contain their own small chromosomes.

Mitochondrial crest: Any of the infoldings of the mitochondrial inner membrane that extend into the matrix.

Mitochondrial DNA (mtDNA): Mitochondrial DNA is found outside the nucleus in a cell and is maternally inherited and therefore not unique to an individual. Because there are thousands of copies of mtDNA per cell and it is less prone to degradation than nuclear DNA. Since it is normally inherited only from mothers, an offspring's mtDNA will be identical to its mother's, unless a mutation has occurred.

Mitosis: Mitosis is a form of eukaryotic cell division that produces two daughter cells with the same genetic component as the parent cell. Chromosomes replicated during the S phase are divided in such a way as to ensure that each daughter cell receives a copy of every chromosome.

Mitotic spindle: The centrosomes organise the production of microtubules that form the spindle fibres that constitute the mitotic spindle.

Moisture: Moisture refers to the presence of a liquid, especially water, often in trace amounts.

Monoepoxide: Any compound containing a single epoxide group, especially when more than one is possible

Monomer: A monomer is a molecule that may bind chemically to other molecules to form a polymer.

Monomeric unit: Unit of polymer molecule containing the same kinds and numbers of atoms as one of the monomers.

Monooxygenases: Monooxygenases are enzymes that incorporate one hydroxyl group into substrates in many metabolic pathways.

Monosaccharide: Monosaccharides are carbohydrates and the simplest form of sugars; can not be decomposed by hydrolysis. A monosaccharide consists of a single sugar molecule. Glucose is an example of a monosaccharide.

Monoterpene: Monoterpenes are a class of terpenes that consist of two isoprene units and have the molecular formula $C_{10}H_{16}$. Monoterpenes may be linear (acyclic) or contain rings.

Monounsaturated fatty acid: A fatty acid missing one pair of hydrogens so that it has one

Monounsaturated: A fatty acid chain is monounsaturated, if it contains one double bond.

Morphogenesis: Morphogenesis is the biological process that causes an organism to develop its shape.

Morphological dormancy: Seed with morphological dormancy have embryos that are not fully developed at the time of dispersal. In other words, the embryo needs time to grow inside the seed before it can germinate.

mRNA transcription: Transcription is the process of making an RNA copy of a gene sequence. This copy, called a messenger RNA (mRNA) molecule, leaves the cell nucleus and enters the cytoplasm, where it directs the synthesis of the protein, which it encodes.

mRNA: Messenger RNA (mRNA) is a large family of RNA molecules that convey genetic information from DNA to the ribosome, where they specify the amino acid sequence of the protein products of gene expression.

Mutagen: A mutagen is a physical or chemical agent that changes the genetic material, usually DNA, of an organism and thus, increases the frequency of mutations above the natural background level.

Mutagenesis: Mutagenesis is a process by which the genetic information of an organism is changed in a stable manner, resulting in a mutation. It may occur spontaneously in nature, or as a result of exposure to mutagens.

Mutant: A mutant is an individual, organism, or new genetic character, arising or resulting from an instance of mutation, which is a base-pair sequence change within the DNA of a gene or chromosome of an organism resulting in the creation of a new character or trait not found in the wild type.

Mutation: A mutation is a permanent change in the DNA sequence of a gene. Mutations in a gene's DNA sequence can alter the amino acid sequence of the protein encoded by the gene.

N

NADH: Nicotinamide adenine dinucleotide

NADP: Nicotinamide adenine dinucleotide phosphate, abbreviated NADP+, is a coenzyme used in anabolic reactions, such as lipid and nucleic acid synthesis, which require NADPH as a reducing agent.

NADPH: NADPH is reduced NADP and is used as a reducing agent. NADPH oxidises to form NADP.

Naphthaleneacetic acid: 1-Naphthaleneacetic acid (NAA) is an organic compound with the formula $C_{10}H_7CH_2CO_2H$. This colourless solid is soluble in organic solvents.

NL: Neutral lipid

Non deep: Associated with afterripening a physiological conditions associated with a loss of dormancy during dry storage.

Non-dormant Seed: A completely non-dormant seed has the capacity to germinate over the widest range of normal physical environmental factors possible for the genotype.

Nonene: Nonene is an alkene with the molecular formula C_9H_{18}.

Non-ionizing radiation: Non-ionizing radiation refers to any type of electromagnetic radiation that does not carry enough energy per quantum to ionize atoms or molecules that is, to completely remove an electron from an atom or molecule.

N-terminus: The N-terminus (also known as the amino-terminus, NH2-terminus, N-terminal end or amine-terminus) refers to the start of a protein or polypeptide terminated by an amino acid with a free amine group ($-NH_2$).

Nucellus: The central portion of an ovule in which the embryo sac develops; the megasporangium.

Nuclear DNA: Nuclear DNA is DNA that is found inside the nucleus of cells, as opposed to mitochondrial DNA [see mitochondrial DNA entry]. Nuclear DNA stores the information necessary for passing genetic attributes to from parents to offspring. Unlike the mtDNA, which is only maternal, nDNA is comprised of genetic coding contributed by both the male and female parents.

Nuclear membrane: A double lipid bilayer membrane which surrounds the genetic material and nucleolus in eukaryotic cells. Also known as the nuclear envelope, nucleolemma.

Nucleic acid: Nucleic acids are complex structures used to maintain the genetic information and found in the nuclei of cell.

Nucleolus: Structure within the nucleus that helps in the synthesis of ribosomes

Nucleopore: Tiny hole within the nuclear membrane that allows nucleic acids and proteins to move into and out of the nucleus.

Nucleoside: The molecule without the phosphate group is called a nucleoside.

Nucleotide: Nucleotides are biological molecules that form the building blocks of nucleic acids (DNA and RNA) and serve to carry packets of energy within the cell (ATP).

Nucleus: Membrane bound structure that contains the cell's hereditary information. The "control center" of the cell, contains the cell's DNA (chromosomes)

O

Octane: Family of alkane molecules, formula C_8H_{18}.

Okazaki fragments: Okazaki fragments are short, newly synthesized DNA fragments that are formed on the lagging template strand during DNA replication.

OL: Open lactone

Oleosin: Oleosins are structural proteins found in vascular plant oil bodies and found in plant cells.

Oligopeptide: A peptide containing a relatively small number of amino acids

Oligosaccharides: An oligosaccharide is a saccharide polymer containing a small number (typically three to ten) of component sugars, also known as simple sugars (monosaccharides).

Organelle: A distinct compartment within a cell, created by the organization of cellular membranes (examples: mitochondrion, chloroplast, endoplasmic reticulum, Golgi apparatus, flagellum).

Organogenesis: Organogenesis entails the regulation of cell division, cell expansion, cell and tissue type differentiation, and patterning of the organ as a whole.

Osmosis: Osmosis is the net movement of water across a selectively permeable membrane driven by a difference in solute concentrations on the two sides of the membrane.

Ovary: Enlarged base of the carpel containing the ovule or ovules. The ovary matures to become a fruit.

Ovule: Located in the ovaries. Carries female gametes. Ovules become seeds on fertilization.

Oxaloacetate: Oxaloacetate is a four-carbon molecule found in the mitochondrion that condenses with acetyl CoA to form citrate in the first reaction of the Krebs cycle.

Oxidation: 1. The combination of oxygen with a molecule with liberation of heat; combustion. 2. Removal of hydrogen from a molecule. 3. Loss of electrons, increasing the positive valence. Oxidation is associated with a loss of energy.

Oxidative damage: Damage to cells caused by reactive oxygen species (ROS); also known as oxidative stress.

Oxidative degradation: The cleavage of a carbon skeleton (often at a C==C double bond) with the introduction of new carbon–oxygen bonds (10.4)

Oxidative phosphorylation: Oxidative phosphorylation is the metabolic pathway in which the mitochondria in cells use their structure, enzymes, and energy released by the oxidation of nutrients to move hydrogen from ADP to phosphate to reform ATP.

Oxygen: Atmospheric air is composed of 79.9 % nitrogen, 20% oxygen and 0.03 % carbon dioxide. Oxygen is required for germination of most of species. If CO_2 concentration is higher than 0.03 %, it retards germination. Respiration increases sharply during seed germination. Since respiration is essentially an oxidative process, an adequate supply of oxygen is a must.

P

Pachytene: Crossing over between pairs of homologous chromosomes to form chiasmata (sing. chiasma).

PAL: Phenylalanine ammonia lyase (PAL).

Panicle: A panicle is a compound raceme or branched cluster of flowers.

PAP: Polyadenylate polymerase

Paradormancy: Dormancy caused by unsuitable environment

PCS: Phytochelatin synthase

Pectin: Pectin is a structural heteropolysaccharide contained in the primary cell walls of terrestrial plants.

Pedicel: A pedicel is a stem that attaches single flowers to the main stem of the inflorescence.

Peduncle: Flower stalk.

Pentose phosphate pathway: The pentose phosphate pathway (also called the phosphogluconate pathway and the hexose monophosphate shunt) is a process that generates NADPH and pentoses (5-carbon sugars).

Pentose: A pentose is a monosaccharide with five carbon atoms.

Peptide bond: A peptide bond is a covalent chemical bond formed between two molecules when the carboxyl group of one molecule reacts with the amino group of the other molecule, causing the release of a molecule of water (H_2O), hence the process is a dehydration synthesis reaction and usually occurs between amino acids.

Pericarp: The ripened and variously modified walls of a plant ovary.

Periderm: The cork cambium and the cork are collectively known as the periderm.

Perisperm: The nutritive tissue outside the sac containing the embryo in some seeds.

Permeable: Allowing passage of motile molecules, *e.g.* a gas or liquid. A seed coat is permeable when water can be absorbed for the seed to imbibe. A barrier may be selective or semipermeable if it allows only passage of small molecules, *e.g.* a polyethylene sheet which is permeable to gas, but impermeable to water.

Peroxidase: A haem enzyme that abstracts two hydrogens from one substrate to reduce hydrogen peroxide, its other substrate to water.

Peroxide scavenging enzyme: Damage to the cell only occurs with an elevated micromolar level of H_2O_2 within the cell, which is prevented by two major scavenging enzymes, alkyl hydroperoxide reductase and catalase. Therefore to increase the level of H_2O_2 in the cell the scavenging enzymes need to be nonfunctional. The scavenging enzymes degrade hydrogen peroxide and alleviate stress on cells.

Peroxisomes: Tiny structures bound by a single membrane that contain enzymes that produce hydrogen peroxide as a by-product. These structures are involved in plant processes such as photorespiration.

Peroxyl radical: Any univalent radical, R-O-O-, derived from a peroxide.

Petal: Located in and above the sepals, often large and colourful, sometimes scented, sometimes producing nectar. Often serve to attract pollinators to the plant.

PGAL: PGAL is a chemical compound that occurs as an intermediate in several central metabolic pathways of all organisms.

Phenylacetic acid: Phenylacetic acid is an organic compound containing a phenyl functional group and a carboxylic acid functional group. It is a white solid with a disagreeable odour.

Phenylalanine: Phenylalanine is an α-amino acid with the formula $C_6H_5CH_2CHCOOH$. This essential amino acid is classified as nonpolar because of the hydrophobic nature of the benzyl side chain.

Phloem loading: Photoassimilates moves from (Source) mesophyll to phloem

Phloem: Relatively soft-walled cells transport organic nutrients in a kind of tissue.

Phosphatidic acid: Phosphatidic acids (PAs) are the acid forms of phosphatidates, a part of common phospholipids, major constituents of cell membranes.

Phosphatidylcholine: Phosphatidylcholines (PC) are a class of phospholipids that incorporate choline as a headgroup.

Phosphatidylethanolamine (PE): Phosphatidylethanolamines are a class of phospholipids found in biological membranes.

Phosphodiester bond: A phosphodiester bond is a group of strong covalent bonds between a phosphate group and two 5-carbon ring carbohydrates (pentoses) over two ester bonds.

Phosphoenolpyruvate (PEP): Phosphoenolpyruvate as the anion, is an important chemical compound in biochemistry. It is involved in the biosynthesis of various aromatic compounds, and in carbon fixation.

Phosphoenolpyruvate carboxykinase: Phosphoenolpyruvate carboxykinase (PEPCK) is an enzyme in the lyase family used in the metabolic pathway of gluconeogenesis. It converts oxaloacetate into phosphoenolpyruvate and carbon dioxide.

Phosphoenolpyruvic acid: Phosphoenolpyruvic acid (PEP), or phosphoenolpyruvate as the anion, is an important chemical compound in biochemistry.

Phosphofructokinase: Phosphofructokinase is a kinase enzyme that phosphorylates fructose 6-phosphate in glycolysis.

Phosphoglucoisomerase: An enzyme that catalyzes the conversion of glucose-6-phosphate into fructose 6-phosphate in the second step of glycolysis.

Phosphoglyceric acid (PGA): Either of two isomeric acid phosphates C3H5O3 (OPO3H2) of glyceric acid that are formed as intermediates in photosynthesis and in carbohydrate metabolism

Phospholipids: Phosphate-containing lipids that are a major component of all cell membranes. Phospholipids are a heterogeneous type of molecule composed of glycerol, phosphate, two fatty acid residues, and 'headgroups' with different chemical properties. The phospholipid bilayer is organized with the fatty acid residues facing the center of the membrane (Hydrophobic or water insoluble) and the headgroups forming the surface of the membrane. As such, bilayers separate water filled compartments and provide an electrically insulating barrier between these two compartments.

Photo – dormancy: Most seeds having photo dormancy germinate only under light conditions. Such seeds are therefore, also called light sensitive.

Photoassimilate: Number of biological compounds is formed by photo assimilation.

Photo-oxidation: This involves an ene reaction between an olefinic centre and singlet oxygen. This reactive form of oxygen is produced by light and a sensitiser such as chlorophyll, erythrosine, rose bengal, methylene blue, etc. The product is a hydroperoxide and reaction occurs with stereomutation and double bond migration. Unlike autoxidation there is no induction period and the reaction is unaffected by antioxidants. It can be inhibited by singlet oxygen quenchers such as carotene. The reaction is quicker than autoxidation.

Photooxygenation: The incorporation of molecular oxygen into a molecular entity. There are three common mechanisms: Type I: the reaction of triplet molecular oxygen with radicals formed photochemically. Type II: the reaction of photochemically produced singlet molecular oxygen with molecular entities to give rise to oxygen containing molecular entities. The third mechanism proceeds by electron transfer producing superoxide anion as the reactive species.

Photophosphorylation: It is the process of converting energy from a light-excited electron into the pyrophosphate bond of an ADP molecule.

Photosynthesis: Photosynthesis is a process used by plants and other organisms to convert the light energy captured from the sun into chemical energy that can be used to fuel the organism's activities.

Photosynthetic organism: Plants and algae that use 'photons' from the sun and chlorophyll (the green stuff in plants) to 'synthesize' or to make their food.

Photosystem: The light-dependent portion of photosynthesis is carried out by two consecutive photosystems (photosystem I and photosystem II) in the thylakoid membrane of the chloroplasts.

Phototropism: Phototropism is a growth movement induced by a light stimulus. Growth towards a source of light is called positive phototropism, while that grows away from the light source is termed negative phototropism. The tips of shoots are usually positively while those of roots are negatively phototropic.

Physiological dormancy: Physiological dormancy (PD) is the most common expression of seed dormancy, and is thought to be caused by a physiological inhibiting mechanism (PIM) of the seed embryo.

Physiomorphic dormancy: Dormancy due to under developed embryo.

Phytin: Phytin is a nutrient supplement, which contains a calcium-magnesium salt of phytic acid.

Phytochrome: Phytochrome is a photoreceptor, a pigment that plants use to detect light.

Phytoene: Phytoene is a 40-carbon intermediate in the biosynthesis of carotenoids.

Phytotoxic: Poisonous to plants

Pi: Inorganic phosphate

Pith: Center of dicot plant stems. In some plants the pith breaks down forming a hollow stem. In older woody plants, the pith is filled with rigid xylem wood fiber.

Plant cells: Eukaryotic cells that differ in several key aspects from the cells of other eukaryotic organisms.

Plant growth regulators: Plant growth regulators are usually defined as organic compounds, other than nutrients, that affect the physiological processes of growth and development in plants when applied in low concentrations.

Plant hormones: The term plant hormones when correctly used is restricted to naturally occurring plant substances.

Plasmodesma: Cytoplasmic channels lined with plasma membrane that connect the protoplasts of adjacent cells across the cell wall.

Plasmodesmata: Plasmodesmata are microscopic channels which traverse the cell walls of plant cells and some algal cells, enabling transport and communication between them.

Plastids: Double membrane bound organelles. It is in plastids that plants make and store food.

Pleiotropy: Pleiotropy occurs when one gene influences multiple phenotypic traits.

Plumule: The area above the cotyledons is thus a miniature shoot and is called the plumule. Primary bud of a plant embryo situated at the apex of the hypocotyl; portion of the seedling axis above the cotyledons, consisting of leaves and an epicotyl, which elongates to form the primary stem.

Point mutation: A point mutation, or single base substitution, is a type of mutation that causes the replacement of a single base nucleotide with another nucleotide of the genetic material, DNA or RNA. The term point mutation also includes insertions or deletions of a single base pair.

Pollen: Grains containing the male gametes. Immature male gametophyte with a protective outer covering.

Polyadenylation: Polyadenylation is the addition of a poly (a) tail to an RNA molecule.

Polyamine: A polyamine is an organic compound having two or more primary amino groups

Polycarpic: Relating to a plant that produces fruit more than once during its lifetime.

Polymerase: A polymerase is an enzyme whose central biological function is the synthesis of polymers of nucleic acids.

Polypeptide chain: A chain of amino acids joined together through peptide bonds.

Polypeptides: Polypeptides are chains of amino acids. Proteins are made up of one or more polypeptide molecules. The amino acids are linked covalently by peptide bonds.

Polyphenols: The most common of the phenolic compounds found in plants - their skins, seeds, and flesh. Polyphenols provide the colour and flavour, as well as acting as antioxidants and neutralizing free radicals. Polyphenols include: flavonoids, lignans, gallic acid, and ellagic acid.

Polyribosome: A cluster of ribosomes linked together by a molecule of messenger RNA and forming the site of protein synthesis.

Polysaccharides: Polysaccharides are compounds consisting of a large number of simple sugars (monosaccharides) linked together by glycosidic bonds. When polysaccharides are composed of a single monosaccharide building block, they are termed homopolysaccharides. Heteropolysaccharides contain two or more different types of monosaccharide. Polysaccharides may have molecular weights of up to several million and are often highly branched. Since they have only the one free anomeric -OH group at the end of a very long chain, polysaccharides aren't reducing sugars and don't show noticeable mutarotation. The most common polysaccharides are cellulose, starch, and glycogen.

Polysome: More than one ribosome can translate an mRNA at one time, making it possible to produce many polypeptides simultaneously from a single mRNA.

Polyunsaturated fatty acid: A fatty acid with two or more double bonds. Common polyunsaturated fatty acids are linoleic, alpha-linolenic, arachidonic, eicosapentaenoic (EPA), and docosahexaenoic (DHA) acids.double bond. A common example is oleic acid.

Porphyrins: Any of various compounds with a structure that consists essentially of four pyrrole rings joined by four =CH- groups ; especially : one (as chlorophyll or hemoglobin) containing a central metal atom and usually having biological activity.

Post dormancy: In this sub phase, a plant organ gradually emerges out from the phase of dormancy. At this stage, it is easy to induce growth respiration.

Precursor: A chemical substance that gives rise to another more important substance

Pre-dormancy: It is the initial sub phase of dormancy in which the organs may be induced to resume the normal course of growth by different treatments.

Primary dormancy: Dormancy, which develops when the seed is attached to the plant and exists when first harvested.

Primary metabolite: A primary metabolite is a kind of metabolite that is directly involved in normal growth, development, and reproduction.

Primer: A primer is a strand of nucleic acid that serves as a starting point for DNA synthesis.

Procambium: The innermost primary meristem - produces cells which will become vascular tissue

Programmed cell death (Apoptosis): Normal benign process of cell suicide, in which the cell shrinks, dissolves its contents, and activates phagocytosis by neighboring cells.

Prokaryotes: Prokaryotic cells are usually independent, while eukaryotic cells are often found in multicellular organisms.

Prolamins: Prolamins are a group of plant storage proteins having a high proline content and found in the all seeds of cereal grains except oats and rice.

Proline: Proline is an α-amino acid, one of the twenty DNA-encoded amino acids.

Prometaphase I: Spindle apparatus formed, and chromosomes attached to spindle fibres by kinetochores.

Propagation: After a free radical has formed, it is capable of propagating the oxidation sequence. This propagation step involves the formation of additional alkyl- or peroxy-radicals and hydroperoxides, which will continue this cycle of radical formation.

Prophase: Prophase occupies over half of mitosis. The nuclear membrane breaks down to form a number of small vesicles and the nucleolus disintegrates.

Prostaglandin: Prostaglandins, are like hormones in that they act as chemical messengers, but do not move to other sites, but work right within the cells where they are synthesized.

Protease (proteinase, proteolytic enzyme): Enzyme such as trypsin that degrades proteins by hydrolyzing some of their peptide bonds.

Protein biosynthesis: The process by which amino acids are linearly arranged into proteins through the involvement of ribosomal RNA, transfer RNA, messenger RNA, and various enzymes.

Proteins: Proteins are organic compounds made of amino acids arranged in a linear chain and folded into a globular form.

Protoderm: The outermost primary meristem - produces cells which will become dermal tissue.

Protoplasm: All the living matter of a cell is called protoplasm.

Protoplasmic respiration: If proteins are used as respiratory substrate that is called as protoplasmic respiration.

Protoplasmic streaming: The movement of solute solution is caused by a combination of diffusion and cytoplasmic streaming.

PTGS: Posttranscriptional gene silencing.

Purine: One of the two types of nitrogenous base compounds that are components of nucleotides. It has a double-ring structure.

Pyrimidine: One of the two types of nitrogenous base compounds that are components of nucleotides. It has a single-ring structure.

Pyruvate carboxylase: Pyruvate carboxylase (PC) is an enzyme of the ligase class that catalyzes the (depending on the species) irreversible carboxylation of pyruvate to form oxaloacetate (OAA).

Pyruvate kinase: Pyruvate kinase is an enzyme involved in glycolysis. It catalyzes the transfer of a phosphate group from phosphoenolpyruvate (PEP) to ADP, yielding one molecule of pyruvate and one molecule of ATP.

Pyruvic acid: An important 3-carbon molecule formed from glucose and glycerol in glycolysis.

Q

Quiescence: Condition in which seed can germinate upon absorption of water. No internal germination barriers exist.

R

Raceme: Usually elongate cluster of flowers along the main stem in which the flowers at the base open first.

Radical: Radicals are atoms, molecules, or ions with unpaired electrons or an open shell.

Radicle protrusion: The definition that a visible protrusion of radicle tip is the completion of germination is not only a definition issue of seed physiologists.

Radicle: The embryonic root, *i.e.* the part of the seed embryo that develops into the primary root. In seeds, the radicle is always facing the micropyle.

Raffinose: A trisaccharide found in many plants, consists of melibiose (galactose and glucose) and fructose. Another plant trisaccharide is gentianose.

Rancidity: A fat/oil is described as rancid on the basis of its undesirable odour and flavour resulting from oxidative deterioration (aldehydes and other short-chain compounds) or from hydrolysis (short-chain and medium-chain acids).

Raphe: A ridge formed on the seed coat if the funiculus is fused with the integument in part of its length in anatropous or campylotropous ovules.

rb locus: The rb mutants decrease the activity of the enzyme ADPglucose pyrophosphorylase and increase its sensitivity to allosteric regulation.

RdRPs: RNA-dependent RNA polymerase

Reactive oxygen species (ROS): Reactive oxygen species are ions or very small molecules that include oxygen ions, free radicals, and peroxides, both inorganic and organic. They are highly reactive due to the presence of unpaired valence shell electrons. ROS form as a natural byproduct of the normal metabolism of oxygen and have important roles in cell signaling. However, during times of environmental stress, ROS levels can increase dramatically, which can result in significant damage to cell structures.

Receptacle: Part of flower stalk bearing the floral organs, at base of flower.

Reduction: A reaction involving the gain of electrons, addition of hydrogen atoms, or loss of oxygen atoms. Reduction is associated with a gain of energy.

Release factor: The new polypeptide is released from the ribosome, aided by proteins called release factors.

Replication fork: A Y-shaped region in a chromosome that serves as the growing site for DNA replication.

Replication: The process of duplicating or producing an exact copy of a polynucleotide strand such as DNA.

Resistant seed coat: In plants like *Amaranthus retroflexus,* the seed coat is hard enough not to allow the radicle to pierce through easily.

Respiration: Respiration is a vital process which occurs in all living cells of the plant but the most actively respiring regions are growing regions like floral and vegetative buds, germinating seedlings, stem and root apices.

Respiratory substrate: The substrates which are broken down in respiration for the release of energy may be carbohydrates, fats or proteins.

Reverse mutation: Mutation from mutant to wild type is called a back mutation or reversion.

Rhizodermis: It is the root epidermis, the outermost primary cell layer of the root.

Ribonuclease: Ribonuclease (commonly abbreviated RNase) is a type of nuclease that catalyzes the degradation of RNA into smaller components.

Ribonucleoside: Portion of an RNA molecule composed of one ribose molecule plus either a purine or a pyrimidine.

Ribose: Ribose is an organic compound with the formula $C_5H_{10}O_5$

Ribosome: The ribosome is a large and complex molecular machine, found within all living cells, that serves as the primary site of biological protein synthesis (translation).

RISC: The RNA-induced silencing complex, or RISC, is a multiprotein complex that incorporates one strand of a small interfering RNA (siRNA) or microRNA (miRNA).

RNA polymerase: RNA polymerase (RNAP or RNApol), also known as DNA-dependent RNA polymerase, is an enzyme that produces RNA.

RNA: Ribonucleic acid (RNA) is a ubiquitous family of large biological molecules that perform multiple vital roles in the coding, decoding, regulation, and expression of genes.

RNAi: RNA interference (RNAi) is a biological process in which RNA molecules inhibit gene expression, typically by causing the destruction of specific mRNA molecules.

Root cap: The outermost protective tip of the root. It consists of cells that are sloughed off as the root grows through the soil. The meristem, the area of cell division, is located right behind the root cap and is protected by it.

Rough endoplasmic reticulum: Endoplasmic reticulum lined with ribosomes

rRNA: Ribosomal ribonucleic acid (rRNA) is the RNA component of the ribosome, and is essential for protein synthesis in all living organisms.

RUBP: Ribulose-1, 5-bisphosphate (RuBP) is an organic substance that is involved in photosynthesis.

Rudimentary embryos: In some members of Orobanchaceae, Orchidaceae, the seeds are shed at a stage when the embryo is immature. Hence, the germination of the seed is postponed till the embryo development is complete.

Rugosus (r) locus: Mutations at the r locus, which encodes a starch-branching enzyme (SBEI or SBEA), exhibit considerable pleiotropy.

S

S Phase: Period of a eucaryotic cell cycle in which DNA is synthesized.

Salicylic acid: Activates genes in some plants that produce chemicals that aid in the defense against pathogenic invaders. Salicylic acid is a monohydroxybenzoic acid, colorless crystalline organic acid is widely used in organic synthesis and functions as a plant hormone.

Saponins: Saponins are a class of chemical compounds, one of many secondary metabolites found in natural sources, with saponins found in particular abundance in various plant species.

Sarcotesta: Soft, fleshy outer layer of a testa.

Saturate: Cause (something) to become thoroughly soaked with liquid so that no more can be absorbed.

Saturated fat: A saturated fat is saturated with hydrogen atoms.

Saturated fatty acid: Saturated fatty acid is fat that consists of triglycerides containing only saturated fatty acid radicals.

Saturated fatty acids: Fatty acids without carbon–carbon unsaturation. For example: lauric (12:0), myristic (14:0), palmitic (16:0) and stearic (18:0) acids.

Scutellum: The scutellum is part of the structure of a barley seed the modified seed leaf.

SDH: Saccharopine dehydrogenase

SDS-PAGE: SDS-PAGE, sodium dodecyl sulfate polyacrylamide gel electrophoresis, describes a technique widely used in biochemistry, forensics, genetics, molecular biology and biotechnology to separate proteins according to their electrophoretic mobility and no other physical feature.

Secondary dormancy: A condition that develops to prevent germination after the seed has been detached from the plant and is exposed to specific unfavorable environmental conditions.

Secondary metabolite: Secondary metabolites are organic compounds that are not directly involved in the normal growth, development, or reproduction of an organism.

Seed coat impermeability to gases: The seed coat of plants like Xanthium is permeable to water, but impermeable to oxygen.

Seed coat impermeability to water: The seed coat in many members of Malvaceae, Fabaceae, Chenopodiaceae, Convolvulaceae and Solanaceae are impermeable to water. When they are buried in the soil, the action of fungi and bacteria also decreases its permeability.

Seed coat induced dormancy: The integuments, during the course of development, undergo thickening and give rise to the seed coat, which protects the embryo. The hardness and impermeability of the seed coat is responsible for inducing seed dormancy.

Seed detrioration: Deteriorative changes occurring with time that increase the seed's vulnerability to external challenges and decrease the ability of the seed to survive.

Seed dormancy: Failure of a viable seed to germinate even when given favorable environmental conditions.

Seed maturation: A process in seed development that occurs after embryogenesis by which a quiescent state is established in a seed. Seed maturation is characterized by storage compound accumulation, acquisition of desiccation tolerance, growth arrest and the entry into a dormancy period of variable length that is broken upon germination.

Seed: Seed is a mature integumented megasporangium or mature ovule consisting of embryonic plants together with store food material covered by a protective coat, the ovary of the carpel contains ovules.

Seed-coat: Protective outer layer(s) on a seed derived from the integuments.

Semi permeable: Partially permeable. Allowing passage of certain, especially small, molecules or ions but acting as a barrier to others. Used of biological and synthetic membranes.

Semi-conservative method: Both the complimentary strands in a DNA molecule separate and each strand functions as a template or mould to prepare its replication.

Semisolid: Intermediate in properties, especially in rigidity, between solids and liquids.

Sepal: Leaf-like structures at flower base, protects young flower bud.

Serine: Serine is an amino acid with the formula $HO_2CCH(NH_2)CH_2OH$. It is one of the proteinogenic amino acids.

Sesquiterpenes: Sesquiterpenes are a class of terpenes that consist of three isoprene units and have the molecular formula $C_{15}H_{24}$.

Sex linked mutation: Mutations occurring in sex chromosomes.

Shikimic acid pathway: The shikimic acid pathway converts simple carbohydrate precursors derived from glycolysis and the pentose phosphate pathway to the aromatic amino acids.

Sieve plates: A perforated wall or part of a wall at the end of one of the individual cells making up a sieve tube.

Sinapine: Sinapine is an alkaloidal amine found in black mustard seeds. It is considered a choline ester of sinapic acid.

Singlet oxygen: Singlet oxygen is a nonradical (does not have an unpaired electron) reactive oxygen species often associated with oxygen free radicals that has strong oxidising activity. Singlet oxygen (1O2) is an electronically excited and mutagenic form of oxygen.

Sink: Sugar is made through photosynthesis, to a sink cell, where the sugar is used.

siRNA: Small interfering RNA (siRNA), sometimes known as short interfering RNA or silencing RNA, is a class of double-stranded RNA molecules, 20-25 base pairs in length.

Sister chromatids: Sister chromatids are two identical copies of a single chromosome that are connected by a centromere.

Skoto dormancy: It is which is induced in seeds that require light for germination when they are imbibed in the dark for extended periods of time.

Smooth endoplasmic reticulum: Endoplasmic reticulum contains no ribosomes.

Soil factor: Soil structure, soil texture and soil temperature influences on seed germination.

Solanad: Basal cell plays only a minor part (or none) in the development of the embryo.

Solute transport: The net flux of solute through a hydrogeologic unit controlled by the flow of subsurface water and transport mechanisms.

Solute: A substance that is dissolved in another substance (a solvent), forming a solution.

Solvent: A substance that can dissolve another substance, or in which another substance is dissolved, forming a solution.

Somatic mutation: Mutations occurring in body cells. These are not transmitted to next generation and hence, termed as non-heritable mutations.

Source: Sugar moves through phloem in plants. It moves from a source cell.

Spherosomes: Spherosomes, also referred to as lipid bodies or oleosomes, are vesicles that stores triacylglycerol in the form of oil (during seed development).

Spindle: The microtubule apparatus that controls chromosome movement during mitosis and meiosis.

Spontaneous mutation: Spontaneous mutations arise from a variety of sources, including errors in DNA replication, spontaneous lesions, and transposable genetic elements.

Stachyose: Stachyose is a tetrasaccharide which consists of the trisaccharide raffinose and a galactose molecule. It is found in plants, for example those in the *Stachys* family, and in soya beans.

Stamen: Male part of the flower, consisting of the anther and filament, makes pollen grains.

Starch synthase: An enzyme that catalyses the transfer of glucose from ADP-glucose to glucose-containing polysaccharides in 1,4-alpha-linkages.

Starch: Starch is a carbohydrate consisting of a large number of glucose units joined by glycosidic bonds. Starch must be broken through digestion before it can be used as glucose sucrose.

Sterols: Sterols, also known as steroid alcohols, are a subgroup of the steroids and an important class of organic molecules.

Stigma: Often sticky top of carpel, serves as a receptive surface for pollen grains.

Stimulus: A factor that causes a response in plant growth

Stoma: A stoma is a pore, found in the leaf and stem epidermis that is used for gaseous exchange.

Stomata: Stomata are minute aperture structures on plants found typically on the outer leaf skin layer, also known as the epidermis.

Stroma: The fluid within the center of the chloroplast is called stroma

Strophiole: Small excrescences arising from various parts of the seed testa *e.g.* the funiculus i anatropous or campylotropous ovules.

Style: The stalk of a carpel, between the stigma and the ovary, through which the pollen tube grows.

Sub apical meristem: The living tissue in which new cells are formed, located below the growing point.

Succinate: Succinate plays a role in the citric acid cycle, an energy-yielding process.

Sucrose: Sucrose or ordinary table sugar, is a disaccharide in which α-D-glucopyranose and β-D-fructofuranose are joined at their anomeric carbons by a glycosidic bond. There are no hemiacetals remaining in the sucrose and therefore sucrose is not a reducing sugar and does not exhibit mutarotation. Sugar is a white crystalline sweet compound found in many plants and extracted from sugar cane and sugar beet. It is used as a sweetening agent in food and drinks. If heated to 200°C, sucrose becomes caramel. When sucrose is hydrolyzed it forms an equimolar mixture of glucose and fructose. This mixture of monosaccharides is called invert sugar. Honeybees have enzymes called invertases that catalyze the hydrolysis of sucrose. Honey, in fact, is primarily a mixture of glucose, fructose, and sucrose.

Superoxide anion: Reduced dioxygen molecule in which the two oxygen molecules are bound by a single covalent bond and carry one unpaired electron with a single electric charge.

Superoxide dismutase: A metal-containing antioxidant enzyme that reduces potentially harmful free radicals of oxygen formed during normal metabolic cell processes to oxygen and hydrogen peroxide -- abbreviation SOD.

Symplast: Movement through plasmodesmata connecting cells.

Synthetic phase: A stage in interphase during which DNA synthesis occurs. At the end of this phase the amount of DNA doubles.

T

TAG: Triacylglycerol

TAM (Tryptamine) pathway: Tryptophan is decarboxylated to form tryptamine (TAM) followed by deamination of the latter resulting in the formation of inodle-3-acetaldehyde (IA1d).

Tannin: A tannin is an astringent, bitter plant polyphenolic compound that binds to and precipitates proteins and various other organic compounds including amino acids and alkaloids.

TCA cycle: Tricarboxylic acid cycle

Telophase: Final stage of mitosis in which the two sets of separated chromosomes condense and become enclosed by nuclear envelopes.

Temperature: The effect on germination can be expressed in terms of cordial temperature *i.e.,* minimum, optimum and maximum temperature. The optimum temperature for most of the seeds is between 15 to 30°C maximum temperature is between 30 to 40°C. Some species will germinate even at freezing point also ex. Alpine.

Template strand: The template strand of DNA is the strand that is actively being transcribed into RNA and will eventually be translated into protein.

Terminal oxidation: Two H-atoms or electrons from the reduced coenzyme ($NADH_2$ or $NADPH_2$) travel through FAD and the cytochromes each with a more positive oxidation -reduction potential and ultimately combine with O_2 molecule to produce one molecule of H_2O. This is called as terminal oxidation.

Termination: The termination step of this oxidation mechanism stops the cycle. The more effective this step becomes, the less oxidation of the lubricant. For this reason, antioxidants are formulated into the lubricants. Antioxidants break into the propagation step of the oxidation mechanism and terminate the process by formation of stable radicals.

Terpenes: Terpenes are a large and diverse class of organic compounds, produced by a variety of plants, particularly conifers, though also by some insects such as termites or swallowtail butterflies, which emit terpenes from their osmeteria.

Terpenoid: The terpenoids, sometimes called isoprenoids, are a large and diverse class of naturally occurring organic chemicals similar to terpenes, derived from five-carbon isoprene units assembled and modified in thousands of ways.

Testa: The outer coat of a seed; usually hard or tough, but may be soft in some species (sarcotesta). Often used as synonymous with seed coat.

Tetraterpene: Tetraterpenes are terpenes consisting of eight isoprene units and have the molecular formula $C_{40}H_{64}$.

TGS: Transcriptional gene silencing.

Thermo - dormancy: Type of dormancy induced by high temperature.

Thiamine: Thiamine is a colorless compound with a chemical formula $C_{12}H_{17}N_4OS$. Thiamine or thiamin or vitamin B1, named as the "thio-vitamine" ("sulfur-containing vitamin") is a water-soluble vitamin of the B complex.

Threonine: Threonine (abbreviated as Thr or T) is an α-amino acid with the chemical formula $HO_2CCH(NH_2)CH(OH)CH_3$. Threonine, an essential amino acid, is a hydrophilic molecule.

3' (Three-prime) end: The 3′-end of a strand is so named due to it terminating at the hydroxyl group of the third carbon in the sugar-ring, and is known as the *tail end*.

Thylakoid disk: Thylakoid disks are disk-shaped membrane structures in chloroplasts that contain chlorophyll.

Thylakoid: The thylakoid is the structural unit of photosynthesis.

Thymine: 5-methyl uracil, also written as 2,6-dioxy, 5-methyl pyrimidine, one of the four nitrogenous bases found in DNA.

Tissue: Tissue is a cellular organizational level intermediate between cells and a complete organism.

Tocopherols: A series of substituted benzopyranols (methyl tocols) occurring in vegetable oils. The four major members are α (5,7,8-trimethyl), β (5,8-dimethyl), γ (7,8-dimethyl) and δ (8-methyl). The C 16 side-chain linked to the pyran ring is saturated in the tocopherols but contains three double bonds in the tocotrienols. α-Tocopherol shows the highest vitamin activity whilst the δ and γ tocopherols are the most active antioxidants. Natural tocopherols are generally used at levels up to 500 ppm along with ascorbyl palmitate (200–500 ppm) which enhances vegetable oils already contain tocols at levels of 200–800 ppm so that added material has only a limited effect. Three types of tocopherol are produced for sale: (i) natural tocopherols isolated from soybean, sunflower or other vegetable oils which are mixtures of the α, β, γ and δ compounds, (ii) natural mixtures which have been methylated to convert the β, γ and δ compounds to α tocopherol (the trimethyl compound) with higher vitamin E activity, and (iii) synthetic α-tocopherol made from trimethylhydroquinone and phytyl bromide. This is entirely the trimethyl compound but a mixture of eight stereoisomers. All three show antioxidant and vitamin E activity but only the last two can be sold as vitamin E.

Tonoplast: A large central vacuole, a water-filled volume enclosed by a membrane.

TPP: Thiamine pyrophosphate

Tracheid: A cell with strengthened walls that functions to transport fluid within plants.

Transcellular streaming: The movement of the particulate and fluid constituents of cytoplasmic through linear files of longitudinally oriented plant cells.

Transcellular transport: It involves the transportation of solutes by a cell through a cell.

Transcription: Transcription is the first step of gene expression, in which a particular segment of DNA is copied into RNA by the enzyme RNA polymerase.

Translation elongation: Translation elongation is simply the ribosome travelling down the message, reading codons and bringing in the proper aminoacyl tRNA's to translate the message out to protein.

Translation: Translation is the construction of an amino acid sequence (polypeptide) from an RNA molecule.

Translocation of Auxin: Auxin moves through the plant by two mechanisms.

It passes in the sap moving through the phloem from where it is synthesized (its "source", usually the shoot) to a "sink" (*e.g.*, the root).

It also passes from cell to cell by the following mechanism.

Auxin can enter the cell by diffusion and also through influx transporters in the plasma membrane. It moves out through efflux transporters – called PIN proteins. Eight different types of PIN proteins have been identified so far. These are transmembrane proteins inserted in localized portions of the plasma membrane.

Translocation: A rearrangement mutation in which part of one chromosome breaks away and joins another.

Transphosphorylation: A reaction involving transfer of a phosphoric group from one compound to another.

Triacylglycerol (triglyceride): Any of a group of lipids that are esters composed of one molecule of glycerol and three molecules of one or more fatty acids.

Triene: An unsaturated hydrocarbon containing three double bonds between carbon atoms.

Triglyceride: A molecule consisting of 3 carbon atoms (glycerol) each with a fatty acid attached to them. They are the storage form of fat.

Triglycerides: A triglyceride (TG, triacylglycerol, TAG, or triacylglyceride) is an ester derived from glycerol and three fatty acids.

Triterpene: Triterpenes are terpenes consisting of six isoprene units and have the molecular formula $C_{30}H_{48}$.

tRNA: Transfer RNAs read the information encoded in the mRNA and transfer the appropriate amino acid to a growing polypeptide chain during protein synthesis.

Tryptophan: Plants and microorganisms commonly synthesize tryptophan from shikimic acid or anthranilate.

β-tubulin: One of the monomeric globular proteins which associate to form the dimer, α, β-tubulin, the basis of microfilaments.

Tubulin: The constituent protein of microtubules of cells which provide a skeleton for maintaining cell shape and is thought to be involved in cell motility.

Tugor pressure: The pressure exerted by water inside the cell against the cell wall.

Tyrosine: Tyrosine or 4-hydroxyphenylalanine, is one of the 22 amino acids that are used by cells to synthesize proteins.

U

UDP: Uridine diphosphate glucose is a nucleotide sugar. It is involved in glycosyltransferase reactions in metabolism.

Ultrastructure: The close detail of a cell, with all organelles visible, that can be seen by using an electron microscope.

Unequal crossing over: Non-reciprocal crossing over caused by mismatching of homologous chromosomes usually occuring in regions of tandem repeats.

Univalent: A single chromosome seen segregating in meiosis.

Unsaturated fat: An unsaturated fat is a fat or fatty acid in which there is at least one double bond within the fatty acid chain.

Unsaturated fatty acid: A term applied to fatty acid to denote the presence of one or more double bonds – the loss of two hydrogens from the fatty acid chain. The more double bonds, the more unsaturated the fatty acid. Two or more double bonds define a fatty acid as polyunsaturated.

Uracil: 2,6-dioxypyrimidine, one of the nitrogeneous bases found in RNA.

UTP: Uridine triphosphate (UTP) is a pyrimidine nucleoside triphosphate, consisting of the organic base uracil linked to the carbon of the ribose sugar, and esterified with tri-phosphoric acid.

V

Vacuole: A vacuole is a membrane-bound organelle which is present in all plant and fungal cells and some protist, animal and bacterial cells.

Vacuoles: Vacuoles are the fluid filled organelles enclosed by a membrane. They can store materials such as food, water, sugar, minerals and waste products.

Valine: Valine is an α-amino acid with the chemical formula $HO_2CCH(NH_2)CH(CH_3)_2$.

Vascular system: This system consists of cells located at the core of the root, whose function it is to transport food and water.

Vascular tissue: Tissue that transports water and nutrients through the plant body in higher plants.

Vitamin C: Vitamin C is water-soluble. It is also called ascorbic acid. Vitamin C is one of the antioxidants. This means it protects cells from damage caused by particular aggressive atoms and molecules called free radicals.

Vitamin E: Vitamin E describes a group of 8 different fat-soluble vitamins. They are antioxidants, which mean they protect cells from damage caused by aggressive types of atom or molecule called free radicals.

Viviparous: The phenomenon of germination of seeds *in situ i.e.,* while still attached to the fruit is called as vivipary. It is commonly seen in Mangrove plants particularly those of the Rhizophoraceae.

W

Warmth: Seeds will not start to germinate, if the temperature is too low. The minimum temperature needed varies with the species of seed.

Water: Water is a basic requirement for germination. It is essential for enzyme activation, breakdown, translocation and use of reserve storage materials.

Wavelength: Wavelength is the distance between identical points in the adjacent cycles of a waveform signal propogated in space or along a wire.

X

Xylem vessels: Xylem vessels are a long straight chain made of tough long dead cells known as vessel elements.

Xylem: Hard cells transport water and mineral nutrients in a kind of tissue.

Z

Zone of elongation: The area immediately behind the tip. Here, cells increase in size through food and water absorption. These enlarging cells push the root through the soil.

Zone of maturation: The region in which the enlarged cells undergo changes to become the specialized tissues of the epidermis, cortex and vascular system.

Zygotene: Homologous chromosomes become closely associated (synapsis) to form pairs of chromosomes (bivalents) consisting of four chromatids (tetrads) or A substage of prophase I of meiosis I during which homologous chromosomes begin to synapse.

✦✦✦